Technology, Society and Sustainability

Lech W. Zacher
Editor

Technology, Society and Sustainability

Selected Concepts, Issues and Cases

Springer

Editor
Lech W. Zacher
Kozminski University
Warsaw
Poland

ISBN 978-3-319-47162-4 ISBN 978-3-319-47164-8 (eBook)
DOI 10.1007/978-3-319-47164-8

Library of Congress Control Number: 2016955523

Printed on acid-free paper

This Springer imprint is published by Springer Nature
The registered company is Springer International Publishing AG
The registered company address is: Gewerbestrasse 11, 6330 Cham, Switzerland

Preface

The book is a result of collaborative efforts uniting various disciplines—from technology studies to philosophy. Relations and interactions between society and technology are scrutinized in a context of sustainability problematique. Including experts from various countries and across different generations makes an interesting contribution to the debate on sustainability. International and intercultural perspectives presented by the authors help to clarify the variety of approaches and contexts in which technologies develop and interact with society and the environment. Better insights into promises and chances, of risks and dangers, especially in the long term, make a shift in consciousness necessary to cope with existing and emerging challenges to individuals, society, and humanity. The authors have tried to present and discuss some of them showing their multidisciplinarity and multiculturality.

The collection consists of eight parts. In Parts I–III devoted to *innovation and technology emerging in and for society*, the authors bring together various technological innovations, not only ICTs, but also their social applications and impacts. It is shown that various technological advances, some of a breakthrough type, create a totally new surrounding for people's activities and lives. Furthermore, they stimulate—via strategies policies, networking, and market-growth and overwhelming technology applications. They are growth- and profit-oriented. Producing many positive impacts, they also create new risks and challenges (e.g., concerning social order, freedom, social justice, legal issues, moral dilemmas of military technologies, technologization of culture, as well as environmental problems). Traditional concerns about social needs, often created by technology and market in the name of progress, should be contested and enriched by sustainability criteria.

In Parts IV–VIII devoted to the relations of *technology, economy, and sustainability*, the authors present TA methods and numerous cases of sustainability-oriented activities and undertakings. Sustainable development is discussed in many areas and contexts (e.g., city development, technology production and applications, organization, business, public finance, consumption, and human life). The road to sustainability is context-dependent; however, strategies, policies, and human behavior changes may in the long run facilitate a timely needed transition. The efforts should be multidimensional and multilevel, local and global, theoretical and practical, and coping both with complexity and diversity of world societies and economies. The authors, presenting multiple ideas, concepts, and real cases (from several countries), have tried to provide inspiring perspectives on sustainable development and to contribute to theoretical backgrounds and transition pathways toward a sustainable future.

The collection is a multidisciplinary and multicultural contribution to the present sustainability discourse. It is focused on two main dimensions of our world: *complexity* and *diversity*. Desirable and urgent transition of socio-technological systems toward a sustainability trajectory of development requires a better understanding of technological trends and social transformations. General advancement of technology does not produce identical changes in various societies, differentiated economically and culturally. Moreover, the abilities to approach sustainable development change over time and space. So there is a constant need for continuing research, analyses, and discussions concerning changing contexts and adequacy of strategies and policies. The authors from 12 countries, of different academic and cultural settings, present their insights, analyses, and recommendations. The collection is focused both on contexts and on activities leading to sustainable trajectories in various domains of economy and social life.

The collection consists of the eight complementing parts. In Part I, there are four papers presenting different dimensions of socio-technological change.

K.I. Pelc constructs a model of innovation diffusion process for a new context—social networking. Broader civilizational aspects of enterprise information infrastructure are discussed by A. Targowski in a perspective of a sustainable future. L.W. Zacher underlines a performative power of technologization of people and their activities. This megatrend is supported by commercialization and marketization, greatly impacting human life and human culture. Technologically pioneering countries are not passive; they conduct policies co-shaping their futures. Information revolution and its overwhelming impacts require such policies. D. Anagnostopoulou describes the digital policies of two important world actors—the European Union and the USA.

In Part II, the authors introduce the issue of technology use in various domains and contexts. A.K. Srivastava and S. Sharma present how social justice can be facilitated by e-policy initiative (in India—the second most populous country in the world). Technology supports politics. ICTs essentially transform our cultural sphere, and media in particular. T. Miczka interestingly reveals the so-called audiovisuality of the third degree, allowing people to move from real space to virtual space. The progression of societies into the future can be accompanied by a concurrent digital preservation of the cultural heritage. So a continuity of culture can be protected (N. Peters, D. Marinova, M. van Faassen, and G. Stasiuk). The innovative services of technology seem countless.

Part III includes in-depth observations—some general, some particular. A. Betlej investigates how the Internet technology makes possible to construct—via an interactive network-based system of virtually embedded relations—an alternative social order in the future. The author tries to answer some crucial questions about the peril and the promise of this technology. J. Miksa continues this discussion, calling for responsibility in the Net. Citizens should feel not only free but also safe (the case of pedophilia is quoted). Threats to civil liberties generated or accompanied by scientific and technological advancements are topics of E. Polak's essay. The balance between citizens' freedom and security is an old challenge in a new technological reality. Legal standards are inevitable to preserve both values, proves M. Skórzewska-Amberg in her paper. W. Lamentowicz focuses on military technologies, on drones in particular. He critically evaluates the legal and moral legitimacy of targeted killing by drones.

Part IV, focused on *technology evaluations and policies,* provides a better understanding of technology's impacts and methods of their evaluation. P. Sienkiewicz and H. Świeboda present a methodological approach for linking TA and systems analysis, while A. Grunwald investigates TA methods and practices (i.e., in German TAB) in the context of sustainable development challenges. Effective translocal TA requires international cooperation. Its empirical case related to Middle and Eastern European countries is developed by G. Banse.

Part V devoted to the *approaches to sustainability* shows that the concept of sustainability and its use may raise controversies. Z. Bisk and P. Boltuc discuss sustainability from a perspective of growth and its limits. Finally, the authors conclude that sustainability can best be attained by technological growth.

The next author, L. Xu, considers sustainability in the context of an uncertain future. Xu introduces a notion of resilience thinking into sustainability discourse. Water resource management is given as an example enhancing socio-ecological resilience.

Sustainability at least as a trajectory is globally crucial in the case of new emerging powers. How to reconcile economic prosperity with sustainability in China?—seek X. Guo, S. Krempl and D. Marinova. Can Confucianism and Taoism help?

Threats to sustainable development in world societies are numerous, tough, and diverse, claims A.P. Wierzbicki. His analysis, based on the project "Poland 2050," evaluates global risks and threats endangering the future of sustainable development worldwide. Part V ends with the chapter by W. Sztumski, who underlines the role of the individual, often underestimated or ignored in sustainability discourse. So he opts for connecting the idea of sustainable development with a concept of "sustainable man" (understood as *Homo sapiens*). Only sustainable people can fully realize a vision of sustainable development, claims the author.

In Part VI devoted to *sustainability and its dimensions and issues.*

E. Stoyanova tries to highlight the relation of corporate social responsibility and practices toward sustainable development. She presents a conceptual framework of embedding sustainable development in organizations through leadership. Many cases are quoted.

Part VII is devoted to practical challenges leading to sustainability. J. Kielin–Maziarz describes the EU Sustainable Consumption and Production

Action Plan in the context of the sustainable development principles. There are more powerful world actors influencing international sustainability. The authors from China (J. Hong and W. Yu) and Australia (D. Marinova, X. Guo, and M. Gollagher) analyze implications for China's resource demand on sustainability in Australia.

Emerging powers like China have their own internal problems due to the ever-expanding cities and their inundation with water. B. Zhao, D. Marinova, and X. Guo outline a sustainability strategy model for urban water management. The next paper by L. Xu and D. Marinova on nano-biotechnology for water sustainability is a good supplement.

Finally, Part VIII, devoted to *sustainability and business,* is oriented toward economic issues. A. Czarczyńska discusses the sustainability challenges in the posttransition countries (like Poland), while analyzing interdependencies of main stakeholders in the sustainable development process worldwide. Stakeholders' strategic thinking is crucial for sustainable development. M. Monaci and M. Magatti postulate a reorientation of companies' engagement in corporate social responsibility (CSR) from mainly adaptive-reactive strategies for legitimization in relation to stakeholders toward a model based on the valorization of social resources and needs. The most desirable will be creating "integrated value"—economic, social, and environmental (some findings of research conducted on a sample of Italian organizations are discussed).

Sustainability characteristics may become the important measures of companies' performance in the present competitive world. They are evaluated by stakeholders as well. L. Soares and C. Chaves propose a concept and methodology suitable to calculate a value of corporate ecological footprint (CEF). To support such an approach, some case studies for the Iberian Peninsula were reported and analyzed.

Sustainability of public finances gains a supranational significance in the time of crisis. Sustainability in this understanding relates not only to finances as such but also to broader issues of economic and social development. N. Gomes investigates in his article the efforts to repair public finances within the EU legal frameworks and programs (the case of Portugal is quoted as an empirical example of a country at financial risk). Needless to prove that social and environmental viability absolutely needs sustainability of public finances.

The collection is a contribution to the present discussion on sustainable development and its challenges for humanity and its organizations. Continuing research and discussion is needed to better understand these challenges and to prepare the appropriate strategies and solutions. Development of socio-technological systems is nowadays very complex; moreover, the world we live in is extremely diverse. Therefore, sustainability discourse must be ongoing, introducing new ideas, concepts, theories, evidence, and experience by various parties—academics, professionals, and practitioners.

The composition of the book and the set of authors emphasize the importance of multi and interdisciplinary and diverse approaches and evaluations, as well as multiculturality. For the purpose of better understanding of present and future challenges, many theoretical and policy-oriented issues are discussed. The borders among them are often fuzzy, some overlap, and some others need further research or action. The authors try to help identify, based on their differentiated knowledge and experience, what is necessary to achieve sustainability in various domains and in various contexts. The technological civilization we live in produces streams (and waves) of innovations. They are generated in societies and should serve them. However, in practice it is neither so simple nor so smooth. The production of innovations and their vast applications are connected with numerous effects and impacts, including negative ones.

Warsaw, Poland Lech W. Zacher

Contents

Contributors

Despina Anagnostopoulou Jean Monnet Chairholder, University of Macedonia, Thessaloniki, Greece

Gerhard Banse Institute for Technology Assessment and Systems Analysis, Karlsruhe Institute of Technology, Karlsruhe, Germany

Alina Betlej The John Paul II Catholic University of Lublin, Lublin, Poland

Tsvi Bisk Center for Strategic Futurist Thinking, Kfar Saba, Israel

Piotr Bołtuć Department of Philosophy, University of Illinois at Springfield, Springfield, IL, USA

Cristina Chaves Faculty of Economics, University of Porto, Porto, Portugal

Anna Czarczyńska Kozminski University, Warsaw, Poland

Margaret Gollagher Curtin University Sustainability Policy Institute, Curtin University, Perth, WA, Australia

Noel Gomes Porto School of Economics, University of Porto, Porto, Portugal

Armin Grunwald Institute for Technology Assessment and Systems Analysis, Karlsruhe Institute of Technology, Karlsruhe, Germany; Office of Technology Assessment at the German Bundestag, Berlin, Germany

Xiumei Guo Curtin University Sustainability Policy Institute, Curtin University, Perth, WA, Australia

Jin Hong School of Management, University of Science and Technology of China, Hefei, Anhui, China

Joanna Kielin-Maziarz Kozminski University, Warsaw, Poland

Sandra Krempl Curtin University Sustainability Policy (CUSP) Institute, Curtin University, Perth, WA, Australia

Wojtek Lamentowicz Academy of Business and Administration in Gdynia, Gdynia, Poland

Mauro Magatti Department of Sociology, Catholic University of the Sacred Heart, Milan, Italy

Dora Marinova Curtin University Sustainability Policy Institute, Curtin University, Perth, WA, Australia

Tadeusz Miczka University of Silesia, Katowice, Poland

Joanna Miksa Ethics Department, University of Lodz, Lodz, Poland

Massimiliano Monaci Department of Sociology, Catholic University of the Sacred Heart, Milan, Italy

Karol I. Pelc School of Business and Economics, Michigan Technological University, Houghton, MI, USA

Nonja Peters Curtin University, Perth, Australia

Ewa Polak Department of Global Economy, Faculty of Economics, Institute of Maritime Transport and Seaborne Trade, University of Gdansk, Sopot, Poland

Talia Raphaely Curtin University, Perth, WA, Australia

Sangeeta Sharma Department of Public Administration, University of Rajasthan, Jaipur, Rajasthan, India

Piotr Sienkiewicz Engineering Institute of Security Systems, National Defence University, Warsaw, Poland

Małgorzata Skórzewska-Amberg School of Law, Kozminski University, Warsaw, Poland

Luisa Soares Faculty of Economics, University of Porto, Porto, Portugal

Anurag Kumar Srivastava Department of Public Administration and Policy, School of Liberal Studies, Pandit Deendayal Petroleum University, Gandhinagar, Gujarat, India

Glen Stasiuk Murdoch University, Murdoch, Australia

Emiliya Stoyanova Economics and Management Department, Varna University of Management, Varna, Bulgaria

Halina Świeboda Engineering Institute of Security Systems, National Defence University, Warsaw, Poland

Wiesław Sztumski Silesian University Katowice, Częstochowa, Poland

Andrew Targowski President Emeritus of International Society for the Comparative Study of Civilizations, Western Michigan University, Kalamazoo, MI, USA

Marijke van Faassen Huygens Instituut voor Nederlandse Geschiedenis (Huygens ING), Amsterdam, The Netherlands

Andrzej P. Wierzbicki National Institute of Telecommunication, The Committee of Future Studies, Polish Academy of Sciences, Warsaw, Poland

Li Xu Curtin University Sustainability Policy Institute, Curtin University, Perth, WA, Australia

Wentao Yu School of Management, University of Science and Technology of China, Hefei, Anhui, China

Lech W. Zacher Center of Impact Assessment Studies and Forecasting, Kozminski University, Warsaw, Poland

Baohui Zhao Curtin University Sustainability Policy Institute, Curtin University, Perth, WA, Australia

Part I
Technological Developments—Civilization—Policies

Chapter 1
Diffusion of Innovation in Social Networking

Karol I. Pelc

Abstract The purpose of this paper was is to present a model of innovation diffusion process applicable for analysis of innovation in social networking. Until recently, the innovation diffusion was measured by the rate of adoption expressed by the number of adopters versus time or by penetration rate within given population. On the basis of literature review and available empirical data, a new, additional dimension called "interactivity level" is proposed in this paper. It is specific to innovations in social networking, where the main purpose of innovation is to improve interactive contacts among actors in a network. The interactivity level reflects amount of active knowledge and effective use of social media/technologies. This dimension has not been systematically studied yet. However, some recent surveys of social media users demonstrate a need and suggest application of that dimension. It is possible to improve evaluation of diffusion of innovation in social networking by introducing a scale of interactivity level, expressed by the mode/scope and intensity of interactive contacts among users of different social media. Such approach implies that adoption of innovation should not be viewed just as an event but rather as a learning process of each adopter. Hence, some phases of that process need to be identified and assessed in terms of dynamics of learning. Six basic phases of this process are proposed in this paper. Comparative study of average interactivity levels of users in different countries is proposed as an indicator of evolution of social media in the global business environment.

Keywords Interactivity level · Innovation adoption · Penetration rate · Diffusion model · Social networking

K.I. Pelc (✉)
School of Business and Economics, Michigan Technological University,
1400 Townsend Dr., Houghton, MI 49931, USA
e-mail: kipelc@mtu.edu

L.W. Zacher (ed.), *Technology, Society and Sustainability*,
DOI 10.1007/978-3-319-47164-8_1

Introduction

Social networking in combination with a number of tools and means facilitating broad interactive communication became one of the important innovations affecting cultural, political and business operations in the twenty-first century. All elements of this broad innovation constitute an implementation of the Web 2.0 platform for exchange of information and opinions as well as for active collaboration among multiple actors participating in a network. Social networking may be viewed as a process of using different social media for achieving certain goals through interactive communication and collaboration. According to Kaplan and Haenlein (2010), "the *social media* is a group of Internet-based applications that build on the ideological and technological foundations of Web 2.0, and that allow the creation and exchange of User Generated Content." Web 2.0 sites allow users to interact and collaborate with each other in a dialog as creators of user-generated content in a virtual community, in contrast to Web sites of Web 1.0 where users are limited to the passive viewing of content that was created for them. The same authors proposed a classification of social media in which they distinguished six types of media. Three of them provide a high level of self-presentation. These are: *blogs, social networking sites* (e.g., Facebook, MySpace, Twitter, Orkut, LinkedIn, QQ and Qzone) and *virtual social worlds* (e.g., Second Life). Other three, *collaborative projects* (e.g., Wikipedia), *content communities* (e.g., YouTube) and *virtual game worlds* (e.g., World of Warcraft), provide low level of self-presentation. Certain amount of confusion arises in the literature due to multiplicity of terms being used by different authors. In addition to those mentioned above, some authors use the term *social technologies* as equivalent of social media seen as software technologies enabling operation of the media (Bughin 2012; Bughin et al. 2011; Brown and Sikes 2012). Others include in their analyses such media as Rich Site Summaries (RSS) and podcasts (Li and Bernoff 2011). More general term of *social networks,* which is sometimes used in this context, refers to certain type of sociological phenomenon that in the current reality is supported by the social media. This concept is subject of both economic and sociological analyses (Benkler 2006; Carlsson 2004; Rainie and Wellman 2012). According to the latter, social networks are facilitated by technological systems: "The internet and mobile phones have facilitated the reshaping of people's social networks, enabling them to be larger and more diverse. And they have reconfigured the way people use their networks to learn, solve problems, make decisions, and provide support to each other." (Rainie and Wellman 2012). Impact of new media on current and future business practices has been emphasized by De Kare-Silver (2011) and strongly supported by Barry Libert in his interview with Knowledge@Wharton (2012). Broader social impact of digital technologies (including social media) in relationship to demographic structure of society has been presented by Tapscott (2009), and multiple perspectives on global networking for innovation supported by those technologies have been described by Pelc (2012).

The main subject of this paper is innovation in social networking as a contemporary phenomenon of growing importance in business and all spheres of social life. That innovation takes place by implementation of several new social media and social technologies. The purpose of the paper is to introduce an additional dimension of innovation diffusion in social networking. That dimension is called "interactivity level." It is intended to characterize the innovation diffusion process not only by the number of adopters (as a function of time) but also by assessment of their mode/depth of adoption, demonstrated by the range and frequency of usage of individual elements or features of the adopted social media/technologies. Empirical data and illustrations concerning the practice of innovation adoption in different countries have been extracted, derived or compiled for purposes of this paper from survey results published by Forrester Research, McKinsey Global Institute and Socialbakers. Importance of social networking for business has been recently explored and discussed in several publications. Brown and Sikes (2012) indicated that social networking is one of the most effective means of digital business. Their survey data suggest that 28 % of executives emphasized benefits for their companies due to this new approach. More particular measurable benefits from using the Web 2.0 media/technologies for increasing the speed of access to knowledge and for reducing communication costs have been confirmed by more than 60 % of respondents in a McKinsey survey of companies (Bughin and Chui 2010). According to another McKinsey study, the most effective social media in support of business processes are blogs, social networking sites and video-sharing when used for scanning of external environment, finding new ideas and managing projects (Bughin 2012). Those data and opinions suggest that an insight into detailed mechanisms of adoption of those innovative tools for social networking is needed.

In the following section, a review is presented of conventional models of innovation diffusion. Then a definition of interactivity level, as a new dimension of innovation diffusion, is proposed together with a suggested way of its application for assessment in social networking. The next section of the paper presents international comparative data on diffusion and interactivity. In the final remarks, we propose some directions of further research.

Rate of Innovation Adoption as a Measure of Diffusion

The diffusion rate, expressed by number of adopters as function of time, is an indicator of adoption of innovation by population of users. It may advance at a different speed depending on access to information about the innovation and on learning capabilities of actors involved in the adoption process. Several models of that process have been presented in the literature (Rogers 2003; Girifalco 1991). Based on the literature review, at least three groups of models can be distinguished. One of them, illustrated by the *Coleman model*, assumes that the source of innovation, and information about it, is open and public, making access to it equally probable for all potential adopters (Skiadas 1985). In this case, diffusion rate is very fast in the early period (initiation) and slows down with time (saturation). Second, clearly defined type, e.g., *Dodd model*, is based on assumption that adoption is driven by imitation process, in which adopters are learning from the earlier adopters and growth of adopters' population is epidemic in its nature (Sharif and Ramanathan 1981). In this, most common case, number of adopters is growing in time according to logistic function. Some authors distinguish such groups of adopters as "innovators," "early adopters," "early majority," "late majority" and "laggards" (Rainie and Wellman 2012).

Third type of models, e.g., *Mahajan–Schoeman model* (Mahajan and Schoeman 1977) combines mechanisms of diffusion of the previous two models, i.e., certain adopters decide on the basis of publicly accessed information, when others follow the imitation route. Depending on proportion between the sizes of two groups of adopters, the rate of diffusion may fluctuate between the two earlier functions of time. In each model, a saturation phase takes place (slow growth) when a large majority of population adopted the innovation. In case of diffusion of innovation in social networking, it is important to consider possibility of different "depth of adoption" by individual adopters. That "depth" is gradually changing during the learning process. As the main functional purpose of social media/technologies is to enable interactive communication involving the user created content (according to Web 2.0 philosophy), the extent of interactivity among adopters is changing with time. Adoption is not just an event; it is a learning process. Individual users, i.e., participants in social networking, operate in different modes depending on phase of learning. It suggests that there is a need to introduce an additional dimension for analysis of this kind of innovations.

Interactivity Level: A New Dimension of Innovation Diffusion

According to Kiousis (2002), *interactivity* can be defined as "the degree to which a communication technology can create a mediated environment in which participants can communicate (one-to-one, one-to-many, and many-to-many), both synchronously and asynchronously, and participate in reciprocal message exchanges (third-order dependency)."

In case of social networking, a measure of interactivity level has to reflect operational capability and effective application of either individual or combined social media for communication in the modes of *one-to-many* and *many-to-many*. Innovation diffusion in this domain may be then characterized by frequency and complexity of those applications. For that purpose, we propose the following definition: *interactivity level* (in social networking) is a number, e.g., on scale of one to six, reflecting complexity and intensity of interactive operation mode of actors involved in social networking. The higher is the number, the more complex operation mode and/or more frequent use of respective social media.

Li and Bernoff (2008, 2011) distinguished interactivity levels of different actors involved in social networking by grouping them into clusters called: creators, critics, collectors, joiners, spectators and inactives. Those groups are identified by surveys of online consumers in different communities and constitute a "social technographics profile" for each of those communities. The authors presented a very useful empirical material, based on Forrester Research surveys, which illustrates practical application of the concept of interactivity level.

The correspondence between interactivity levels and operational modes in subsequent innovation adoption phases on one part and tasks analyzed in the studies presented by Li and Bernoff on the other is shown in Table 1.1. The following innovation adoption phases (levels of interactivity) are included: initial learning, field exploration, focused observation, selection and classification,

Table 1.1 Interactivity levels, interactivity modes/phases and corresponding interactivity tasks by actor groups in surveys conducted by the Forrester Research

Interactivity level (1)	Interactivity mode and innovation adoption phase (2)	Interactivity tasks[a] (according to Forrester Research surveys) (3)
1	Initial learning –Learning about existence of social media	No tasks
2	Field exploration –Random observations of different user created contents –Acquaintance with some features and communication potential of social media	Passive tasks Reading social media presentations, e.g., blogs, online forums, customer reviews and ratings Listening to podcasts Watching videos (YouTube)
3	Focused observation –Observation of social networking sites: testing own capability to use them –Learning how to use social networking effectively	Visiting social networking sites (e.g., Facebook. Twitter, LinkedIn) Maintaining profile on a social networking site
4	Selection and classification –Limited use of social media for communication –Classifying different forms of social media	Applying tags for classification of Web pages, photographs, etc. Online voting for Web sites Using RSS feeds
5	Evaluation –Experimenting with interactive communication through selected media –Contributing to contents created by other users	Comments on logs of other authors Rating products/services Editing wiki texts
6	Design and implementation –Interactive broad range networking and communication with other users of social media –Creating own contents and publishing in a broad range of social media	Web page design and publication Blog design and publication Photograph/video creation and publication Music creation and publication Texts creation and posting

[a]Lists of tasks presented in column 3 has been adopted from the book: Li and Bernoff (2008)

evaluation, design and implementation. They are reflecting the learning process of interactive communication in social networking. All adopters practice interactivity by gradually moving from Level 1 of initial learning about existence of social media to Level 6 of design and implementation of interactivity by the user created contents of messages.

Global Aspects of Innovation Diffusion in Social Networking

Diffusion of innovation is dependent on many factors: economic, social, technological, cultural, etc. In case of social networking, that process has been relatively fast in both the adoption rate by general population and the adoption rate by businesses, customers, suppliers, etc. In this section, we focus on diffusion of innovation in social networking in different countries as characterized by the overall rate of adoption of social media in a population and by changes in the level of interactivity in social networking among the online consumers. The adoption of social network sites by general population is illustrated by statistics on the use of Facebook, based on data published by Socialbakers (2013). This company is using the term of "penetration of population" as equivalent of the adoption rate of innovation, i.e., percentage of population using the social networking site at a given point of time. In Tables 1.2, 1.3 and 1.4, the penetration by Facebook in the early months of 2013 is shown for three groups of countries.

Table 1.2 presents the top ten countries in terms of the level of Facebook penetration (% of total population of the country). It is interesting to note that the

Table 1.2 Highest Facebook penetration in population of countries

Rank order (top 10)	Country/continent	Facebook penetration rate % of population (2013)
1	Monaco/Europe	107.93[a]
2	Qatar/Asia	78.66
3	Aruba/South America	74.73
4	Falkland Islands/South America	74.30
5	Gibraltar/Europe	73.88
6	Turks & Caicos Islands/North America	72.19
7	Iceland/Europe	71.86
8	United Arab Emirates/Asia	67.75
9	Singapore/Asia	58.66
10	Brunei/Asia	56.09

Percentage of population using Facebook in respective countries in 2013. Data derived from: Socialbakers "Facebook Statistics by Country," http://www.socialbakers.com/, retrieved 03/01/2013

[a] Penetration rate above 100 % suggests that some users own more than on Facebook account

Table 1.3 Lowest Facebook penetration in population of countries

Ranked from the lowest end[a] (BOTTOM 10)	Country/continent	Facebook penetration rate % of population (2013)
1	Niger/Africa	0.38
2	Tajikistan/Asia	0.56
3	Uzbekistan/Asia	0.58
4	Guinea/Africa	0.67
5	Burkina Faso/ Africa	0.88
6	Somalia/Africa	1.11
7	Malawi/Africa	1.17
8	Madagascar/Africa	1.34
9	Afghanistan/Asia	1.45
10	Mali/Africa	1.58

Percentage of population using Facebook in respective countries in the early months of 2013. Data derived from: Socialbakers "Facebook Statistics by Country," http://www.socialbakers.com/, retrieved 03/01/2013

[a]China in not listed in this table. The lowest penetration of Facebook (0.04 %) in that country is mainly due to intermittent blocking of facebook by government rules, and due to popularity of other social networking media e.g. QQ or Qzone

Table 1.4 Facebook penetration in population of selected developed countries

Country/continent	Facebook penetration rate % of population (2013)
Canada/North America	53.45
US/North America	52.56
UK/Europe	51.61
France/Europe	39.07
Germany/Europe	30.64
Poland/Europe	26.63
S. Korea/Asia	19.59
Japan/Asia	10.85

Percentage of population using Facebook in respective countries in the early months of 2013. Data derived from: Socialbakers "Facebook Statistics by Country," http://www.socialbakers.com/, retrieved 03/01/2013

leader of this "elite group" is Monaco with penetration rate over 100 %, meaning that there are several people owning more than one Facebook account. Only two other European nations qualified into the group of global top ten: Gibraltar and Iceland with over 70 % penetration of population. There is only one country of North America among them: Turks and Caicos Islands (on the Caribbean Sea). Majority in this group is constituted by countries of small total population, mostly islands or Arab countries with relatively small population and high level of GNP per capita such as Qatar (population 1.9 million, GDP/capita $98.000) and United Arab Emirates (population 5.4 million, GDP/capita $67.000).

Table 1.3 presents the bottom ten countries ordered by the lowest percentage of Facebook penetration. All countries on the list of lowest penetration are located in Africa or Asia with penetration levels ranging from around 0.4–1.6 %.

Table 1.4 presents Facebook penetration rates of selected developed countries of North America, i.e., Canada and the USA, of Europe: the UK, France, Germany and Poland, and of Asia: Hong Kong, Japan and South Korea. It may be seen that the highest Facebook penetration in this group of countries is 54.31 % in Hong Kong, when in Canada, the USA and UK it is almost the same in the range between 52 and 53 %. In terms of total number of Facebook users, the USA is still at the global top with over 163 million users.

All those data, extracted from the Socialbakers Web site http://www.socialbakers.com/facebook-statistics/, present just one point on the timescale (early months of 2013). To analyze dynamics of innovation diffusion, one would have to build time series of data for certain period of time, e.g., ten years. The Socialbakers offer such possibility, not only for Facebook but also for Twitter, YouTube, Google+ and LinkedIn, considered to be the top five social media in terms of scale of their penetration.

As mentioned in the previous section of the paper, we suggest applying interactivity level as an important dimension of innovation diffusion in social networking. In Table 1.5, data have been compared for interactivity levels in selected six developed countries between 2007 and 2010 (or 2009 for some countries). They indicate changes in levels of interactivity in each country

Table 1.5 Changes in interactivity levels of social media users in selected countries between 2007 and 2010 (or 2009)

	USA		UK		Germany		Japan		S. Korea	
Interactivity level	2007 (%)	2013 (%)	2007 (%)	2013 (%)	2007 (%)	2013 (%)	2007 (%)	2013 (%)	2007 (%)	2013 (%)
1. Initial learning	44	18	54	37	49	52	26	23	36	9
2. Field exploration	48	73	37	50	44	38	70	69	39	76
3. Focused observation	25	51	21	38	12	21	22	26	41	48
4. Selection & classification	12	21	5	6	12	4	6	11	14	19
5. Evaluation	25	37	16	21	22	12	36	30	27	46
6. Design & implementation	18	24	9	15	8	9	22	34	38	49

Average % of online consumers involved in interactive social networking at different interactivity levels

Data compiled from: Forrester Research surveys presented by Li and Bernoff (2008) and in http://empowered.forrester.com/tool_consumer.html (2014)

reflecting dynamics of adoption process among online consumers participating in the surveys conducted by Forrester Research (reported by Li, C. and Bernoff, J. in two subsequent editions of their book). Those changes illustrate phases of innovation adoption process according to the concept presented in the earlier part of this paper.

From data of Table 1.5, it is possible to recognize that the most dynamic changes in interactivity levels (adoption phases of innovation) took place in South Korea, where percentage of online consumers actively involved in social networking increased most visibly at each level of interactivity higher than Level 1. For instance percentage of users communicating interactively at Level 6 changed from 38 to 49 % between 2007 and 2013. It means that 49 % of online consumers performed fully interactive communication using broad range of social media in 2013 (see Table 1.1 for description). It is the highest percentage of interaction at that level achieved in all five countries under consideration. At the same time, percentage of Level 1 interactivity (initial online consumers involved in social networking still remained at the lowest level of interactivity learning) decreased in that country from 36 % in 2007 to 9 % in 2013. It means that only 9 % of consumers did not advance to higher level of interactivity in 2013.

For comparison, data of the same Table 1.5 indicate that the least dynamic changes in interactivity level took place in Germany, where percentage of online consumers actively involved in social networking at several levels of interactivity has changed very little. For instance percentage of users communicating interactively at Level 6 remain almost the same 8 and 9 % between 2007 and 2013. At the same time, percentage of online consumers operating at the lowest Level 1 (initial learning) remained also almost unchanged (from 49 to 52 %). Both of these data indicate a very low dynamics of adoption of interactivity in social networking by online consumers in that country. Accuracy, representativeness and verifiability of all those data are limited due to very nature of surveys. However, they demonstrate a rather clear pattern of different learning processes taking place in respective countries. These differences are probably driven by cultural and social factors in each case.

Final Remarks

Results of study reported in this paper support two initial concepts. The first of them concerns the possibility of applying conventional models of innovation diffusion to analysis of social networking (Pelc 2013). The second, more original concept, concerns the need of new dimension in that analysis, i.e., interactivity level. In case of social networking, that dimension reflects the very nature of innovation diffusion. Introducing of interactivity level as an indicator of adoption is based on observation that the decision on innovation adoption in social networking is not a single event. To the contrary, it is a process, during which the adopter is learning experientially how to communicate interactively through social media. We suggested that six phases of this process can be distinguished. The proposed dimension of interactivity level should be explored further by collecting data and time series in a longer period of observations. Distinction of just six levels of interactivity within the adoption process of social networking may also require verification. The global aspect of social networking implementation could be a subject of further research encompassing an enlarged set of countries and their cultural, economic, social and technological capabilities. That research could reveal new ways of diffusion of social networking in global business environment and suggest new methods for accelerating learning processes with all their multidimensional consequences. The central goal of research would be to explore prospects for interactive communication through social networking among global partners in business and in all other spheres of social, economic, cultural and political activity.

References

Benkler Y (2006) The wealth of networks: how social production transforms markets and freedom. Yale University Press, New Haven and London

Brown B, Sikes J (2012) Minding your digital business: McKinsey Global Survey results. McKinsey Q. https://www.mckinseyquarterly.com/article=2975. Accessed 17 Aug 2012

Bughin J, Chui M (2010) The rise of the networked enterprise: Web 2.0 finds its payday. McKinsey Q. https://www.mckinseyquarterly.com/article=2716. Accessed 17 Aug 2012

Bughin J, Byers AH, Chui M (2011) How social technologies are extending the organization. McKinsey Q November 2011,10

Bughin J (2012) Wiring the open-source enterprise. McKinsey Q January 2012, 4

Carlsson B (2004) The digital economy: What is new and what is not? Struct Change Econ Dyn 15:245–264

De Kare-Silver M (2011) E-shock 2020: how the digital technology revolution is changing business and all our lives. Palgrave Macmillan, Basingstoke and New York

Girifalco LA (1991) Dynamics of technological change. Van Nostrand Reinhold, New York

Kaplan AM, Haenlein M (2010) Users of the world, unite! The challenges and opportunities of Social Media. Bus Horiz 53:59–68

Kiousis S (2002) Interactivity: a concept explication. New Media Soc 3:355–383

Knowledge@Wharton (2012) Barry Libert's four steps for harnessing the power of social media. http://knowledge.wharton.upenn.edu/articleID=3054

Li C, Bernoff J (2008) Groundswell: winning in a world transformed by social technologies. Harvard Business Press, Boston

Li C, Bernoff J (2011) Groundswell, expanded and revised edition: winning in a world transformed by Social technologies. Harvard Business Press, Boston

Mahajan V, Schoeman M (1977) Generalized model for the time pattern of the diffusion process. IEEE Trans Eng Manage 24(1):12–18

Pelc KI (2012) A model of global network environment for innovation. In: Proceedings of the 2012 global business & international management conference. Orlando, FL, January 15–17, 2012. ISSN 2155-1219: 118–129

Pelc KI (2013) Global dynamics of innovation in social networking. In: ASBBS proceedings 2013, 20th annual conference of the American society of business and behavioral sciences, Las Vegas Feb. 21–24, 2013, 20:340–350. http://asbbs.org/files/ASBBS2013V1/Index.htm

Rainie L, Wellman B (2012) Networked: the new social operating system. The MIT Press, Cambridge, MA

Rogers EM (2003) Diffusion of innovations, 5th edn. Free Press, New York
Sharif MN, Ramanathan K (1981) Binomial innovation diffusion models with dynamic potential adopter population. Technol Forecast Soc Chang 20(1):63–87
Skiadas C (1985) Two generalized rational models for forecasting innovation diffusion. Technol Forecast Soc Chang 27(1):39–61
Socialbakers (2013) Facebook statistics by country. http://www.socialbakers.com/. Accessed 01 Mar 2013
Tapscott D (2009) Grown up digital: how the net generation is changing our world. McGraw-Hill, New York

Chapter 2
The Architecture of Sustainability-oriented Enterprise and Civilization

Andrew Targowski

Abstract The evolution of the Classic Enterprise Information Infrastructure into Sustainability and Global Enterprise Information Infrastructure is defined. However, it is not the end of evolution. Since the SE operates within larger entities, such as Local, National, Global Information Infrastructures, these ones create the Civilization Information Infrastructure. The latter is the foundation for modern civilizations and furthermore is the foundation for the emerging Global Civilization with its repercussions for lower-level infrastructures as well as for the World Civilization. If such civilization wants to survive, it must be able to monitor and predict its sustainability in relationship with enterprises. In conclusion some recommendations will be addressed for the pathways to a sustainable future.

Keywords Enterprise information infrastructure · Enterprise systems · Business intelligence · Sustainability intelligence · Global intelligence · Civilization intelligence · Sustainability key performance indicators · Management dashboard · Civilization monitoring systems

Introduction

The purpose of this study is to define a concept how to transform a classic enterprise into sustainability and global-oriented enterprise, which will be economically vital, environmentally accountable, and socially responsible. Furthermore, such enterprise's intelligence system should be integrated with national and civilizational levels of the Monitoring and Predicting Systems. The approach to solve these issues is based on graphic modeling the mentioned systems. As a result of this study, the pathways to a sustainable future of an enterprise and civilization are offered.

A. Targowski (✉)
President Emeritus of International Society for the Comparative Study of Civilizations, Western Michigan University, Kalamazoo, MI, USA
e-mail: andrew.targowski@wmich.edu

L.W. Zacher (ed.), *Technology, Society and Sustainability*,
DOI 10.1007/978-3-319-47164-8_2

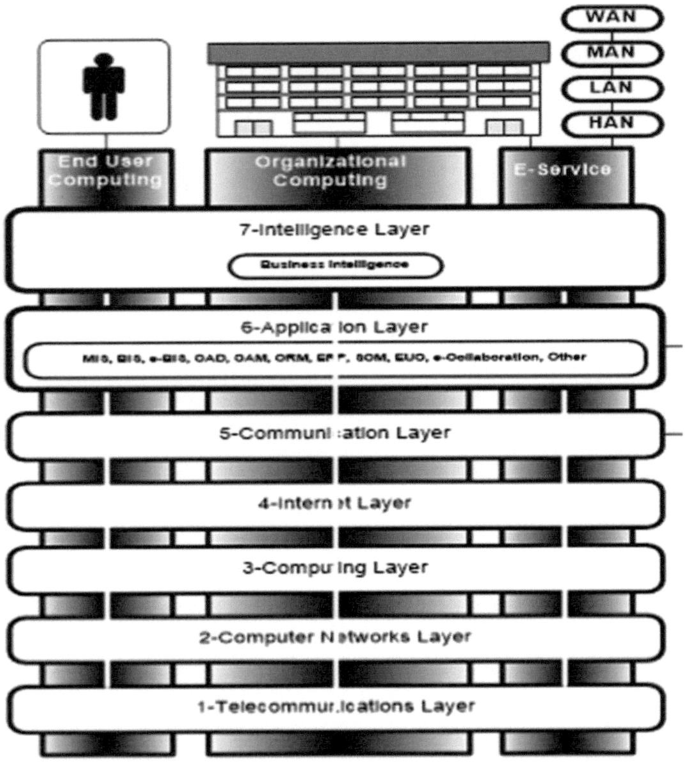

Fig. 2.1 Classic Enterprise Information Infrastructure architecture

Classic Enterprise Information Infrastructure

The Classic Enterprise Information Infrastructure (C-EII) is illustrated in Fig. 2.1. It contains 7 specialized layers, where the 6th and 7th Layer is the most visible for the end users. A set of applications is evolving along with the development of IT concepts and business needs. In the 2000, it is based upon work from the office via in building, local, metropolitan, and national networks/infrastructures (LAN, LII, MII, NII) and from a home via a home network/infrastructure (HII) for telework. The 7th Intelligence Layer is also an application layer, which specializes in managing of the whole enterprise with the support of Knowledge Management System, composed of Enterprise Data warehouse, Data Mining, Knowledge Database, and Management Dashboard, also known as business intelligence.

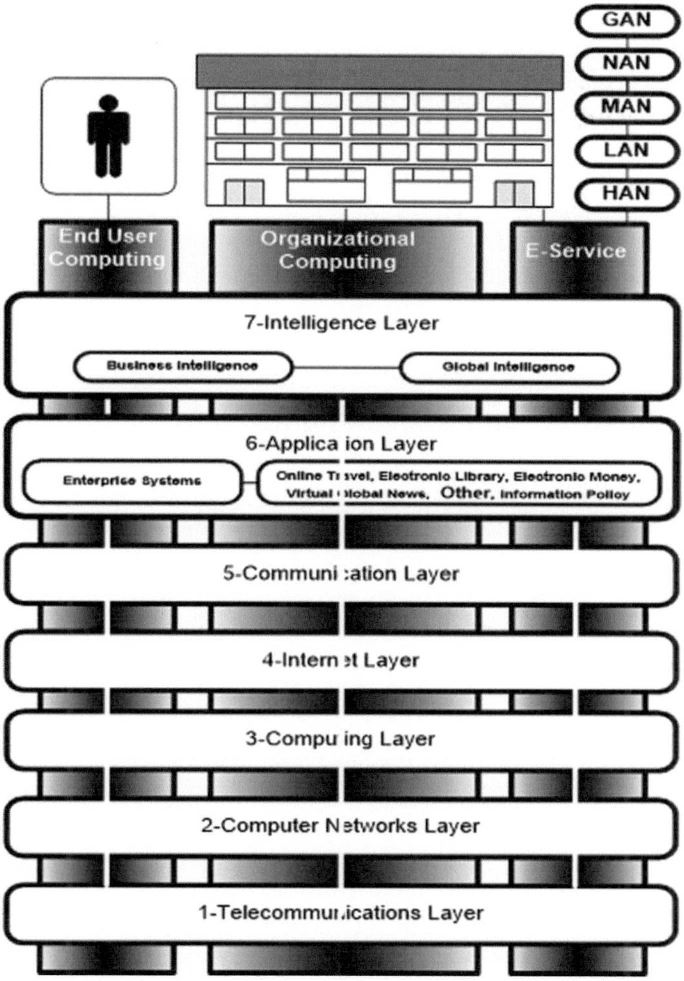

Fig. 2.2 Global Information Infrastructure architecture

Global Enterprise Information Infrastructure

The Global Enterprise Information Infrastructure (G-EII) is the extension of the C-EII through the Global networks/infrastructure, as it is illustrated in Fig. 2.2. The user-visible Layer 6 has applications more complex than applications of a classic enterprise, since they have to cover that enterprise geographic presence around the Globe and must comply with a given set of nations' legal rules. This requirement is particularly important in the Human Resources Applications, which must comply with each country's rules. In the G-EII, new applications are in demand, such as e-Collaboration, e-CAD/CAM. The e-Collaboration allows for simultaneous team or teams work in a virtual space, including virtual reality, saving on the costs of trips to be in meetings. E-CAD/CAM is particularly applicable in off-shore outsourcing of manufacturing processes. Also e-Library is a convenient application, particularly for remotely located users, without a limited access to good libraries.

The 7th Intelligence Layer contains two intelligence-oriented systems: the classic business intelligence and global intelligence. The difference between BI and GI is in their content; the latter analyzes business processes in the global platform and traces the Globalization Index to be aware of a given enterprise's global operations.

Global Civilization and Global Enterprise

The current third wave of globalization takes place on the threshold of the third millennium and is the most extensive to date. Globalization refers to a multidimensional set of social processes that create, multiply, stretch, and intensify worldwide social interdependencies and exchanges while at the same time fostering in people a growing awareness of deepening connections between the local and distant (Steger 2003). The world is shrinking fast and comes together as a Global Civilization, which shapes our lives and changes politics, work, and families.

A model of Global Civilization is shown in Fig. 2.3. This model indicates that the Global Civilization is possible because of the Internet and global transportation systems and is driven by market forces only, which in many opinions are driven by stateless corporations' greed and unregulated policies, since their strategies and operations are very difficult to regulate by international organizations. In other words,

Fig. 2.3 The Solar Model of Global Civilization in the twenty-first century

the Global Civilization is getting out of social control, and while it can be stopped, it cannot be the only solution for the World Civilization's progress and survival.

The notion of "globalization" and its universality is perceived by many as a Western value only. According to United Nations statistics, most of the people in the world do not have running water, most are illiterate, most have less than a high school education, and many are malnourished. Similarly, the "Silicon Valleys" of the "Third World," in places such as Bangalore are sensationally displayed as further evidence of this globalism, when just a few blocks away from the internet cafes and computer shops in Bangalore (which themselves occupy only a few blocks), rural India in all its traditional manifestations resumes its predominance. Thus, with the exception of the Group of Eight industrialized countries (G8)—all of which except one are Western—the majority of people on this globe do not truly and meaningfully benefit from, nor form a crucial part of, that globalization.

Through the time from 4000 B.C. to 1800 A.D., our civilization was growing 3 % per 1000 years, and the budgeting of strategic resources was not the issue (Maddison 2001). Since the Industrial Revolution in the nineteenth century, civilization was in the Accelerated Growth and in the twenty-first century it entered the Growth Trap period, when the Accelerated Growth is even intensified by the growth of population and managerial/global or even super capitalism, which looks for tremendous growth in executive benefits and replaces voters by lobbyists.

We used to think and act in terms of a local community, nation, region, even a group of nations, but now we need to take these considerations in a broader—planetary context, if we want to sustain our social life. The planet is so large for every individual but for the population is becoming smaller and smaller. In the last 200 years, the population grown from 300 million to 6.7 billion and is still growing. We have about 4.7 acres of available footprint, but we use 5.4 acres in terms of calculated resources. "We are living beyond our ecological means. The planet is shrinking, because we are running out of resources. We are using the planet with such intensity that it is unable to restore itself" (Steffen 2008: 16).

The global-oriented enterprises are mostly stateless corporations which promote the following business practices:

1. The sky is the limit in business? Really, what about depleting resources and inequality?
2. Growth-centered business? *Starbucks* around each corner?
3. Enlarge the market share? 200 new *Wal-Mart* stores every year in the USA?
4. Efficiency obsessed business? What about the environmental destruction—200,000 acres of cropland under a single manager? But smaller farms produce much more food per acre (in tons, calories, and in dollars).
5. Only business effectiveness? Only minimized cost is the most important factor? Neglecting environment and community costs.
6. Getting business moving? But where business is moving is less important, at $2.5 billion/day foreign trade deficit, and exporting debts is it the American business direction?

7. Globalization is better than localization? To satisfy stateless corporations. Perhaps the truth is vice versa.
8. If you "do not fit it is your fault, re-skill." Government advises, go to community colleges and be another craftsmen? What about university graduates?
9. We teach "information and knowledge" but what about teaching "wisdom"? Can we differentiate knowledge from wisdom?
10. We teach short-term decision making? Long-term sounds like central planning? No vision is plus?
11. Human resources of the 1960–1970s versus of the 2000th. Workers are strategic resources (past) versus disposable commodity (today).
12. Anti-Fordism, factories without workers? It is possible but necessary?
13. Social cohesion and economic forces are splitting apart, but we are irrelevant in teaching about it.
14. The "football strategy"—the leader (CEO) takes all? What about the other stakeholders?
15. From individualism to super-individualism (there is no such thing as "society," only individuals and families (M. Thatcher)?
16. Only leadership is important in *autistic* business of isolated individuals?
17. Stay at home and virtualize? Since it is possible but is it necessary?
18. An urban core giving way to an urban prairie. It is what we strive in advanced civilizations?
19. We in Western Civilization cannot compete with others; therefore, we have to accept the decline of our level of living? It is not true, since we compete with sweatshops and this is not fair competition.
20. Other.

These questionable business practices lead to the necessity of developing a sustainable and global enterprise which will (Princen 2007):

1. Mitigate the super-consumerism by promoting "cautious consuming".
2. Promote the rational principle of "sufficiency" in the context of the strategic resources depletion, since the economy cannot operate as if there is never enough and never too much. Sufficiency is contrary to modern society's dominant principle efficiency.

Sustainable and Global Enterprise Information Infrastructure

The footprint methodology determines humanity's total impact on the planet in terms of being within or exceeding the earth's biocapacity. Today, the earth has just over 15 hectares of bioproductive land per person, but the average per capita footprint is 22 hectares. It means that each person on earth (on average) is using about 6 more hectares of productive land than is available. This is known as overshoot. Since this extra land is not technically available, the figure of 6 hectares is a way to represent the fact that human demands on the environment are greater than the earth can support.

The opposite of overshoot is "sustainability"—living within the limits of the environment. Sustainability it is "progress that meets the needs of the present without compromising the ability of future generations to meet their needs" (The Brundtland 1987). In broad terms, business sustainable practices mean that the current generation's moral obligation is to ensure that future generations enjoy at least as good a quality of life as the current generation has now (Pezzey 1989). In order to do so, the current generation must apply the following strategies:

- Economic vitality
- Environmental accountability
- Social responsibility

According to Dow Jones, measuring business sustainability or corporate sustainability is a business approach that creates long-term shareholder value by embracing opportunities and managing risk deriving from economic, environmental and social developments. Several types of sustainability indexes measure a given corporation's ability to run a sustainable business. The architecture of the Sustainable and Global EII is shown in Fig. 2.4.

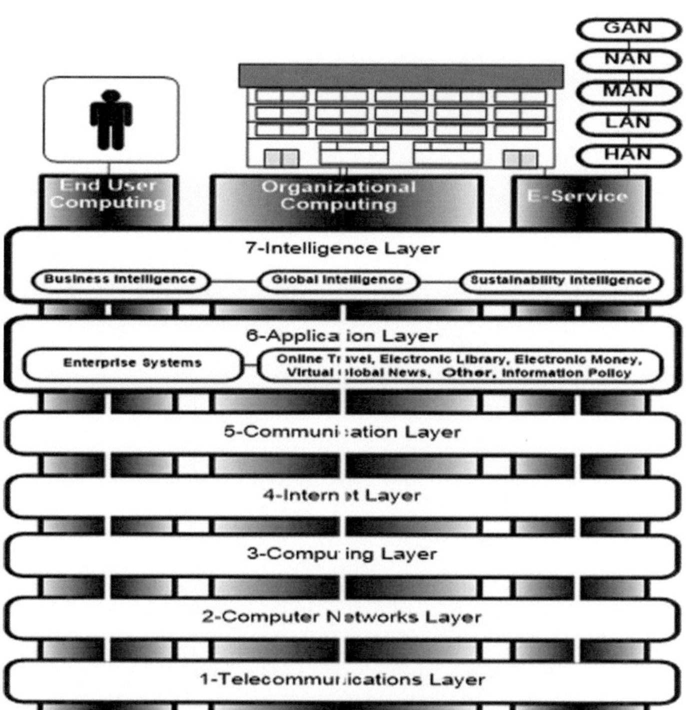

Fig. 2.4 Sustainable, Global Enterprise Information Infrastructure architecture

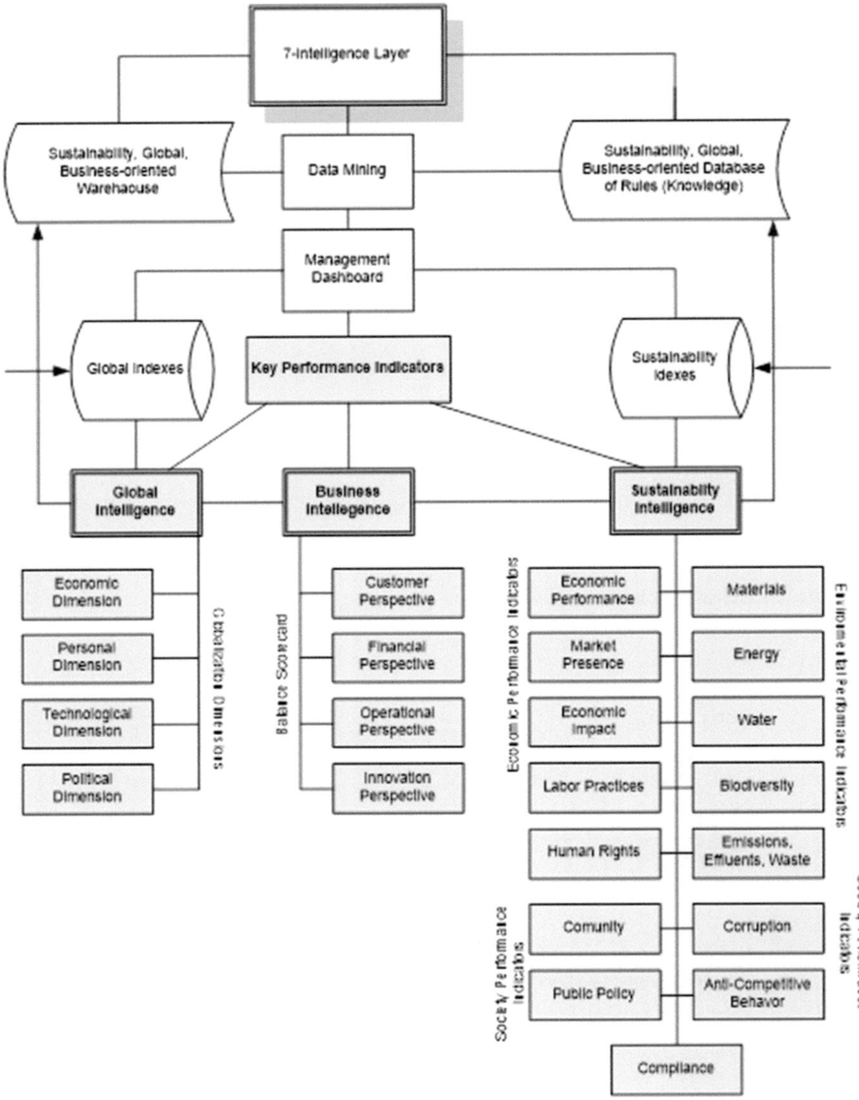

Fig. 2.5 Architecture of the 7th Intelligence Layer (list of Key performance indicators is limited to their

Figure 2.5 defines the content of the 7th Intelligence Layer. The main component of the 7th Layer is a set of key performance indicators (KPI) monitored by the systems of Business Intelligence (BI), Global Intelligence (GI), and Sustainability Intelligence (SI). The BI's Key Performance Indicators belong to the well-established Balance Scorecard.

The GI is based on key performance indicators published and updated by the A. T. Kearney (www.atKearney.com/main). The SI is based on indexes published and updated by Dow Jones and Global Reporting Initiative (GRI) in Amsterdam.

The sustainable, global enterprise is managed by a set of three kinds of KPIs. The business KPIs are applied every day, week, months, quarter, year, and so forth. The GI and SI Key Performance Indicators may be applied every quarter or every six months, at least every year. These three sets of intelligence should be coordinated into one composite Management Dashboard.

The sustainable, global enterprises operate within the civilization boundaries and either protect or destroy civilization. Therefore its place must be noticed in the civilization monitoring system, and vice versa.

Monitoring and Predicting Systems of Sustainable Global Civilization

The current level of the world affairs on the one hand is tending to deemphasize the meaning of a state for the sake of larger organizations such as EU or NAFTA; on the other hand, more and more the political and economic competition in the world is taking place at the level of a civilization. Huntington (1993) argues that the clash of civilizations characterizes a new world order after the Cold War (1945–1991). Civilization is a larger in space and time entity composed of humans (society), their culture and infrastructure. In the last 6000 years, we had about 30 civilizations, which nowadays are: Western, Eastern, Chinese, Japanese, Hindu, Islamic, Buddhist, and African, oriented by religion and emerging Global Civilization driven by business (Targowski 2009).

The future of World Civilization is bleak, since the combination of Population Bomb, Ecological Bomb, and Depletion of Strategic Resources Bomb creates the Death Triangle of Civilization which about 2050 will be very evident Targowski 2009: 404). Its first symptoms are evident nowadays, under the form of the overcrowded Planet, deforestation and land degradation, greenhouse effect, floods, drought, shrinking strategic resources, and so forth. The financial and economic crisis in 2008–2009, which was triggered mostly by the global, stateless corporations, shows the relationships among the enterprise and civilization levels must be established for the sake of the humans' survival, even in the perceived future.

The relationships among Intelligence Systems of Sustainable, Global Enterprises and a given civilization such one as the Western Civilization and the World Civilization are depicted in Fig. 2.6. The Civilization Monitoring and Predicting System (CMPS) is the set of the following components:

- Aggregated KPIs of the Enterprises (sustainability, globalization, and business),
- Population Index
- Living Planet Index
- Biophysical Index

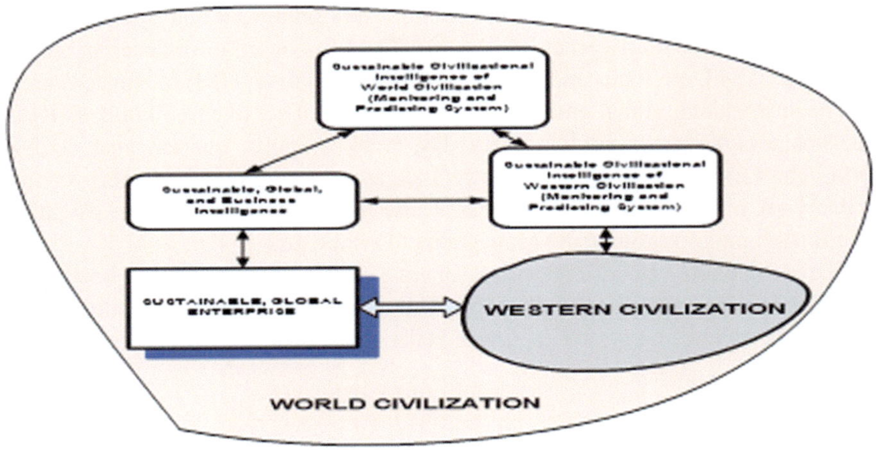

Fig. 2.6 Monitoring and redirecting systems of civilizations (an example limited to the western civilization)

- Wars and Conflicts Index
- Well-being of Nations
- Eco-efficiency Index
- Resources Index
- Societal Index
- Globalization Index
- Genuine Progress Indicator
- Other
- Aggregated Index of a given civilization
- Aggregated Index of World Civilization

There are about 20 different indexes which measure the dynamics of civilization but there is no one aggregated index which could easily monitor and predict the Planet's status and impact of human behavior and develop sustainability spirit among institutions involve in developmental activities.

Conclusion

The pathways to a sustainable future of civilization depend upon the following activities:

1. Expanding the number of sustainable, global enterprises (SGE)
2. Establishing the relationships among the SGE level and national, and civilizational levels
3. Embedding strategies of sustainability in development-oriented organizations

Fig. 2.7 The Solar Model of Sustainable, Global Civilization in the twenty-first century

4. Strengthening national and civilizational coordination, leading to:
   a. Improving livelihoods on fragile lands
   b. Transforming policies associated with the use of land, water, energy
   c. Getting the best from cities
   d. Other
5. Strengthening the institutions to solve global problems
6. Other

Figure 2.7 defines a model of the Sustainable, Global Civilization, which perhaps has chances for more rational development.

References

Brundtland Commission (1987) Our common future. In: The world commission on environment and development, Geneva. Retrieved 23 Jan 2009. http://geneva-international.org/GVA/WelcomeKit/En-vironment/chap_5.E.html

Prahald CK, Porter ME (2003) Harvard Business Review on Corporate Responsibility. Harvard Business School Press, Boston

Huntington S (1993) Clash of civilizations. Foreign Aff Summer 72(3):22–49

Laszlo Ch (2003) The sustainable company. Island Press, Washington

Maddison A (2001) The world economy, a millennial perspective. OECD, Paris

Pezzey J (1989) Economic analysis of sustainable growth and sustainable development. World Bank Environment Department, Washington. Working Paper 15

Princen Th (2007) The Logic of sufficiency. The MIT Press, Boston

Targowski A (2009) Information technology and societal development. Premier Reference Source, Harshey, PA

Bibliography

Anderson J, Ikenberry GJ, Risse Th (2008) The end of the west? Cornell University Press, Ithaca
Braudel F (1993) A history of civilization. Penguin Books, New York
Brown LR (2001) Eco-economy. W.W. Norton & Company, New York
Dernbach JC (2009) Agenda for a sustainable America. ELI Press, Washington
Friedman ThL (2008) Hot, flat, and crowded. FARRAR, STRAUTS and GIROUX, New York
Jamieson D (2008) Ethics and environment. Cambridge University Press, New York
Krugman P (2008) The return of depression economics and the crisis of 2008. W. W. Norton & Company, New York
McKibben B (2007) Deep economy. Times Books, New York
Melko M (1969) The nature of civilization. Porter Sergeant Publisher, Boston
Millennium Ecosystem Assessment Board (2005) Ecosystems and human well-being. Island Press, Washington
Sachs JD (2008) Common wealth. The Penguin Press, New York
Steingart G (2008) The war for wealth. McGraw Hill, New York
Soros G (2008) The new paradigm for financial markets, the credit crisis of 2008 and what it means. PublicAffairs, New York
Toynbee A (1995) A study of history. Barnes & Noble, New York
Turow LC (1996) The future of capitalism. William Morrow and Company, New York
World Bank (2003) Sustainable development in a dynamic world. The World Bank and Oxford Press, Washington
World Bank (2008a) Global monitoring report. The World Bank, Washington
World Bank (2008b) The growth report, strategies for sustain growth and inclusive development. The World Bank, Washington
Zuboff Sh, Maxim J (2002) The support economy. Viking, New York

Chapter 3
Technologization of Man and Marketization of His Activities and Culture of the Future

Lech W. Zacher

Abstract Technologization of man and human activities has a long history. The industrial revolution of the eighteenth century and the scientific and technological revolution of the twentieth century accelerated this process. Its present forms as the information revolution, biorevolution, and nanorevolution created a new reality. However, technology has been a subject of increasing commercialization and marketization what has a detrimental influence on culture. Dominating (also in the cyberspace) the mass pop culture is oriented mostly to entertainment and consumption of technological gadgets. Could this trend be modified or reoriented? Anyway the complex relations and interactions of technology and culture should be investigated in an interdisciplinary and systemic way which can be instrumental for positive actions and changes.

Keywords Technologization · Commercialization · Marketization · Mass pop culture · Entertainment · Consumption

Technologization of man and his activities has been progressing since the beginning of the human civilization. The industrial revolution of the eighteenth century and the scientific and technological revolution of the twentieth century have made this process even accelerated. Its present form is the information revolution and the new social space, i.e., cyberspace. Technology nowadays is a subject of increasing commercialization and marketization what has a significant influence on its "shape." Technology is a creation of men and their cultures, however, not rarely, is dominating and impacting culture, its development, directions, and products. Mutual relations of technology and culture are very complex and multidimensional. Their future is hard to predict.

L.W. Zacher (✉)
Center of Impact Assessment Studies and Forecasting, Kozminski University, Warsaw, Poland
e-mail: lzacher@kozminski.edu.pl

L.W. Zacher (ed.), *Technology, Society and Sustainability*,
DOI 10.1007/978-3-319-47164-8_3

Premises of Technologization

Technologization of man and his environment and his activities is a long process occurring at the dawn of the human civilization. Basically, it can be treated as historically evolutionary process, however, with some discontinuities which were connected with technological breakthroughs. The eighteenth century was the beginning of the industrial revolution, which introduced many novelties such as massification of production, its machinization, new types of power generation, and also new skills and competencies of people (workers, technicians, managers). The twentieth century was proclaimed—by J. D. Bernal—the era of the scientific and technological revolution. This revolution overlapped to some extent previous industrial advances. However, its basis was funded on scientification of technology, dynamic development of research and education, diffusion of scientific thinking, and technological rationality. So the end of the twentieth century and the beginning of twenty-first century was a time of domination of new directions (or "developmental channels") of this revolution—from plastics to composites and then to the information revolution and biotechnological revolution and, still in its embryonic state, the nanotechnological revolution. These revolutions created a new—rather artificial—environment for humans and their life, not to mention that they changed the man himself. The transformed environment is not only artificial but also—thanks to technology—intelligent. In spite of the fact that these revolutionary processes were performed by societies—by their scientific investigations, innovations, and their applications, by institutions and organizations, strategies, and policies, and by their behavior, attitudes, and imaginations—their results were only partially planned, desirable, expected, and positive. These revolutions generated serious risks, dangers, and negative side effects (immediate or long term). However, they proved also a great *transformative ability* practically in all domains of people's life. Moreover, they transformed the people themselves, which were their creators and agents of technological change. Radical in scale and depth, transformations occurred in the R & D sphere, education, industry and services, medicine, management, and also in institutions and organizations, in people's consciousness, their imagination, lifestyles, behavior, politics, and culture (not solely technological), as well as in media, entertainment, sport, and last but not least in the military sphere.

Technologization of people was growing and making useful—in scientific discourse—such terms as *technological man*, *technological civilization*, and *technological culture* (technopoly). In the present period of technological development, it seems that important role is played by the information revolution (which is also called the computer, microprocessor, telematic, and digital one). Its effects are expressed as informalization and networking of the human world. Moreover, this revolution generated a new—an additional—social space: cyberspace. So the space of human functioning was enlarged and became practically borderless; communication became immediate, *online*, and interactive. Cyberspace means to at least some extent a separatedness and voluntary—or not—exclusion. Any way a territory or rather space available for man (as a symbol of *homo sapiens*, not in meaning of sex or gender) and his functioning was fundamentally enlarged. So the world of people became significantly technologized and (virtually) enlarged and at the same time more complicated and complex, also very differentiated (various gaps, recently—digital divide) and diverse (that is why we use the term *civilization of diversity* as opposed to formerly expected universalistic civilization). Differentiation, variety, and diversity are expressed not only in a form of multilevel and multitrajectory development (what seems somewhat natural since the so-called take-off moment of the development of countries was historically and geographically uneven, as well as its pace and directions), but also in diversity of its positive and negative effects and impacts generated by technology. These effects were determined not only by technology per se, but rather were dependent on a *cultural ability* to create technology and use it in practice.

Globalization is a significant dimension of development of technology and its overwhelming influence. Globalization as a visible process begun from Columbus and attained its climax at present—thanks to new means of transportation (ships, airplanes, transcontinental trains, etc.) and telecommunications and media (telephones, radio and TV, cables, satellites, the Internet). Global mediatization and the global auditorium enable a hegemonization of cultural influence and imposition of life patterns (Westernization, Americanization). However, such a situation can change in the multi- or post-polar world, in which the so-called new powers (as China or India) find their proper place. So hegemonism can be weaker, while multiculturalism becomes meaningful and stronger. Moreover, the "inner structure" of globalization may be heavily impacted by racial and religious changes, not to mention the growing pressure from the part of non-Western immigration. Do some pose the question whether the "Christian, Western culture of the white man" will be dislodged by the culture of Islam, by Asiatic and African natives, and whether "clashes of civilizations" (to recall Huntington) will be intensified and more dangerous? Do the new powers become not only the new poles of growth, but also new centers of advanced technologies?

Overcoming the Misleading Generalizations

Generalizations concerning technology happen to be misleading. Technology is globalized now, but it does not mean that all countries, societies, or companies have the same technology, on the same level, equally modern and efficient and that they all can develop a *high-tech economy* based on *cutting-edge* technologies. Access and performance have many limitations. Many factors and favorable conditions are necessary—for example, aspirations, competences, technological culture, skilled workforce, markets, and of course capital. It was not commonly understood by politicians, aid organizations, Western companies, and even scientists. E. F. Schumacher a few decades ago had tried to change such thinking and policies, introducing such terms and concepts as *intermediate technology* and *appropriate technology*. Recently, similar ideas were stimulated by the movement called Critical Technical Practice, often in the context of such programs as Information Communication Technologies for Development (ICT4D) and One Laptop Per Child (OLPC). However, the mainstream thinking on technology, its design, transfer, and applications was determined through decades by the world leaders and their scientific communities. So, a high-tech narrative is now predominant, technological progress is considered a goal, technology becomes an idol, and societies' goal becomes to achieve high productivity, military power, and regional hegemony. Such ideology drives technology transfer and the arms race even in poor societies at the expense of other goals.

So the world scientific and technological leaders (often based on network cooperation and brain drain) create novelties that require usually expensive and long-term research, while the rest of the world should modernize itself as much as possible using such means as foreign capital inflow (FDI, joint ventures), technology import, imitation of development and consumption patterns, technical cooperation, and so on. Nowadays, modernization is costly, and economic, organizational, and personnel requirements are difficult to fulfill. These who are delayed (backward) have to rely on transfer of technology, knowledge, and skills. Sometimes the latecomer's chance emerges, and the latecomer is able to "jump over" some levels of technology (China tried to do it, but the first time failed; at present, seems successful). However, the price is high and its negative side effect is *technological dualization* of economy and people's lives. Anyway modernization narrative aims at de facto all countries—less developed, transitional economies (for example, new members of EU), and even leading countries where there are usually some neglected areas or types of economic activity that are delayed or obsolete. Technical modernization has, of course, its political and cultural dimensions.

Co-evolution of technological systems and social systems—actually the latter seems predominant—determines the trend of change. However, in the very differentiated world, there is in fact multilevel and multitrajectory development of technology and its applications, effects, and impacts as well. It is so in spite of all systems–network linkages in research, innovations, strategies and policies, and implementation activities. Diversity remains since conditions and abilities vary. Science and research are *drivers* of technology development. Creativity is oriented not really on human needs (needs and desires are now created, often artificially by marketing, fashion, ads), but on generation of demand, on facilitation of human life, on cumulating gadgets, and on making customers dependent. So such a creativity strongly connected with the market enables meeting people's real and artificial needs and obtaining profit at the same time (quite often government takes risk, if an expected result is not certain; there are also government orders, tax reductions, etc.). But culture—broadly understood—remains a condition, limitation, also a stimulus and favorable context of the co-evolution of technological and social systems and of creativity.

Technology and Culture

Culture—in necessary feedback with technology—also is being technologized and industrialized up to the stage of the so-called technopoly (to recall Postman), which means technology domination over culture. Quite frequent theoretical division between civilization as material potential of humanity and culture as immaterial (symbolic, spiritual) looses its explanatory power. Technology, various technologies create new possibilities for culture, they determine significantly its directions, its "products" and its social assimilation, its emotional perception, not to mention diffusion, possibilities of reception (not possible without technical devices—such as a radio and TV set, video, mp3, movie apparatus, home movie). The creation of various "cultural products" is now not possible without photograph and film cameras, recording devices, light systems, computers, and also studios. Taking into consideration the so-called cyber-culture shows that border between technology and culture—in many of its dimensions—is more and more fuzzy. In the long-term perspectives of trans- and post-humanism, we can expect the emergence of a new entirely technical culture. Such culture created and performed by machines, computers, and culture could be then called a *post-culture*, a form possible thanks to historical trends (technologization, virtualization) and to new practices of artificial intelligence, automata, man-machine systems, robots, and cyborgs that are quite imaginable.

For the time being, culture is significantly supported—in its expression and access—by technology. Thanks to media, especially electronic, the contemporary "products" of culture reach the global auditorium. Audio and video productions and their diffusion would not be possible at all without technology. Present fascination with pictorial forms of culture and communication ("pictorial civilization") drives transformations of *homo sapiens* to *homo videns* (to use G. Sartori's term—Sartori 1999). Pictorial bias is responsible for underestimation of the "audio element." Therefore, sounds of music attack us from everywhere—in hotels, in railway stations, in the supermarket, in the car and airplane, and while walking (young people have earphones in their ears almost permanently). Music is present in film productions, theater performances, and operas and music halls, not to mention street musicians and noisy neighbors listening to music or dancing. Music recording, CDs, stereo systems, and music players constitute a large, profitable industry. Music piracy and bootleg music CDs are now technically easy. Anyway sound is still fundamental, and technology makes it happen—its generating (i.e., electronic music) and amplifying (amplifiers are at present a must for pop music stars performing for very big audiences as in the case of Michel Jarre or Elton John who use complicated electronic devices and lasers). Live rock or pop concerts nowadays have audiences of a million or more listeners. Such a scale was never possible before the ICT revolution. On the Internet, sound and picture are intertwined and complementary. Such pop stars as Lady Gaga and Justin Bieber have more than a billion views on YouTube.

Man is multisensoric; however, a sense of sight seems the most important. Probably that is why technology development is going in the direction of screen, visualization, pictoriality, and iconic presentation. *Homo videns* likes visual expression and spectacles. It was noticed by G. Debord when he wrote on "society of spectacle," and by E. Goffman when he used "theatre of life" as an important metaphor useful to describe social behavior. Immense mass spectacles are more frequent now—the bigger the spectacle, the bigger the profit. They are transmitted by media additionally. In some sense, tournaments of computer games or other Internet events can be treated as mass events. Contemporary culture is more and more close to entertainment and becomes mass and popular in its nature. It is evident that such characteristics as feedback drive technologies and profitable business practices are useful and functional to this end. That is why we talk about business of culture and culture industries as the music industry, media industry, audiovisual industry, publishing industry, computer games industry, industry of mass events as festivals, museum exhibitions, entertainment parks, etc. These industries—important for youth, for digital natives—function both in real life and virtual life. Hypertext, multimedia, and hyperreality are new technical means and new spaces generated by ICTs.

Mass production, inexpensiveness, and access to culture (and entertainment often connected with it) construct its specific "structure." In the past culture, in particular the so-called high culture, its creation and consumption were mostly a domain of elites (aristocratic, intelligentsia, of persons educated and having their own generational cultural heritage). At present, the situation is different—technologization, dissemination among the masses, and price moderation—all due to market competition and the dominant development trajectory toward *market society* (which profile is increasingly consumption—entertainment) cause upward movement of not culturally refined vast masses (or "the people masses" as it was called sometimes). So the most accessible and common type of culture is a primitive culture or plebeian pop culture, which is increasingly commercialized and light and often amateurish. Such a type of culture is now in demand on a large scale. It does not require any particular theoretical knowledge, competences, refined taste, and skills of wise selection of offers of cultural industries and media. It does not need any generational cultural heritage. Cultural resources are treated as commodities in the world supermarket. What is sufficient is emotions, feelings, pleasure, and to be a fan (the last attitude is the main expression on Twitter and Facebook with the ability to "Like" things in that social realm). Cyberspace is a paradise for amateurs who can go global. Cultural capital presently changes its structure with the help of technology. Some rather simple—for youth, in particular—skills and competencies are needed for using audio and video systems, home movie, computers, tablets, various music players, iPads, and iPhones. So technological culture especially in its practical, operational form becomes a significant part of the cultural capital. Of course, growing technological culture is a positive asset; it makes out of us a technological man.

Cyberspace being a new social space is also a space for creation, dissemination, and consumption of culture. The consumption can transform itself into a participation, not only amateurish. However, in this space mass culture, primitive and simplistic culture seems to be the most expansionistic and predominant. In the social portals are really the masses (in Facebook with more than 1 billion participants), so almost *ex definitione* it is not an aggregation of highly educated and sophisticated people being cultural elite. In not so advanced societies (e.g., in Poland), there are portals gathering "colleagues from elementary school" (*Our class*) or from prison (*Our cell*). All these show that the Internet culture in great part is a rather primitive pop culture that transforms itself into entertainment (so it can be called an entertainment culture); it is a culture of fan attitude (fan pages) and of permanently ongoing plebiscite ("I like it"). On the other side, it is market culture, culture of supply and demand, and culture of customer, client, and provider. Technology is able to provide such culture and to co-shape it. Internet groups or network tribes (they are not societies or communities in the traditional sociological understanding though they are often called so) are *pop-market human aggregates*. However, they have other significant characteristics, not rarely contradictory. There are, for example, transparency, openness, freedom (no censorship), transborderness, and even questioning property rights. Postulates are also formulated to treat the Internet as a *global public good* and an access to it as a *human right* (demands of Pirates Parties are even more extreme). Immanent for cyberspace activities is de-hierarchization and equal chances for all connected—everybody can be a broadcaster or film maker (for YouTube) or Wikipedia editor or network artist. There are no limitations for self-expression and self-promotion (in bloggers' cases) and for communication and interactions. Moreover, the cyberspace makes it possible not only to communicate personally and interact but also to manifest political opinions, to protest on a large, even international scale, to exert pressure on politicians, business, media, to participate in public debates, and last but not least to perform participatory democracy having a technical possibility of permanent referenda online.

ICTs' Impacts

The aforementioned technology-driven possibilities that are still anew will shape the culture broadly understood, not really symbolic culture or art per se, but culture of communication, behavior, political culture, culture of co-operation, and the like. It is difficult to predict how these various dimensions of culture will look in the future. Will this new neo-culture dominate the old culture based on other, pre-electronic technologies? Will there be people excluded from it? Nowadays, even the uneducated and poor can—at least to some extent—consume the old culture. In this new phase called *informationalization* (to recall Castells term)—besides netocratic *Weltanschauung* and netocratic ethic—will some new cultural hierarchies, some opposition groups, some virtual opponents, and revolutionaries emerge? And, if so, will they influence the nature and characteristics of the future culture? It is difficult to predict how strong the pressure will be of new technologies creating new possibilities, *ergo* new applications. New technology may be ahead of culture ascribed to the older technological generations. Development of technology practically has no limits (except for funding and sometimes cognitive barriers, which is normal); it is semiautonomous (however, it is not fully beyond the existing culture) and notoriously is ahead of culture (according to W. Ogburn's principle of the technology—society relations). It is so because culture is as a rule less dynamic, more conditioned by the past and by previous values, and also institutions. It is not only creative but also adaptive. So it is possibly a continuation of the present revolutionary type of technological change. Thus, it seems probable that technology can dominate entirely over culture, which will be at the same time a premise for its further technologization and transition toward trans- and post-humanism.

Emotions, feelings, visuality, compulsive communication, being always *online*—all these are provided or facilitated, even excessively, by ICTs. This excess of possibilities, information, contacts, reactions, expectations, and options seems to be a problem. Additionally, the future will be connected with a new culture of creativity of one's own individual virtual worlds (thanks to the VR technology that will be more and more developed, cheap and accessible). This culture may be the culture of virtual escape from reality. Will the new world will resemble *Second Life* with avatars and perhaps advanced robots? Under such conditions, the traditional meaning and discourse on culture may be totally inadequate. This challenge will be faced—if at all—by the future generations of information society (i.e., by fully digital natives, *born digital*). Will they be totally limited by technological imagination or will they try to transgress it? Culture seems to be less and less a kind of "steering mission" and "producer of history" (by the way, it is doubtful if ever it played such a role effectively). Culture stops to be a way of maintaining some sort of equilibrium and common world building. Under the conditions symbolized by such terms as *civilization of diversity*, *globalization(s)*, *network individualization,* and *privatization of the future,* culture becomes more in concert with technological rationality, technological opportunities, and tempting offers—promoted via media and consumed via market. Such a situation yields a *liquidity*. According to Bauman, this liquidity of reality is imminent and implied by self-intensifying, compulsive, and obsessive modernization (Bauman 2000). And such modernization is technology based and is not reflexive. The more new technology, the bigger demands for it and bigger its influence, also on culture. Thus, the pressure of technology will not be weaker in the future as well.

Changing Roles and Meanings

The academic-humanistic perception of the role and meaning of culture has often been a burden of wishful thinking and often underestimates or overlooks influence and impacts of technology. Predominance of technology seems to be indispensable in the face of environmental and demographic challenges and is fortified by the cognitive mission of technology, which is connected with research and experiments, not to mention its crucial utilitarian function for the military, industry, media, and households. So technology has its powerful agents everywhere. In the space of technology development, there are various actors, stakeholders, and users, who are also detrimentally impacted and excluded. Distribution of "technological fruits" as well as their costs and risks is not equal in societies and through generations. In the practice of development, the technological and economic rationalities are approaching each other. The so-called technoscience is significantly practically profiled. Innovations are commercialized and marketed. A part of costs of their development and risk is often covered by government funds and agencies (e.g., the Advanced Technology Program in the USA). So it is difficult to win a battle against technology or to get an advantage over it. Simply technology pays—in spite of costs, risks, and catastrophes. A return to a stage of primitive ("barbaric") technology seems rather improbable.

The situation of culture is fundamentally different. Culture is a "high value," not a practical need necessary to survive. For this reason, its primitivization seems quite possible (and such a process is de facto ongoing). This process is multicausal. In the past, the high culture was represented by higher classes—aristocracy, noble class, and intelligentsia. They inherited traditions, cultural needs, and "cultural stuff" (books, paintings, pianos, etc.) and passed them to next generations. Mass dissemination of culture, its "democratization" means downward direction to rather primitive tastes and amateurship (circus culture, street culture, tawdry culture, break dance, low-level literature). Revolution of the masses, which were often of countryside origin and uneducated, has broken or radically transformed and downsized the process of historical cultural accumulation and of respecting and inheriting high culture and cultural competences as well. This can be countered by the argument concerned with the emancipatory mission of technology. Technology is a key of development at large. However, nowadays culture, especially high in spite of fantastic technological possibilities (as hi-fi recording, global broadcasting, etc.), is not a *sine qua non condition* of anything—business profit, professional career, or social relations. In the present civilization called "civilization of impatience and haste," the dominating (in media) and overwhelming (the youth) pop culture is basically a *culture of events*. Privatized human life, prevalent strive for consumption and entertainment, and economic instability (mass unemployment or its danger in many countries) all of these are reasons and conditions in which it is a daily life that predominantly shapes culture, not as previously when culture had a superior position. Availability of technological devices and access to the Internet auditorium lead not to mass creativity and artistic values, but to spreading rather simplistic amateurship and rubbish-type culture. In a such culture, the most important thing is a number of fans, not professional evaluations and criticism. Cultural style of life is determined to a significant extent by the Internet gurus (e.g., Zuckerberg, Wales, Assange) and popular celebrities. Of course, mass popularization of culture, its possible interactivity, its immediate messages, and opportunity for creativity are undoubtedly positive phenomena.

However, it is not sure whether they will outweigh mediocracy in the future culture.

There is a growing evidence that culture terminates step by step its role as a driver of development of societies that were shaping them fully in the past. Now culture becomes rather a *context* that has much less formative influence and that is subordinated to various feedbacks, for example in relation with technology. In market societies, culture defends itself by raising its quality and allowing commercialization at the same time—however, with support from enlightened governments and citizens (gifts, foundations, grants, and state budget financing). In culturally advanced societies that at the same time have a numerous population, the high culture has a good chance in their big markets. But is this true for the whole world and for the whole of mankind? There are some hopes in globalization, and also in the possibility of transfer and imitation of cultural leaders' patterns. So hopes are located de facto in technology, which enables this but not automatically, not equally, and not immediately.

Clash of generations—traditional vs. digital—is also manifested in culture, cultural activities, and cultural values. The digital world promotes a different approach to culture—culture should be free, accessible for all, as well as knowledge and information; the Internet should be a public good, and access to it should be treated as a human right. These are rather revolutionary postulates. But still we live in the hybrid world (real and virtual). That is why it is difficult to predict the future of culture (its all forms—symbolic, material, technological, spiritual, etc.) and its interactions with technology.

Modeling the Future Development

Both culture and technology are domains of human behavior and activities, and of creativity and risks, and they are sources of profits and possible negative side effects. These domains, at least until now, connect or even integrate with man, with the *homo sapiens* species; they are—interactively acting—driving forces of its evolution. Mankind develops thanks to technology (this means a co-evolution) and is in a dialectical relation with the surrounding environment. This environment is partly—still—natural (and excessively exploited and devastated by technology, production, and other human activities) and partly man-made, artificial, and built mainly with a help of technology. This *eco-artificial environment* influences and limits man and his activities, but also generates the new developmental occasions. In the evolution of man and his world of spirit and of practice, culture plays a significant role in all aspects and dimensions, and at all levels. Culture impacts and shapes people (in the generational sequence in particular) and also co-shapes technology and its practical applications, and the environment. But it is also transformed itself—for example, it is increasingly technologized, electronized, virtualized, disseminated, and used. A systems approach and socio-cybernetics can be instrumental in the analysis of such systems as technology, culture, environment, and society. These systems can be complementarily investigated as networks (Web theory is useful). Both aforementioned approaches together unmask their complexity and multiplicity of relations, which are quite difficult to recognize and to analyze and to present in schematic figures. To make a figure readable, a far-reaching schematization is necessary. Figure 3.1 (see Annex) is an attempt of presenting an ideal model of influences and interactions in the evolutionary processes of the human world of life (in regard to technology in particular).

The model can be presented at least in three versions:

- as a model of diffusion (one way and of different intensity—in time and space),
- as a model of interactive impacts (multidirectional, both spontaneous and steered with differentiated strength and effectiveness),
- as a model of one-directional domination (i.e., asymmetry) in which technology, culture, and environment are dominators (single or in coalition)—that had its place in history (e.g., in the form of total dependence on nature a long time ago or nowadays while we depend on technopoly and what can happen as well in the future).

While estimating the great role of culture in driving and humanizing the technoeconomy development, it seems rational to demand—as in the case of the idea of sustainable development—also a *sustainability of culture*. It can be pursued—in a similar way—in the form of concepts, strategies, and policies derived from normative and empirical premises and convictions, maintaining that perseverance of culture and its sustainability is a *sine qua non* condition of perseverance of humanity.

Cultural sustainability is important in itself, but it is also necessary for a whole sustainability of society. It may play a special guiding role in societal choices—for example, in consumption, environment protection, relation between men–animals, alternative energy, eco-innovations and eco-design, and ecological security. Social culture and cultural values are essential for reaching and maintaining a trajectory of sustainable development. In the case of failure to accomplish this purpose, the culture may degenerate and become a culture of fear and opposition (to technology and growth). Cultural change is always lagged, always behind technological change (Ogburn's principle). Sustainability requires more prospective orientation of cultural change and focusing on risky aspects of technological advances. For this purpose, cultural sustainability should rely on humanistic values since the future of humanity is at stake.

In the long perspective (50–100 years and more), transformations of man and his world are able to create evolutionarily a new reality, more technological in substance than biological. A world of self-replicating robots and cyborgs and self-organizing systems of machines, the world strongly virtualized will certainly produce new (evolutionary?) mechanisms and new (technological?) principles of game for the coming Transhuman Era and then for the Post-human Era. This can revert the present proportions of the biological and the technological and lead to "the world without us." *Nota bene* the cosmologists predict in a very long future the end of the human species. The aforementioned trends and events may radically accelerate this end. The beginnings of transhuman culture are already visible. However, it seems extremely hard to imagine a culture of a post-human man or a culture of machines. Perhaps technology, per se, and its abilities of simulation and creation could greatly help.

Annex: Approaches and Conceptualizations

To better understand a complexity of all interrelationships, conditions, and reciprocal influences of technology and culture—in the context of development of societies and human individuals and of various processes reconfiguring them—several schemes together with interpretative description are presented below.

Figure 3.1 illustrates graphically components (sectors, spheres) of the human world of life. It shows its complexity and multidimensionality and existing types of impacts (connected with market, politics, strategies, and behavior).

Figure 3.2 illustrates mechanisms and processes of development. Systems, strategies, policies, and behavior shape and modify development. Processes of development should be investigated in the context of relations of knowledge, market, and culture what allows for etiquetting societies according to their characteristics and dominated features (Fig. 3.3). Societies of the world are very diverse as to these characteristics and to their levels and intensity. Generalizations are hard, though some trends seem to be quite clear.

In the past decades, there were many efforts to give new names to changing societies. Social change was predominantly shaped or driven by technology,

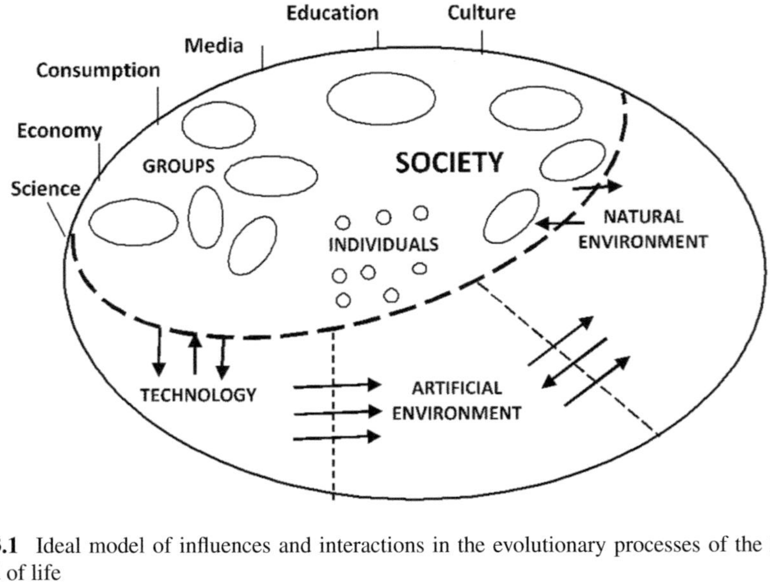

Fig. 3.1 Ideal model of influences and interactions in the evolutionary processes of the human world of life

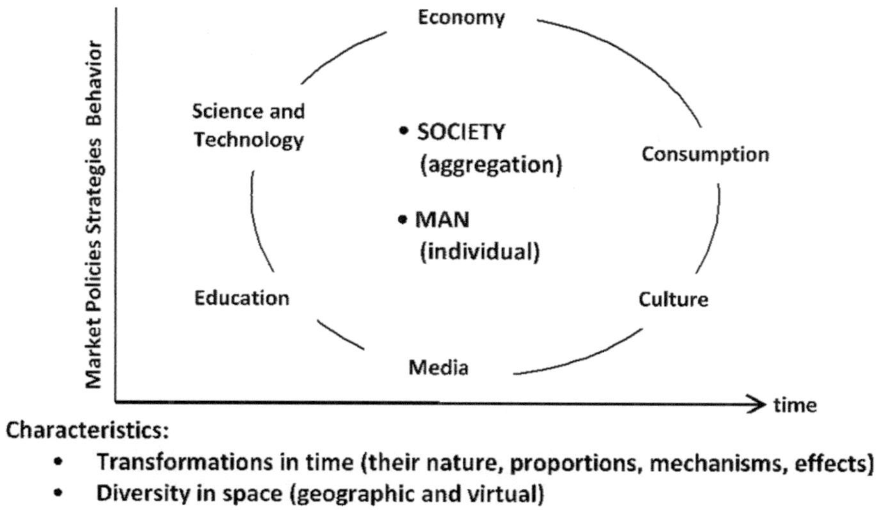

Fig. 3.2 Components (sectors, spheres) of the human world of life

market, and culture. These domains determine a fabric of social change in general (Fig. 3.4).

In the real cases, the etiquettes could overlap. Usually a society could be characterized by a set (a mix) of etiquettes. Social change was a change in such a set regarding its elements, their meanings, proportions, relations, and importance. Unsuccessful change was often connected with existing gaps of various sorts (see Fig. 3.5). Recognizing gaps enables a better understanding of development and its barriers (diverse through societies and regions).

These gaps can be assigned to countries, societies, and organizations. There are, of course, many more gaps and divides as information, organizational, management, digital. They are often interconnected and mutually stimulating each other. Many gaps are inherited and difficult to overcome. Identification and recognition of gaps and their various constellations are important not only theoretically but also for strategies and policies elaboration and their implementation. Evidently, all details of gap classification are subject of interpretation.

The next scheme (Fig. 3.6) is, in fact, the listing of processes characterizing the present changes in societies (as a rule making possible by technology and stimulated by technology) and challenges (not to confuse them with goals), challenges connected both with a global survival and development and with issues associated with market, politics, and multiculturalism. The scheme below presents a complex picture of sometimes really negative trends in societies (differentiated throughout space and time). It is designed also to define and explain various uncertainties, risks, and dangers. They should be examined in depth and should constitute a basis for new proper policies and strategies, and for research and societal education as well.

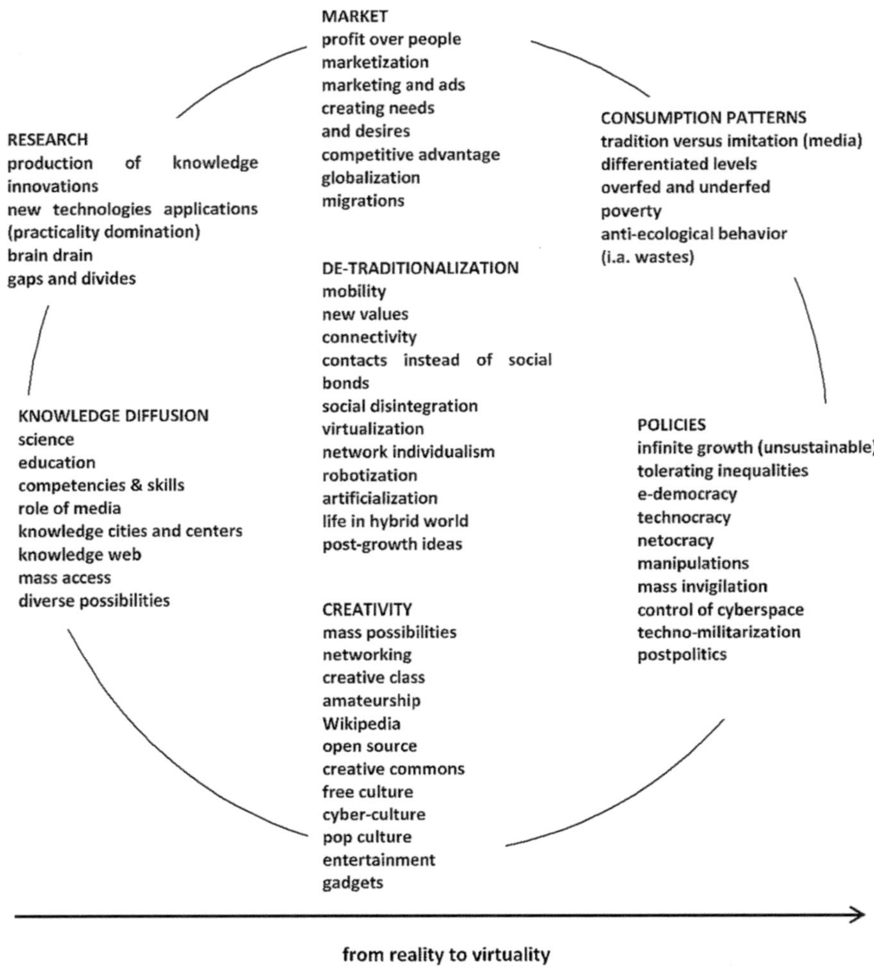

Fig. 3.3 Mechanisms and development processes. Selected systems, strategies, policies, and behavior in the context of development, knowledge, and culture

Global processes and challenges are located in structures and networks (or their "mixes"), so are their mechanisms and developmental processes and phenomena, as well as their effects and consequences.

Figure 3.7 (below) illustrates the failure of the concepts (and predictions)—often associated with theory of modernization based on expectation of common imitation of development patterns (technological, economic, political, cultural, and so forth). Such imitation had to concern a consumption. A wide distance between leading countries of the world and the rest does not allow for really common and effective imitation and—what's more important—achievement of similar results. It is so in spite of the ongoing processes of integration, globalization, technology transfer, migrations and international networking, and co-operation. Despite

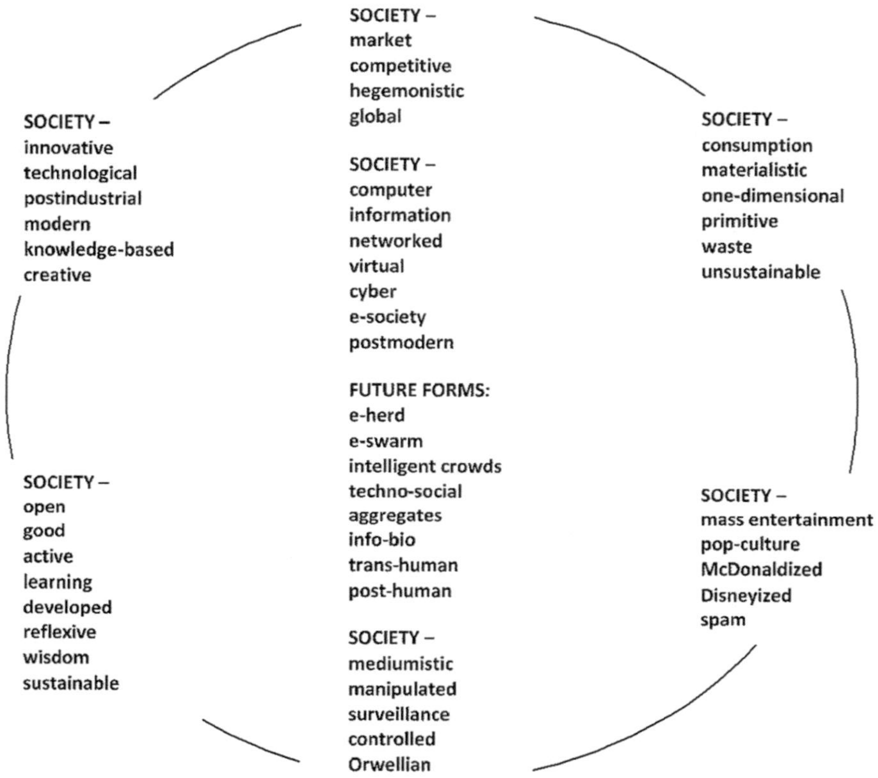

Fig. 3.4 Etiquettes for societies (knowledge–market–culture)

similarities, the predominant feature seems to be a diversity—not of declared goals and policies but of real actions and their effects. The reason of it is on one side a large differentiation of national developmental potentials and on the other side not equal cultural ability to creation and application of technology in various domains of human life. This should be emphasized in policies for development and international aid and cooperation.

In the subsequent stages of human civilization and its development, significantly co-shaped, and stimulated by technology (discoveries, inventions, innovations, their applications, and diffusion), various "profiles" of cultural challenges associated with these stages emerged. Figure 3.7 is a proposal of ordering the existing types of these challenges in relation to the stages of development (from prehistory to Post-human Era).

This classification of the types of culture prevailing in the subsequent stages of development of human civilization is arbitrary. Needless to say that these stages and the corresponding cultures sometimes overlapped or had "fuzzy" borders. In some cases, the transition from one to other was evolutionary, and in other cases revolutionary; sometimes changes were inductive, and sometimes they were indigenous (Fig. 3.8).

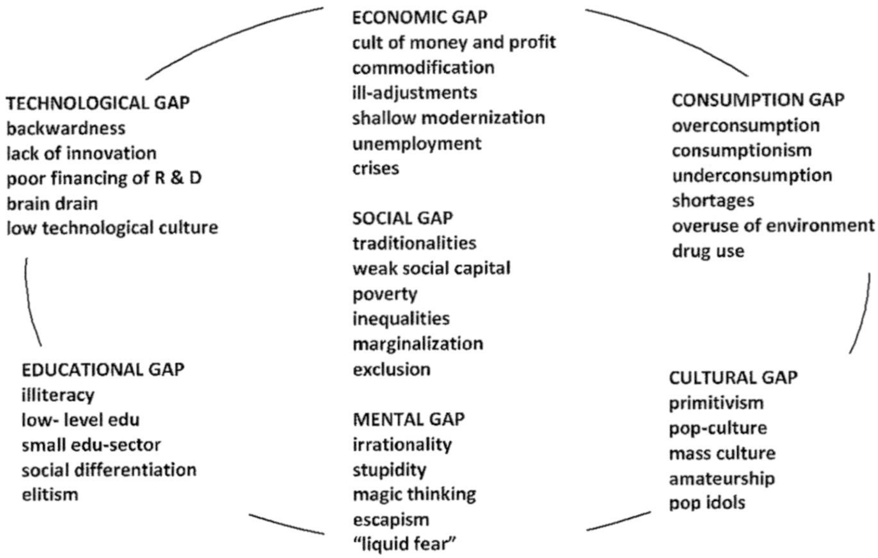

Fig. 3.5 Components and characteristics of development gaps

General processes: / **Areas and impacts**

technologization .... artificialization
globalization .... reconfigurations
McDonaldization .... uniformization
Disneyization .... entertainment
mass reach .... "democratization", pop-culture, cult of amateur
infantilization .... pop-fashion entertainment
barbarization .... primitivism and vulgarity
exhibitionism........ permissiveness, Internet behaviors (stalking, bullying, etc.)
idiotization .... advertisement, media sensationalization, tabloids, esoterics

General global challenges (not goals)

limits to growth, uncertainties, risk and dangers, survival, development and progress } demography (overpopulation of Earth); resources (exhaustion, increasing costs); environmental devastation (worsening quality); crises (ecological, political, social); conflicts and wars (over resources and hegemony); catastrophes (natural, technological); poverty and its impacts (health, social, political)

- competitiveness – dependence – hegemony
- marginalization – exclusion
- diversity and hybridization of the world:
  poor – rich
  low tech – high tech
  less sustainable – more sustainable
- global mobility (migrations)
- ageing of industrial (advanced) and democratic societies
- change of the national and world proportions and strength of races, religions, cultures, etc.

Fig. 3.6 General processes and global challenges

Civilization of diversity
multitrajectory and
multilevel development
of societies and of technology and its
influences
effects (also detrimental side effects)
interactions

Role of technology – cultural ability to

transformative	creation
networking	adaptation
globalizing	effective use
transfer of knowledge	coping with risks and negative side effects
and technologies	achieving synergy
international cooperation	modification

Fig. 3.7 Failure of universalism and its ideology

- "struggle over fire" (prehistoric stage) – culture of survival
- "construction" (infrastructure, industry, struggling with nature) – industrial culture
- building artificial environment (on large scale) – culture of denaturalization continued
- scientific and technological revolution (research paradigms, nuclear energy; space exploration, ICT, biotechnology, nanotechnology) – scientific and technological culture, culture of technologization and artificialization, cyber-culture emergence
- uncertainties, risks, dangers and undesirable detrimental side effects of technology and development – culture of risk evaluation, and culture of consequentionalism
- new struggle for survival (crises, demography, finances, economy, environment, politics, conflicts and wars, mafia, terror) – culture of global survival
- emerging new concept, strategies and trajectory of sustainability – culture of sustainability
- future-oriented civilization – culture of transition and change (trans- and posthumanism in perspective)
- "world without us" – posthuman culture, machinekind culture (?)

Fig. 3.8 Stages of civilization development and technology—types of culture

Essential features of all de facto processes of development are uncertainties and feedbacks (both positive and negative). So analysis of the change in human setting (or generally speaking—of civilizational context) in connection with change in man (as individual) and in society (or various forms of human collectivities) is crucial for an understanding of the past and also of the present world. Individual and societal transformations are both "products" and "generators" of broadly understood culture. However, such analysis does not mean the recognition of the future, which is an effect ("resultant," "sum") of all types of changes. The future, the long term in particular, seems mostly unknown and open ended. All the more, we should think about it in a scientific way (using forecasts, simulations, scenarios, strategy elaboration, long-term policies, planning, computer modeling, etc.), drawing our attention to the instrumental and determining role of technology and its relations with culture.

Figure 3.9 is a graphic presentation of such issues as feedbacks and uncertainties in development.

Analysis of feedbacks is typical for a systems approach. Uncertainties are investigated with risk analysis methods. Technological and social changes can be also analyzed with the aid of Web theory. However, for this reason the types of networks should be distinguished. The exemplary classification can be as follows.

Technocultural changes always have—and will have in the future—their proponents and opponents, both in theory and in practice. However, it goes without saying that the most important are agents of change or in other words carriers of change. Various subjects designed to conduct change or to slow it down can be engaged. This is illustrated in Fig. 3.10.

In conclusion, it is worthwhile to return to the idea of social assessment of technology (TA in short), from which its advanced concepts, methodologies, and procedures enable also a recognition of various relations, influences, interactions, and impacts—both advantageous and disadvantageous for people and environment (Fig. 3.11). They make it possible to formulate essential research questions addressing at the same time political and cultural issues and controversies. Some exemplary research questions politically and culturally bounded are presented in the table below Fig. (3.12).

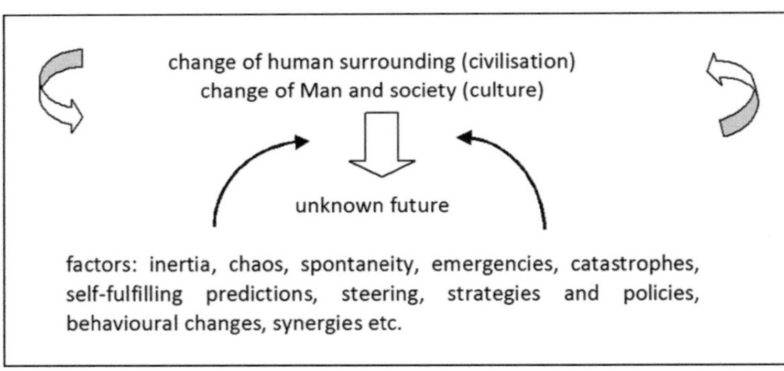

Fig. 3.9 Feedbacks and uncertainties in development process

Fig. 3.10 Exemplary enumeration of traditional and teleinformatic networks

Exemplary domains:

Exemplary network types: bonds – societal
linkages – economic
contacts – human
influences – cultural
impacts – technological
power - political
manipulations – mediumistic

Subjects in question:
various structures, networks, organizations, institutions, groups, individuals
- who wants to maintain *status quo* (as good)
  because of inertia and tradition
  vested interests
  business as usual attitude
- who resists changes (as risky and detrimental) in the name of ideology, values (e.g., greens, anarchists, anti-globalists, anti-system movements etc.) and to promote their political position
- who proposes and creates alternatives (to ameliorate) (e.g., innovators, visionaries, intellectuals, alter-globalists, prospective thinkers, proponents of sustainable development)

Such subjects are present virtually in all structures of societies, also on international and global scale. However, their proportions, strength, and reach determine "shape" of change, its direction, and transformative power.

Fig. 3.11 Agents and opponents to technocultural changes (basic classification)

Questions and issues:	
• what technology	when and where
– has done so far	for whom
– is doing now	at what costs
– can do in the future	whose costs
– cannot do	
• how to study technology	time horizon (short, long term)
– assess its impacts	positive and negative
– modify its features and impacts	innovations and redistribution
• who is a subject of	political mode:
– choice	economocracy (business, market)
	technocracy (technicians, techno-structure)
– decisions	democracy (democratic procedures, negotiations, deliberation, consensus seeking, participation, public good)
– evaluation	netocracy (networking of business and political decision-makers with mass-media)
	autocracy (political autocratism, plutocracy)

Fig. 3.12 Technology evaluation—selected research questions

In parallel in a similar way, some questions addressed to various forms and dimensions of culture of the future can be formulated. Such exemplary questions can be as follows:

- What is culture nowadays?
- What are its types and dimensions?
- What are the directions of its transformations and change?
- How are they caused or supported by technology?
- What are new domains of culture and what happens in them (e.g., in cyberspace)?
- What is the present influence of culture on the "shape" of world and the lives of people?
- What are the short- and long-term consequences of cultural transformation in the world?
- What will the culture of the future be like, also from the perspective of trans- and post-humanism?

To have a more complete view and understanding of the ambiguous relations and interactions of technology and culture, it is worthwhile to confront both lists of questions. And even more important seems to be confronting the answers to these questions.

References

Adas M (1989) Machines as the measure of men: science, technology, and ideologies of western dominance. Cornell University Press, Ithaca

Adas M (2006) Dominance by design: technological imperatives and America's civilizing mission. Belknap Press, Cambridge

Aronovitz S et al (eds) (1996) Technoscience and cyberculture. Routledge, New York

Batteau AW (2010) Technology and culture. Waveland Press, Long Grove

Bauman Z (2000) Liquid modernity. Polity Press, Cambridge

Berger P, Huntington SP (2002) Many globalizations: cultural diversity in the contemporary world. Oxford University Press, New York

Bernal JD (1939) The social function of science. Macmillan, London

Brand R (2005) Synchronizing science and technology with human behavior. Earthscan, London

Breton Ph (2011) The culture of the internet and the internet as cult: social fears and religious fantasies. Litwin Books, Duluth

Cartelli A (ed) (2012) Current trends and future practices for digital literacy and competence. IGI Global, Hershey

Castells M (2000) The rise of the network society, vol 1, 2nd edn. Blackwell, Oxford

Castells M (ed) (2004) The network society. a cross cultural perspective. Elgar, Cheltenham

Debord G (1992) La Société du Spectacle. Gallimard, Paris

Dinello D (2005) Technophobia! Science fiction visions of posthuman technology. University of Texas Press, Austin

Downing JDH (2000) Radical media: rebellious communication and social movements. Sage, Thousand Oaks

Featherstone M (2000) Technologies of post-human development and the potential for global citizenship. In: Pieterse JN (ed) Global futures: shaping globalization. Zed Books, London-New York

Fisher D, Wright L (2001) On Utopias and Dystopias: towards an understanding of the discourse surrounding the internet. J Comput Mediat Commun. doi:10.1111/j.1083-6101.2001.tb00115

Fox MF, Johnson DG, Rosser SV (eds) (2006) Women, gender, and technology. University of Illinois Press, Urbana

Friedel R (2007) A culture of improvement technology and the western millennium. MIT Press, Cambridge
Fuller M, Goffey A (2012) Evil media. MIT Press, Cambridge
Fuller S (2011) Humanity 2.0: what it means to be human past, present and future. Palgrave Macmillan, London
Goffman E (1959) The presentation of self in everyday life. Anchor Books, New York
Hanks C (ed) (2010) Technology and values—essential readings. Wiley, Blackwell, Malden and Oxford
Harbers H (ed) (2005) Inside the politics of technology: agency and normativity in the co-production of technology and society. Amsterdam University Press, Amsterdam
Hughes TP (2004) Human-built world: how to think about technology and culture. Chicago University Press, Chicago
Huntington SP (1996) The clash of civilizations and the remaking of world order. Simon & Schuster, New York
James J (2008) Re-estimating the difficulty of closing the digital divide. J Am Soc Inf Sci Technol. doi:10.1002/asi.2089759
Johnson DG, Wetmore JM (eds) (2009) Technology and society—Building our sociotechnical future. MIT Press, Cambridge
Keen A (2007) Cult of the amateur: how today's Internet is killing our culture. Doubleday, New York
Kelty ChM (2008) Two bits: the cultural significance of free software. Duke University Press, Durham
Kirkup G et al (2000) The gendered cyborg: a reader. Routledge, London
Kurzweil R (2005) The singularity is near. Viking, New York
Lie M, Sørensen KH (eds) (1996) Making technology our own? Domestication technology into everyday life. Scandinavian University Press, Boston
MacCarthy E, Kelty Ch (2010) Responsibility and nanotechnology. Soc Stud Sci. doi:10.1177/0306312709351762
Maines RP (2009) Hedonizing technologies: path to pleasures in hobbies and leisure. Johns Hopkins University Press, Baltimore
Mariner RB, Piehler GK (eds) (2009) The atomic bomb and American society: new perspectives. University of Tennessee Press, Knoxville
McInerney P-B (2009) Technology movements and the politics of free/open source software. Sci Technol Human Values. doi:10.1177/0162243907309852
Miller VA (2011) Understanding digital culture. Sage, London
Morgan K et al (2004) Human perspectives in the internet society: culture, psychology and gender. WITPress, Southhampton
Nayar PK (ed) (2010) New media and cybercultures anthology. Wiley, Malden
Nye DE (2006) Technology matters: questions to live with. MIT Press, Cambridge
Oudshoorn N, Pinch TJ (eds) (2005) How users matter: The co-construction of users and technology. MIT Press, Cambridge
Philbrick M, Barandiaran J (2009) The national citizens' technology forum: lessons for the future. Science and Public Policy
Plant S (1992) Zeroes and Ones: Digital Women and the New Technoculture. Doubleday, New York
Postman N (1992) Technopoly. Vintage, New York
Rheingold H (2002) Smart Mobs—The Next Social Revolution—Transforming Culture and Communities in the Age of Instant Access. Basic Books, Cambridge
Rosen BC (1998) Winners and losers of the information revolution: psychological change and its discontents. Praeger Publishers, Westport
Rycroft RW, Kash DE (1999) The complexity challenge: technological innovation for the 21st century. Pinter, London
Sartori G (1999) *Homo videns*. Roma: Gius, Laterza & Figli

Schroeder R (2007) Rethinking science, technology, and social change. Stanford University Press, Stanford
Schumacher EF (1973) Small is beatiful: economics as if people mattered. Harper&Row, New York
Seidensticker B (2006) Future hype: the myths of technology change. Bernett-Koehler Publishers, San Francisco
Silver D, Massanari A (eds) (2006) Critical cyberculture studies. New York University Press, Cambridge, MA
Smith MR, Marx L (1994) Does technology drive history? The dilemma of technological determinism. MIT Press, Cambridge
Spiller N (ed) (2002) Cyber reader: critical writings for the digital era. Phaidon, London
Stivers R (2004) Shades of loneliness: pathologies of a technological society. Rowman & Littlefield, Lannan
Sussan R (2005) Les utopies posthumaines: contre-culture, cyberculture, culture du chaos. Omniscience, Sophia Antipolis
Taylor PA, Harris JL (2005) Digital matters-theory and culture of the matrix. Routledge, London
Tredinnick L (2008) Digital information culture: the individual and society in the digital age. Chandos Publishing, Oxford
Turkle S (1995) Life on the screen: identity in the age of the internet. Simon and Schuster, New York
Vanderburg WH (2005) Living in the labyrinth of technology. University of Toronto Press, Torobto
Vanderburg WH (2007) Technology and the law: Who rules? Bull Sci Technol Soc. doi:10.1177/0270467607302688
Wajcman J (2004) Technofeminism. Polity/Blackwell, London
Wolfe C (2010) What is Posthumanism?. University of Minnesota Press, Minneapolis
Zacher LW (2007). E-Transformations of societies. In: Anttiroiko A-V (Ed.) Electronic government: concepts, methodologies, tools and applications, information science reference (IGI Global), vol. 6, Chapt. 8.5. Hershey—New York: Hershey
Zacher LW (2008) Some repetitive reflections on visioning the future. In: Sharma S, Sharma PK (eds) Transformative pathways: attainable Utopias. Prateekasha Publications, Jaipur
Zacher LW (2009) Information society discourse. In: Encyclopedia of information science and technology. Hershey: IGI Global
Zacher LW (2012a) Society, market and technology nexus as contexts of ICT policies and applications: some issues and reflexions. Int J Inf Commun Technol Hum Dev. doi:10.4018/jicthd.2012070103
Zacher LW (2012b) Toward democratization of science and technology spheres. Some opportunities and problems. In: Bammé A et al (eds) Yearbook 2011 of the IAS-STS. Profil, München
Zacher LW (2013) Human and Societal Potentials for Transcending the Crisis of Civilization. In: Targowski A, Celiński MJ (eds) Spirituality and Civilization Sustainability in the 21st Century. Nova Science Publishers, New York

Chapter 4
The Digital Agenda of the European Union and the Digital Policies of the USA

Despina Anagnostopoulou

Abstract The contribution reviews the digital policies and legislation on e-commerce and e-government in both the EU and the USA: In order to protect business and consumers the EU has issued specific Directives on e-signature, pre-contractual information, certification of e-service providers, consumer sales, consumer rights, an optional Common European Sales Law, data protection, VAT facilitation (MOSS) and SEPA modernization in order to exploit all the potential of the Single Market in terms of e-commerce. The USA, where the Internet was born, has early launched digital policies and is the leader in the e-commerce and the digital economy, thanks to its contribution in Internet governance and the freedom the US allow to the innovating firms.

Keywords US digital policy · EU Digital Agenda · E-commerce in EU and USA · Digital Single Market

Introduction

The Increasing Importance of E-Commerce

Electronic communications started in USA in 1958. It was the research on the US ARPA project which increased technological progress in networks, protocols and software and led to the ARPAnet (1969). More inventions followed such as the EDI[1]

[1]Electronic Data Interchange.

D. Anagnostopoulou (✉)
Jean Monnet Chairholder, University of Macedonia,
Egnatia 156, 54006 Thessaloniki, Greece
e-mail: danag@uom.gr

L.W. Zacher (ed.), *Technology, Society and Sustainability*,
DOI 10.1007/978-3-319-47164-8_4

(1972) and the WWW[2] (1992, CERN) which was created as a management system of multimedia resources that permits data exchange between computers (Tziva 2007:26). The development of e-commerce was fast: in the period 1994–1997 many business firms created a web page on the Internet without electronic transactions (e.g. amazon.com, e-bay). In the period 1997–2000 the first electronic transactions were made possible, and since 2000 e-commerce is a widely considered way to increase profit. In the EU, the turnover increased from 10 billion euros (2001) to 70 billion euros (2008). In 2008, 33 % of EU consumers used e-commerce for their purchases mainly from e-shops established in their own state, while only 7 % trusted cross-border transactions (Anagnostopoulou 2011:295–297).

The Concept and Forms of Electronic Commerce at the European and International Level

E-commerce is generally defined as the occurrence of transactions via Internet, other networks (intranet, extranet) or other applications (electronic press, e-radio, e-television, etc.). These transactions may be between: Business to Consumer (B2C), Business to Business (B2B), Consumer to Consumers (C2C or User to User) and lately Consumer to Government (C2G).

International texts define e-commerce as applying "to any kind of information in the form of a data message used in the context of commercial activities"[3] or "conducted over computer-mediated networks" (OECD 2002:89). E-commerce is the commerce conducted through electronic data transmission which includes image, sound and/or texts. The electronic commerce covers *"the production, advertisement, sale and distribution* of products"[4] "or s*ervices, whether between businesses, households, individuals, governments and other public or private organizations conducted over the Internet"* (OECD 2002:89).

International Law and Soft Law on Electronic Commerce

The United Nations

It was first the International Covenant on Civil and Political Rights (1966) which provided that freedom of expression could include "other media" besides the oral,

[2]European Centre for Nuclear Research (1954); see http://public.web.cern.ch/public/en/About/About-en.html. All websites were accessed for the last confirmation on 25 March 2013.

[3]UNCITRAL, *Model Law on Electronic Commerce*, article 1.

[4]European Parliament Resolution of 5 February 2009 on international trade and the internet, (2008/2204(INI)), OJ C 67E, 18.3.2010, p. 112.

written, printed or artistic form, thus covering electronic transmission of information.[5]

UNCITRAL[6] has elaborated the Model Law on Electronic Commerce (1996) in order to encourage and promote the electronic forms of communication. The Model Law was the first international text to declare that electronic texts are equivalent to written texts (handwritten or typed) and accepted as originals and that e-signature is valid if the identity of the person can be recognized. In addition, it established a framework of rules for electronic transactions (risk transfer, liability in case of fraud or wrong transmission, time of dispatch, etc.).[7] The Model Law was signed by 67 states, among which the USA, Canada, China, Australia, New Zealand, and India. From the EU member states only the UK, Ireland and France signed it.[8]

In 2000, the United Nations Centre for Trade Facilitation and Electronic Business (Boss and Kilian 2008; Faria 2006:689) adopted the Model E-Commerce Agreement[9] (Igglezakis 2009:223).

In 2001, the United Nations General Assembly adopted the UNCITRAL Model Law on Electronic Signatures which established the legal force of electronic signatures under certain conditions and requirements for their credibility (as reliable as appropriate) independently of the means used (neutrality).[10] The Model Law was signed by 32 states including China, Qatar and India. Neither the US nor the EU member states have signed this Model Law, except for an overseas territory of the UK, Montserrat.[11]

In 2005 the United Nations urged for the signature of the UNCITRAL Convention on the Use of Electronic Communications in International Contracts. The Convention covers only e-commerce B2B, acknowledges electronic messages and signatures as equivalent to handwritten under specific conditions. It establishes the place and time of dispatch and receipt of electronic communications for the conclusion of e-contracts in a digital environment. The Convention was signed by 20 states, e.g. China, Russia, Iran, and it is ratified by only seven states.[12] It

[5]International Covenant on Civil and Political Rights, UN Treaty Series, Vol. 68, p. 17 and 999,1-14668 art. 19 par. 2; see http://treaties.un.org/doc/Publication/UNTS/Volume%20999/volume-999-I-14668-English.pdf.

[6]United Nations Commission on International Trade Law.

[7]See http://www.uncitral.org/uncitral/en/uncitral_texts/electronic_commerce/1996Model.html.

[8]See http://www.uncitral.org/uncitral/en/uncitral_texts/electronic_commerce/1996Model_status.html.

[9]General Assembly Resolution 51/162 of 16 December 1996. UNCITRAL, 1985—Recommendation on the Legal Value of Computer Records; see http://www.uncitral.org/uncitral/en/uncitral_texts/electronic_commerce/1985Recommendation.html.

[10]http://www.uncitral.org/uncitral/en/uncitral_texts/electronic_commerce/2001Model_signatures.html.

[11]http://www.uncitral.org/uncitral/en/uncitral_texts/electronic_commerce/2001Model_status.html.

[12]The states that ratified the UN Convention are Honduras, Dominican Republic and Singapore. See http://www.uncitral.org/uncit-ral/en/uncitral_texts/electronic_commerce/2005Convention_status.html.

entered into force on 1 March 2013.[13] The Electronic Communications Convention builds upon earlier instruments drafted by the Commission and, in particular, the UNCITRAL Model Law on Electronic Commerce and the UNCITRAL Model Law on Electronic Signatures. These instruments are widely considered standard legislative texts setting forth the three fundamental principles of electronic commerce legislation, which the Convention incorporates, namely non-discrimination, technological neutrality and functional equivalence (UNCITRAL 2008).

Other International or Regional Organizations and Associations

The Council of Europe has adopted two Conventions: the Convention for the Protection of Individuals with regard to Automatic Processing of Personal Data (1981) and the Convention on Cybercrime (2001),[14] as well as its Additional Protocol concerning the criminalization of acts of a racist and xenophobic nature committed through computer systems (2003).[15]

The OECD has published guidelines concerning the protection of privacy on the Internet (1980),[16] cryptography (1997),[17] the protection of the e-consumer (1999),[18] spamming (2006) etc.[19] It has also adopted the Declaration of Seoul for the Future of E-economy (2008).[20] In addition, the OECD Directorate for Science, Technology and Industry (STI) analyses how information and communication

[13]According to Art. 23 the Convention will enter into force on "the first day of the month following the expiration of 6 months after the deposit of the third instrument of ratification, acceptance, approval or accession. http://www.uncitral.org/pdf/english/texts/electcom/06-57452_Ebook.pdf.

[14]Council of Europe, Convention on Cybercrime, Budapest, 23.XI.2001, ETS No 185. See http://conventions.coe.int/Treaty/ EN/ Treaties/html/185.htm.

[15]Additional Protocol to the Convention on cybercrime, concerning the criminalization of acts of a racist and xenophobic nature committed through computer systems, Strasbourg, 28.I.2003; see http://conventions.coe.int/Treaty/en/Treaties/html/189.htm.

[16]The OECD Guidelines on the Protection of Privacy and Transborder Flows of Personal Data; see http://www.oecd.org/document/20/0,3343,en_2649_34255_15589524_1_1_1_1,00.html. Grammatikaki-Alexiou A.: *E-commerce*, op.cit., p. 143 (in Greek).

[17]Recommendation of the Council concerning Guidelines for Cryptography Policy, 17 March 1997; see http://www.oecd.org/document/11/0,2340,en_2649_34255_1814731_119820_1_1_1,00.html.

[18]Guidelines for e-Consumer Protection (1999); see http://www.oecd.org/document/51/0,3746,en_21571361_43348316_1824435_1_1_1_1,00.html.

[19]See www.oecd-antispam.org, 2006 OECD recommendation on cross-border enforcement of laws against spam and http://www.oecd.org/department/0,3355,en_2649_22555297_1_1_1_1_1,00.html.

[20]*The Seoul Declaration for the Future of the Internet Economy; see* http://www.oecd.org/dataoecd/49/28/40839436.pdf.

technologies (ICTs) contribute to sustainable economic growth. The OECD Digital Economy Papers cover ICT-related issues and policy reports.[21]

Since its establishment (1995), the WTO has shown interest on Trade in Information Technology Product (13 December 1996) and on Global Commerce (1998). However, the Hong Kong Ministerial Declaration on the Doha Work Programme has deleted the issue of e-commerce from the Doha Agenda.[22] According to the WTO, e-commerce is defined as "the production, advertising, sale and distribution of products via telecommunications networks", though a distinction may be drawn between a) the delivery of content on physical carrier media and b) the content digitally encoded and transmitted electronically over the Internet and thus independently from physical carrier media, over fixed and wireless networks.[23]

The ICC has published the ICC eTerms 2004 as model contract terms to be used in electronic cross-border business.[24] It has also introduced the "Business Action" to support the information society and participates the Internet Governance Forum.

The "Commission on the Digital Economy" seeks to realize the full potential of e-commerce by developing policy and practical tools that encourage competition, growth, predictability, compliance and the secured, free flow of information in cross-border trade, via the Internet and other information and communication technologies. The Commission on the Digital Economy currently informs the Internet Corporation for Assigned Names and Numbers (ICANN), the European Parliament and the Commission of the European Union (EU), the Council of Europe, the Asia-Pacific Economic Cooperation (APEC), the World Trade Organization (WTO), the World Bank, the G8/G20, the International Telecommunication Union (ITU), the United Nations Conference on International Trade Law (UNCITRAL), the UN Educational, Scientific and Organization (UNESCO), the UN Conference on Trade and Development (UNCTAD), the UN ICT-related agencies and initiatives and the Organization for Economic Cooperation and Development (OECD).[25]

---

[21]http://www.oecd-ilibrary.org/science-and-technology/oecd-digital-economy-papers_20716826.

[22]Trade in Information Technology Products (also known as Information Technology Agreement (ITA)) of the Singapore Ministerial Declaration of the First Session of the Ministerial Conference of the WTO, 13.12.1996, pt. 18, Geneva Ministerial Declaration on Global Electronic Commerce of the Second Session of the Ministerial Conference of the WTO (20 May 1998), Hong Kong Ministerial Declaration on the Doha Work Programme of the Sixth Session of the Ministerial Conference of the WTO, adopted on 18 December 2005, point 46 on E-commerce.

[23]European Parliament Resolution of 5 February 2009 on International Trade and the Internet (2008/2204(INI)), OJ C 67E, 18.3.2010, pp. 112–120.

[24]http://www.tradelex.com/component/option,com_docman/task,doc_view/gid,100/Itemid,28/lang,es_ES.

[25]http://www.iccwbo.org/about-icc/policy-commissions/digital-economy/.

The EU Framework and Policy for E-Commerce

The EU Framework Up to 2010

The European framework comprised mostly Directives on protection of personal data (1995),[26] the Information society (1998),[27] E-signatures (1999),[28] E-commerce (2000),[29] E-financial services (2002),[30] E-tendering of public contracts (2004), E-services (2006),[31] E-billing and VAT (2006),[32] E-payments in SEPA (2007).[33] The main drive was the initiative e-Europe approved by the European Council in March 2000 as a part of the Lisbon process.[34] Its goal was the promotion of the information society by the liberalization of telecommunications, the establishment of a clear legal framework for e-commerce building on consumer trust and the support for the industry and R&D by introducing cheap and fast Internet access. It was followed by other initiatives such as the Europe2002 initiative aiming at increasing the number of Internet connections in EU, opening all communication networks up to competition and stimulating Internet use by promoting training and consumer protection;[35] the GoDigital initiative, aiming at raising awareness and spreading best practice on e-commerce among SMEs;[36] the e-business legal portal (2002) managed by a network of 16 Euro Info Centres; the e-Europe2005 initiative, aiming at stimulating the development of services, applications and content while speeding up the deployment of

---

[26]Directive 95/46/EC, OJ L 281 of 23.11.1995, p. 31, amended by Directive 2007/64/EC, OJ L 319 of 5.12.2007, p. 1. See also Directive 2002/58/EC of 12.7.2002, OJ L 201 of 31.7.2002, p. 37.

[27]Directive 98/34/EC, OJ L 204 of 21.7.1998, p. 37 amended by Directive 98/48/EC, OJ L 217 of 5.8.1998, p. 18 and Directive 2006/96/EC, OJ L 363 of 20.12.06, p. 81.

[28]Directive 1999/93/EC, OJ L 13 of 19.1.2000, p. 12.

[29]Directive 2000/31/EC, OJ L 178 of 17.7.2000, p. 1.

[30]Directive 2002/65/EC, OJ L 271 of 9.10.2002, p. 16.

[31]Directive 2006/123/EC, OJ L 376 of 27.12.2006, p. 36.

[32]Directive 2006/112/EC, OJ L 347 of 11.12.2006, p. 1 amended by Directive 2006/138/EC, OJ L 384 of 29.12.2006, p. 92.

[33]Directive 2007/64/EC, OJ L 319 of 5.12.2007, p. 1.

[34]Commission Communication on a Commission initiative for the special European Council of Lisbon, 23 and 24 March 2000—eEurope—An information society for all, COM (1999) 687.

[35]Commission Communication "eEurope 2002: Impact and Priorities, a communication to the Spring European Council in Stockholm", 23–24 March 2001, COM (2001) 140.

[36]Commission Communication "Helping SMEs to Go Digital", COM (2001) 136, www.europa.eu.int/godigital.

secure broadband Internet access;[37] the i2010 strategy, defining broad policy guidelines for the information society and the media,[38] etc.[39]

The Directive 2000/31/EC on Certain Issues of E-Commerce

The Directive on e-commerce aims at the protection of free movement of goods and services in the information society while safeguarding the protection of privacy. Its goals are the equivalence of e-contracts, the protection of e-consumers with pre-contractual information, the definition of the place of establishment independently of the place of the server as well as the obligations of certification providers. There are four main principles of the Directive:

(a) The principle of the country of origin means that the e-service provider must comply with the national provisions of the member state of its own establishment (art.3§1). Restrictions on the free movement by the host member state are permitted under conditions based on public policy, public security and public health, as well as general interest, e.g. the protection of consumer (art.3§4),
(b) The principle of prohibiting prior authorization by the host member state (art.4§1),
(c) The principle of the respect of professional rules of the country of origin (art.8 par.1 and art. 10 par. 2) and d) The principle of no general liability of intermediaries (e.g. Google) for monitoring the information which they transmit or store nor a general obligation actively to seek facts or circumstances indicating illegal activity (art. 15).[40]

The Protection of the E-Consumer Through Various EU Directives

Obligatory provision of pre-contractual information to the consumer (e.g. the identity of the provider and geographical address, the cost of product or service, VAT number

---

[37]Commission Communication "eEurope 2005 Action Plan: An information society for all", COM (2002) 263.

[38]Commission Communication "i-2010—A European Information Society for growth and employment", COM (2005) 229.

[39]Commission Communication, "Implementing the Community Lisbon Programme—Modern SME Policy for Growth and Employment", COM (2005) 551. European Parliament resolution of 5 February 2009 on enhancing the role of European SMEs in international trade (2008/2205(INI)), PE_TA(2009)0048.

[40]CJEU Joined cases C-236/08 to C-238/08, Google France SARL and Google Inc. v Louis Vuitton Malletier SA, European Court Reports 2010, p. I-2417.

and Professional Affiliation) is regulated by Directives 2000/31/EC,[41] 2006/123/EC[42] and 2011/83/EU.[43] Provision of information concerning the terms and conditions in online conventions is regulated by Directive 93/13/EC.[44] In addition, e-consumers have the right of withdrawal,[45] the right to be protected from spamming (opt-in approach)[46] and the right to be protected by the consumer's home rules.[47]

The E-Signature

The Directive 99/93/EC has established the fundamental EU framework for e-signature in order to facilitate its legal recognition. The e-signature has two functions: a) the confirmative function meaning that the e-signature confirms the identity of the sender and b) the confidential function meaning that only the receiver can read the mail sent (protection from undesirable intruders).

The definition of the e-signature is "data in electronic form, which are attached to or logically associated with other electronic data and serve as a method of authentication". The signatory is only a natural person.

The advanced e-signature is based on a recognized certificate and created by a secure system. The "certification service provider" is an entity or a legal or natural person who issues certificates or provides other services related to e-signatures (art. 2 § 11). The advanced encrypted signature is based on public and private keys used (algorithms or devices) and:

(a) Is uniquely linked to the signatory;
(b) Is capable of identifying the signatory;
(c) Is created using means that the signatory can maintain under his sole control; and
(d) Is linked to the data to which it relates in such a manner that any subsequent change of the data is detectable.

However, certain problems appeared because of the use of e-signature: first, there are different national regulations regarding the process of certification;

[41]Directive 2000/31/EC of the European Parliament and of the Council of 8 June 2000 on certain legal aspects of information society services, in particular electronic commerce, in the Internal Market ("Directive on electronic commerce"), OJ L 178, 17.7.2000, p. 1.

[42]Directive 2006/123/EC, OJ L 376 of 27.12.2006, p. 36.

[43]Directive 2011/83/EU amending Directives 93/13/EEC, 1999/44/EC, and repealing Directive 85/577/EEC and 97/7/EC, OJ L 304, 22.11.2011, p. 64.

[44]Directive 93/13/EEC, OJ 1993, 95, p. 29. See CJEU Case C-451/12, Josune Esteban García, OJ C 399, 22.12.2012, p. 12.

[45]Directive 97/7/EC, OJ L 144 of 4.6.1997, p. 19, as amended by Directive 2007/64/EC, OJ L 319 of 5.12.2007, p. 1.

[46]Directive 2002/58/EC of 12.7.2002, OJ L 201 of 31.7.2002, p. 37.

[47]Regulation (EC) 593/2008, OJ L 177 of 4.7.2008, p. 6.

second, there is limited use of advanced e-signature; third, the user is not familiarized with the e-signature, and last but not least, there are interoperability problems between e-signature systems.[48]

Therefore, the European Commission has declared the need to reform electronic authentication and electronic identity for businesses, in order to enable e-invoicing and e-government transactions in its Single Market Act.[49]

Domain Names

The domain names provide the identity of the user (e.g. business firm) in the Internet. The Internet Corporation for Assigned Names and Numbers (ICANN-1998)[50] is responsible for the coordination of the global Internet system of unique identifiers and for its stable and secure operation. It has established the Uniform Domain Name Dispute Resolution Policy and Rules (Karadimitriou 2008:50–51).

Domain names in the EU are assigned according to the .eu Top Level Domain (Reg.733/2002).[51] Public policy rules concerning the implementation and functions of the .eu Top Level Domain and the principles governing registration are provided for by the Reg. (EC) No 874/2004.[52] The Dispute Resolution Procedure is established according to the recommendations by WIPO (Tountopoulos and Hatzopoulos 2001:57).

Conclusions on the Situation Up to 2010 for the EU

UNCITRAL has played a major role for the development of e-commerce by providing a framework for the legal force of e-signatures; however, neither the EU nor the majority of its member states have signed the model laws. The reason is that the EU legislation is more detailed and more binding than the international rules regarding e-commerce issues. The main goals of the EU so far have been firstly, to

[48]EESSI standards—European Validation Authorities Gateway to ensure cross-border interoperability for e- signatures.

[49]Commission Communication on the Single Market Act, COM(2010) 608 final; see below. Commission Staff Working Document On Line Services, including E-Commerce in the Single Market, SEC (2011) 1641 final, Accompanying Document to COM (2011) 942, A Coherent Framework to boost Confidence in the Digital Single Market of E-commerce and other online services, 11.1.2012, p. 84–86.

[50]Internet Corporation for Assigned Names and Numbers; see http://www.icann.org/en/about.

[51]Regulation (EC) 733/2002, OJ L 113 of 30.4.2002, p. 1, as amended by Regulation (EC) 1137/2008, OJ L 311 of 21.11.2008, p. 1.

[52]Regulation (EC) 874/2004, *OJ L 162 of 30.4.2004, p. 40.*

achieve a balance between the EU and the international efforts concerning e-commerce and secondly, to establish a stable framework to regulate e-commerce and to protect at the same time the electronic consumer. However, this strategy was not enough to boost e-commerce and promote growth in the EU.

The New EU Policy on E-Commerce and Growth

Relauncing the Single Market, Mario Monti's Report and the Single Market Act

Mario Monti's report "A new Strategy for the Single Market"[53] underlined the importance of e-commerce together with innovative services and eco-industries as holding the largest growth and employment potential. It was on the invitation of Commission President Barroso that Mario Monti, former member of the European Commission, prepared a report on the relaunch of the Single Market, a key strategic objective of the EU on May 9, 2010.[54] In addition, the European Parliament has issued a resolution recalling the importance of e-commerce on global trade.[55] The European Council and the European Parliament have since called for further strengthening the European digital leadership and completing the Digital Single Market by 2015.[56] The efforts have led to the adoption of the Single Market Act and the Digital Agenda.[57]

The EU 2020 Strategy and the Flagship Initiative "A Digital Agenda for Europe"

In 2010 the European Commission developed its EU 2020 Strategy by establishing three development priorities on smart, sustainable and inclusive growth.[58] In addition, the EU 2020 strategy proposed seven flagship initiatives:

---

[53]Report by Mario Monti to the President of the European Commission: "A new strategy for the single Market" of 9 June 2010, page 9, http://ec.europa.eu/bepa/pdf/monti_report_final_10_05_2010_en.pdf.

[54]http://ec.europa.eu/internal_market/smn/smn58/docs/monti_en.pdf.

[55]European Parliament Resolution of 5 February 2009 on International Trade and the Internet (2008/2204(INI)), OJ C 67E, 18.3.2010, p. 112.

[56]European Council conclusions of 28/29 June 2012 conclusions of 1/2 March 2012. European Parliament resolution of 21 September 2010 on completing the internal market for e-commerce (2010/2012(INI)) P7_TA(2010)0320, OJ C 50E, 21.2.2012, p. 1.

[57]Commission Communication on the Single Market Act, COM (2010) 608 final.

[58]Commission Communication, EUROPE 2020: A strategy for smart, sustainable and inclusive growth, COM (2010) 2020 final.

(i) An innovation Union,
(ii) Youth on the move,
(iii) A Digital Agenda for Europe,
(iv) A resource-efficient Europe,
(v) An industrial policy for the globalization era,
(vi) An agenda for new skills and jobs and
(vii) A European platform to tackle poverty.

Thus, the Digital Agenda for Europe[59] was adopted in 2010, as an integral part of the Europe 2020 strategy, to stimulate the digital economy and address societal challenges through ICT. It is a complete set of actions which are designed ultimately to establish the Digital Single Market. Services are a crucial sector for the EU's economic recovery, accounting for over 70 % of all jobs and all net job creation in the single market. Access to fast and ultra-fast broadband and financing the infrastructure throughout the whole EU and in particular for geographically disadvantaged regions are also very important actions. The European Commission has also set specific arithmetic goals for the future. In particular, by 2013 the EU should have achieved a comprehensive coverage of basic broadband networks (30 megabytes/second or higher-speed broadband services by 2020). By 2015, the proportion of EU Internet users from the current 60 % should have risen to 85 %, online shopping should have increased to 50 %, cross-border shopping should have increased to 20 % and e-procurement to 50 %.[60]However, the above aims were not achieved by the end of 2015. While 53 % of citizens shop online, only 16 % engage in cross-border eCommerce according to the Digital Agenda Scoreboard 2016.

In addition, the diversity between member states is significant.[61]

The Digital Agenda is based on seven pillars: I. Achieving the Digital Single Market; II. Enhancing interoperability and standards by adopting an EU Interoperability Strategy and Framework;[62] III. Strengthening Online Trust and Security by adopting a Reinforced Pan—European Network and Information Security Policy, Cyber-Security Platforms and Pan-European Computer

[59]In the late 1980s, the US Digital Agenda at WIPO was to persuade the international community to use copyright law as the principal form of legal protection for computer programs; see Samuelson, P.: The U.S. Digital Agenda At The World Intellectual Property Organisation, pp. 1–3, at http://people.ischool.berkeley.edu/~pam/courses/cyberlaw97/docs/wipo.pdf. The Australians have used the word Digital Agenda when they modified their 1968 law on Copyright and related rights in 2001, Copyright Amendment (Digital Agenda) Act, no 110/2000; see http://www.wipo.int/wipolex/en/details.jsp?id=297.

[60]See also European Commission: Broadband coverage in Europe in 2011, Mapping progress towards the coverage objectives of the Digital Agenda (2011).

[61]See http://ec.europa.eu/digital-agenda/en/country/greece.

[62]Commission Communication "Towards interoperability for European public services", 16.12.2010, COM (2010) 744 final.

Emergency Response teams;[63] IV. Promoting fast and ultra-fast Internet access needed for video conferencing by adopting a European Radio Spectrum Policy.[64] Broadband will bring better opportunities for communities that are more isolated and will boost the EU economy since 10 percentage points more broadband means 1–1.5 % more GDP growth;[65] V. Research and Innovation with increased private and public ICT investment in order to develop a new generation of Web-based applications and services;[66] VI. Enhancing digital literacy, skills and inclusion, their recognition and evaluation and increasing the participation of vulnerable groups[67] and VII. The internationalization of "Internet governance".[68]

The main tool for coordinating the above pillars is the *Digital Agenda Scoreboard* which assesses the progress achieved regarding the targets set out in the Digital Agenda. In addition, it provides analysis and detailed data on all the policy areas covered by the Digital Agenda.[69]

In order to finance the development of information and communication structures provided for by its Digital Agenda the European Commission has established the "Connecting Europe Facility",[70] with guidelines for the preparation of broadband infrastructure projects. In addition, in the context of the European Cohesion Fund, the European Commission has drafted guidelines for smart specialization strategies which must be implemented in order to benefit from regional funding. Moreover, in the context of the regulatory framework on electronic communications, the European Commission has adopted a recommendation on access-pricing schemes in the wholesale market in order to stimulate investment in fibre deployment, and reviewed the 2009 guidelines on state aid for broadband networks. Last but not least, the European Commission adopted a guide on cost reduction techniques for construction works with the aim of reducing them by half.

In 2012, the European Commission has updated its Digital Adenda by adopting seven new priorities in order to confront the patchy pan-European policy

[63]See http://ec.europa.eu/digital-agenda/en/our-goals/pillar-iii-trust-security.

[64]See http://ec.europa.eu/digital-agenda/en/our-goals/pillar-iv-fast-and-ultra-fast-internet-access.

[65]Kroes, N., Vice-President of the European Commission responsible for the Digital Agenda, SPEECH/13/263 of 26.3.2013, Getting every home and town in Europe "broadband ready". Available at http://europa.eu/rapid/press-release_SPEECH-13-263_en.htm.

[66]See http://ec.europa.eu/digital-agenda/en/our-goals/pillar-v-research-and-innovation.

[67]See http://ec.europa.eu/digital-agenda/en/our-goals/pillar-vi-enhancing-digital-literacy-skills-and-inclusion.

[68]See http://ec.europa.eu/digital-agenda/en/our-goals/international.

[69]See http://ec.europa.eu/digital-agenda/en/scoreboard.

[70]Commission Staff Working Document On Line Services, including E-Commerce in the Single Market, SEC (2011) 1641 final, Accompanying Document to COM (2011) 942, A Coherent Framework to boost Confidence in the Digital Single Market of E-commerce and other online services, 11 January 2012, p. 106.

framework and maximize *"the digital sector's contribution to Europe's recovery"* based on the conclusions of the Digital Agenda Assembly 2012.[71] These new priorities are[72]:

(a) A new *regulatory framework for broadband networks* on stronger non-discriminatory network access and new costing methodology for wholesale access to broadband networks, net neutrality, universal service and mechanisms for reducing the civil engineering costs of *broadband roll-out* leading to New State Aid Guidelines,
(b) New public *digital service infrastructures*, especially for *cross-border interoperability* in eIDs and eSignatures, business mobility, eJustice, electronic health records and cultural platforms,
(c) A Grand Coalition on *Digital Skills and Jobs* in order to increase *IT training placements,* support Web entrepreneurs and make business "start-up friendly",
(d) A new EU *cybersecurity strategy* and a Directive to establish a common minimum level including an online platform to prevent and counter cross-border cyber incidents, and incident reporting requirements,
(e) An update of *EU's Copyright Framework* via a structured stakeholder dialogue in order to submit a proposal,[73]
(f) An acceleration of *cloud computing* through public sector buying power, dismantling current national "fortresses" and negative consumer perceptions[74] and
(g) A new micro- and nano*electronics industrial strategy*, to increase Europe's attractiveness for investments in design and production as well as to increase its global market share.

The Digital Single Market Policy and Actions

The Digital Single Market is the first pillar of the Digital Agenda. According to the definition of the European Commission in 2015, the Digital Single Market is "one in which the free movement of persons, services and capital is ensured and where the individuals and businesses can seamlessly access and exercise online activities under conditions of fair competition, and a high level of consumer and personal

---

[71]European Commission: IP/12/10 of 11 January 2012 "Stimulating growth and employment: an action plan for doubling the volume of e-commerce in Europe by 2015"; see http://europa.eu/rapid/press-release_IP-12-1389_en.htm.

[72]Op.cit.

[73]MEMO/12/950 of 5 December 2012, Commission agrees way forward for modernizing copyright in the digital economy.

[74]European Commission: IP/12/1225.

data protection, irrespective of their nationality or place of residence" (https://ec.europa.eu/digital-single-market/en/digital-single-market). With its initial 20 envisaged actions, it aims at the removal of barriers that block the free flow of legal digital content and online services across national borders and boost the music download business, establish a single area for online payments and further protect EU consumers in cyberspace. The Digital Agenda will update the EU Single Market rules for the digital era.[75] Its goals are to set up new, globalized businesses, while consolidating existing businesses. It will thus contribute in a rapid increase in jobs based on European culture and expertise and will restore balance in the single market by anchoring it on three pillars:

(a) An economic strand to support the growth of businesses; development of an electronic identity of each business
(b) A social strand to restore the confidence of the EU citizens such as the Code of Rights of the users of online services and
(c) An enhanced governance strand which will promote growth and jobs in the Single Market by establishing a better dialogue with civil society as a whole; a close partnership with the various market participants; an efficient provision of information for citizens and enterprises; and closer monitoring of the application of single market legislation.

In 2010, the Single Market Act[76] actions included simplifying the pan-European licensing for online works and simplifying distribution of creative content, opening up public data resources for reuse, protecting IP rights on line against counterfeiting and piracy, improving Single European payment and e-invoicing and establishing Alternative Online Dispute Resolution[77] by a Directive on alternative dispute resolution for consumer disputes and a Regulation on online out-of-court dispute resolution for consumer disputes as well as improving the framework for civil law proceedings.[78] According to Proposal No 2, the Commission was to submit a proposal for a framework Directive on the management of copyrights, with the aim of opening up access to online content by improving the governance, transparency and electronic management of copyright. The Commission would also propose a Directive on orphan works.[79] According to Proposal No 5, the

[75]http://ec.europa.eu/digital-agenda/en/our-goals/pillar-i-digital-single-market.

[76]Commission Communication "Towards a Single Market Act For a highly competitive social market economy 50 proposals for improving our work, business and exchanges with one another", COM (2010) 608 final.

[77]See http://ec.europa.eu/digital-agenda/en/pillar-i-digital-single-market/action-13-complementing-consumer-rights-directive.

[78]Commission Proposal for a Directive on alternative dispute resolution for consumer disputes and amending Regulation (EC) No 2006/2004 and Directive 2009/22/EC (Directive on consumer ADR), COM (2011) 793 and Commission Proposal for a Regulation on online dispute resolution for consumer disputes, COM (2011) 794 final.

[79]European Parliament Resolution of 12 May 2011 on unlocking the potential of cultural and creative industries (2010/2156(INI)), OJ C 377E, 7.12.2012, p. 142.

Commission was to take initiatives to develop electronic commerce in the internal market in order to confront problems faced by consumers in the digital economy, e.g. guidelines for the member states to guarantee the effective application of provisions in the Services Directive to combat discrimination against recipients of services because of their nationality or place of residence.[80] According to Proposal No 46 the Commission would submit an initiative on the use of alternative dispute resolution in the EU and a Recommendation on the network of alternative dispute resolution systems for financial services as well as a European system for the settlement of online disputes for digital transactions.

In October 2012, the European Commission adopted the Single Market Act II, putting forward twelve key actions concentrated on four main drivers for growth, employment and confidence: (a) integrated networks, (b) cross-border mobility of citizens and businesses, (c) *the digital economy* and (d) actions that reinforce cohesion and consumer benefits. Working towards the completion of the digital single market by 2015, the Commission proposed to: (i) facilitate e-commerce in the EU by making *payment services* easier to use, more trustworthy and competitive, (ii) address a key underlying cause of *lack of investment in high-speed broadband connection*, i.e. its civil engineering costs, and (iii) adopt electronic invoicing standard in public procurement procedures.[81]In May 2015, the Digital Single Market Strategy was adopted by the European Commission. It is built on three pillars: (1) better access for consumers and businesses to digital goods and services across Europe; (2) creating the right conditions and a level playing field for digital networks and innovative services to flourish; (3) maximising the growth potential of the digital economy. The European Commission will forward several new proposals e.g. on e-commerce of digital content, better parcel delivery, ending unjustified geo-blocking, analysing the role of online platforms, ensuring cybersecurity, apply smart metering etc. It will also review legislation on copyright law, telecoms rules, audiovisual media framework etc. (European Commission, IP 15-4919).

Specific New Tools for the Promotion of E-Commerce

The Code for EU Online Rights

As EU citizens were becoming increasingly disenchanted with the single market there was an urgent need to establish a coherent framework to boost

[80]See also proposals no 6 on standardization reform, no 22 on mutual recognition of e-identification and online e-authentication across the EU and no 28 on a European Radio Spectrum Action Programme. Communication from the Commission on the Single Market Act, Towards a Single Market Act For a highly competitive social market economy 50 proposals for improving our work, business and exchanges with one another, COM (2010) 608 final.

[81]IP/12/1054, Single Market Act II: twelve priority actions for new growth (3 Oct 2012).

confidence in the Digital Single Market. Thus, in December 2012 the European Commission published the "Code of EU online rights", which compiles the basic set of rights and principles in order to protect citizens when acceding and using online networks and services. These rights and principles are not truly new and not always easy to grasp, because their use is not exclusive to the digital environment. They had already been scattered, e.g., across various Directives, Regulations and Conventions in the areas of e-communications, e-commerce and consumer protection. In addition, some member states might have introduced higher levels of protection than those stipulated by EU minimum harmonization directives when transposing them into national law. This pertains specifically to some parts of the consumer protection legislation.[82]

The European Consumer Agenda and the Directive 2011/83/EU on Consumer Rights in Distance and Off-Premises Contracts

In the Digital Single Market, consumer confidence (and purchasing) is still at a very low level. Too often, consumers are prevented from buying products through e-commerce or are discriminated against on the grounds of their nationality or place of residence by undertakings in other member states. However, consumers who shop online have twice as much choice in the domestic market and 16 times more choice in the single market.

Therefore, the Directive 2011/83/EU[83] repeals previous directives on distance and off premises contracts (85/577/EEC and 97/7/EC) and (re)establishes consumer confidence by (re)adopting consumer rights though this time on the basis of full harmonization.

Before the consumer places his order, the trader is obligated to inform the consumer in a clear and prominent manner, for all details pertaining to his name and address, fiscal identity, etc. (pre-contractual information).

After the consumer places his order, the consumer may electronically fill in and submit the model withdrawal form or any other unequivocal statement on the trader's website.

New harmonized rules on the transfer of risk in sales contracts and the default time-limit for the delivery of goods as well as a ban on hidden charges, on pre-ticked boxes which impose surcharges higher than the trader's actual costs for the use of a certain payment means (e.g. credit cards) and on charges for telephone hotlines higher than the standard telephone rate for calls. Contracts on the provision of utilities (water, gas, electricity) and contracts concerning digital content will also be covered by the new rules. However, certain areas such as healthcare services, passenger transport and gambling are excluded.

[82]See https://ec.europa.eu/digital-agenda/sites/digital-agenda/files/Code%20EU%20online%20rights%20EN%20final%202.pdf.

[83]Directive 2011/83/EU on consumer rights will change the requirements for pre-contractual information.

Despite all these provisions, the Directive 2011/83/EU still presents certain disadvantages: the fragmentation of national laws on remedies (guarantees) and unfair contract terms still remains since national laws still apply on the validity, termination or enforceability of a contract, transfer of ownership of goods, as well as gambling and healthcare services (Preamble par. 14 and art. 1 par. 3 of the Directive). In addition, member states may impose linguistic requirements on consumer contracts (art. 6 par. 7). Last but not least, there is only a limited right of withdrawal for contracts for the supply of digital content (computer programs, applications, games or music).

In addition, the New Consumer Agenda is based on an integrated approach and with the impact of the digital revolution on consumer behaviour as one of its cornerstones (COM(2012) 225 final, par. 3.2.). It aims at extending the Internal Market Information System (IMI) and the Consumer Protection Cooperation Network (CPC).

Draft Common European Sales Law (CESL)

As Vivian Reding states: "For the simple purchase of a [product], several national contract laws may apply at the same time, leading to a substantial degree of legal uncertainty for businesses as well as for consumers". Then questions arise: "Which law applies", "Can I trust the legal systems of other countries" and "Where and at what price can I get swift and reliable legal advice?", "Should I get involved in these countries and risk being sued abroad?" and then answers like this "Sorry, the product you have requested cannot be purchased by you as you are a non-resident[84]".

The draft CESL[85] was presented by the European Commission in 2011 and aimed to tackle all these problems by establishing an optional instrument for a Common European Sales Law. It has passed through scrutiny by the Council (December 2011) and the European Parliament (March 2012), but the proposal was not approved.[86] In case of approval, CESL would have been an optional second regime within each member state's law. The CESL would have been both national law and EU law. Where the parties would have validly agreed to use the CESL for a contract, only the CESL would govern the matters addressed in its rules. Its aim was to complement the EU Consumers Rights Directive 2011/83/EU (Fleischer 2012:235–252). In May 2015, in order to remedy the legal

[84]Reding V.: Keynote Speech at the Conference *"Towards a European Contract Law"* co-organized by the Study Centre for Consumer Law of the Catholic University of Leuven and the Centre for European Private Law of the University of Münster.

[85]Commission Proposal for a Regulation of the European Parliament and of the Council on a Common European Sales Law, COM (2011) 635—SEC (2011) 1165, SEC (2011) 1166.

[86]European Parliament resolution of 8 June 2011 on policy options for progress towards a European Contract Law for consumers and businesses (2011/2013(INI)), OJ C 380E, 11.12.2012, pp. 59–66.

fragmentation and adopt certain fully harmonised rules, the European Commission has submitted a proposal of a Directive on certain aspects concerning contracts for the online and other distance sales of goods (COM(2015) 635 final).

Data Protection Reform

The data protection proposal will open up the EU's Digital Single Market with one single legislative act for data protection. The implementation of the current Data Protection Directive (1995) is fragmented and complicated, and therefore, it will be replaced by a Regulation according to art. 16 TFEU.[87] Indeed, the Regulation (EU) 679/2016 (L 119/1) will replace the old Directive 95/46/EC in May 2018 and will establish a uniform and simplified set of rules across the EU. It will facilitate SMEs who will address their requests to just one single data protection authority (one-stop-shop). It will also provide for better cooperation between data protection authorities on cases with a wider European impact. In addition, the European Union Court of Justice has issued many judgments safeguarding data protection as a fundamental right in the EU and privacy (e.g. of the consumers) as an integral part of human dignity and personal freedom.[88] In this way, the same rules will apply to all businesses, including non-European ones, while at the same time consumers are ensured that their fundamental right to privacy is protected in a world of total connectivity, "where people exchange their personal data for digital services".[89]

The eIDAS Regulation

The Regulation (EU) 910/2014 (L 257/73) on electronic identification and trust services for electronic transactions in the internal market (eIDAS Regulation) enables secure and seamless electronic interactions between businesses, citizens and public authorities across the EU in "just one click". It has replaced Directive 99/93/EC since June 30, 2016 and will establish interoperability standards between national electronic identification schemes (eIDs) for submitting tax declarations, enrolling in a foreign university, remotely opening a bank account, setting up a business in another Member State, authenticating for internet payments, bidding to on line call for tender, etc. (https://ec.europa.eu/digital-single-market/en/trust-services-and-eid).

---

87 Reding, V.: *EU Data Protection rules: Better for business, better for citizens*, SPEECH/13/269, 26/03/2013. See also Commission Communication, "*A comprehensive approach on data protection in the European Union*", COM(2010) 609, 04 Nov 2010.

88 EU Charter of Fundamental Rights, art. 7 and 8.

89 Reding V., *EU Data Protection rules: Better for business, better for citizens*, SPEECH/13/269, 26/03/2013.

The Copyright Reform

In 2012 the EU has adopted Directive 2012/28/EU on certain permitted uses of orphan works (L 299/5) in order to facilitate the digitisation and dissemination of cultural heritage. In 2014, the EU has adopted 2014/26/EU on collective rights management and multi-territorial licensing of rights in musical works for online uses (L 84/72). The Directive will facilitate the multi-territorial licensing by collective management organisations. In addition, the European Commission has issued a legislative proposal on cross-border portability of online content services which aims at ensuring that consumers who buy or subscribe to films, sport broadcasts, music, e-books and games can access them when they travel in other EU member states (COM(2015) 627).It has also submitted proposals in order to allow for wider online availability of content across the EU and to adapt exceptions and limitations to the digital world (COM(2015) 626, see https://ec.europa.eu/digital-single-market/en/copyright).

VAT Rules for E-Commerce and MOSS

Current VAT rules do not contribute to the Digital Agenda and their complexity dissuades SMEs from online transactions.[90] Beyond a certain threshold, set by each member state at EUR 35 000 or EUR 100 000, *the seller of products must register in the member states where he sells.* The seller is therefore deterred from cross-border activities since he is obliged to know the thresholds of the host member states and fulfil the VAT obligations (registration for VAT, completing VAT returns and payment of the VAT due) in each member state where sales exceed the threshold. On the other hand, *the EU supplier of services* to final consumers in the EU is taxed in the member state where that supplier is established. Consequently, businesses could take advantage by establishing their company seat in those member states applying a lower VAT rate.

Directive 2008/8/EC of 12 February 2008[91] has set that as from 2015, electronic services provided by an EU supplier to a non-taxable person (e.g. final consumer) will also be taxable at the place where the customer is established and the supplier himself will be responsible for collecting the VAT paid in the member state of the customer. If the customer is established outside the EU, no VAT will be due in the EU.[92]

---

[90]Commission Communication "*Towards a simpler, more robust and efficient VAT system tailored to the single market*", COM (2011) 851.

[91]Council Directive 2008/8/EC of 12 February 2008 amending Directive 2006/112/EC as regards the place of supply of services, OJ L 44, of 20.02.2008; p. 11, available at: http://eur-lex.europa.eu/LexUriServ/LexUriServ.do?uri=OJ:L:2008:044:0011:0022:EN:PDF.

[92]Before Directive 2008/8/EC, VAT for electronic services provided by non-EU suppliers is currently collected and paid in the member state where the customer is established.

In addition, the European Commission has proposed the introduction of an optional Mini One Stop Shop mechanism (MOSS)[93] that would help to alleviate administrative burdens for businesses, assist in VAT identification and declaration as they will be able to fulfil all their obligations in their own member state. It will first cover suppliers of telecommunications, broadcasting television and electronic services provided to final consumers (since 2015) and afterwards, when so agreed by the member states, all suppliers of services will be covered.[94]

Payments and SEPA

Though e-commerce is performed by distance electronic payments, almost 35 % of Internet users have doubts concerning security of payment and additional charges, as well as the cost and the quality of cross-border delivery services. The Single Euro Payments Area (SEPA) is now considered to be the "acquis of the European Union" on which to develop a strategy on new pan-European payment services respecting the right to data protection and confidential communications. This aim will be achieved by removing the barriers to entry and competition on these markets, ensuring the transparency of payment services for both consumers and sellers, improving and accelerating the standardization and interoperability of payments by card, Internet or mobile phone, increasing the level of security of payments and data protection.[95]

Conclusions on the EU Policy for Promoting E-Commerce

According to the European Commission, "E-commerce is a key twenty-first century marketplace for the EU, that may contribute to knowledge economy and its competitiveness", as well as have a significant impact on jobs and growth. "It is a vital force and an important catalyst for achieving the aims of the EU 2020 strategy by unlocking the full potential of the EU single market". The "relaunch of the

---

[93]Commission Proposal for a Council Directive amending Directive 77/388/EEC with a view to simplifying value added tax obligations, COM (2004) 728 final, 29.10.2004. Bibliographic notice available at: http://eur-lex.europa.eu/Notice.do?val=390797:cs&lang=en&list=390798:cs,390488:cs,390797:cs,&pos=3&page=1&nbl=3&pgs=10&hwords=.

[94]Commission Staff Working Document On Line Services, including E-Commerce in the Single Market, SEC (2011) 1641 final, Accompanying Document to COM (2011) 942, A Coherent Framework to boost Confidence in the Digital Single Market of E-commerce and other online services, 11 January 2012, pp. 102–103.

[95]Commission Green Paper "Towards an integrated European market for card, internet and mobile payments", COM (2011) 941.

single market is therefore an essential element of the EU 2020 strategy" since "certain elements of the flagship initiatives will help to structure the operation of the single market" in particular the Digital Agenda for Europe, the Union of innovation and an modern industrial policy, increasing "synergies between the various flagship initiatives" and exploiting "the considerable opportunities for growth presented by the digital economy". Otherwise, the cost of the failure to complete the Digital Single Market is estimated to be at least 4.1 % of GDP between 2010 and 2020, i.e. EUR 500 billion.[96]

The EU Digital Agenda Policy aims to overcome the patchwork of different laws, rules, standards and practices, often with little or no interoperability, to strengthen consumer rights and skills and to change the legal framework to cross-border transactions (payments, deliveries, dispute resolution, risk of abuse).

The EU faces national legislative fragmentation using either optional instruments based on the principle of subsidiarity (e.g. CESL, Alternative Dispute Resolution) or Regulations based on the new provisions of the Treaty of Lisbon (Data protection, SEPA, etc.) or full harmonization Directives (Consumer Rights Directive). Relaunching the Single Market means not only adopting new measures but also reviewing existing mechanisms such as the Directives on e-commerce, e-signatures, as well as the intellectual property and the EU Data protection system. However, national opposition in the cumbersome EU legislative procedures for adopting the Digital Agenda proposals does not help in furthering e-commerce. Proposals take as much as 2–3 years to be adopted and may take another 2 years to be implemented (e.g. Consumer Rights Directive, etc.). This is the *cost of non-integration* that the EU has to pay in comparison with its competitors in the international trade, e.g. the USA that the use of e-commerce amounts to even 66 %.

In addition, increasing e-commerce and services in the EU *requires firm and concerted action in line with the Digital Agenda for Europe*. Indeed, the Digital Agenda is the cross section of many, if not all, EU policies headed by different Commissioners. Therefore, besides monitoring the measures proposed, the European Commission is vigilant to ensure the *consistency and convergence of initiatives* and the avoidance of any *newly created obstacles* to the development of online services, even by its own legislation.

The global dimension of e-commerce and of online services is of high importance for the EU and entails more active participation in the work of existing international organizations and bodies (UNCITRAL, Council of Europe, OECD, ICC) and sufficient coordination. The ultimate aim is to facilitate e-commerce of European business firms operating both inside and outside the EU and of business

[96]Commission Communication "Towards a Single Market Act For a highly competitive social market economy: 50 proposals for improving our work, business and exchanges with one another", Brussels, 27.10.2010, p. 4, COM (2010) 608 final, available at http://eur-lex.europa.eu/LexUriServ/LexUriServ.do?uri=COM:2010:0608:FIN:en:PDF.

firms from third countries established or not in the EU. The significance of e-commerce to global trade is shown also by the fact that the 2013 Public Forum, the WTO's largest annual outreach event for civil society, academia, business, the media, governments, parliamentarians and intergovernmental organizations, took place on 1–3 October 2013 in Geneva, with the theme "Expanding Trade through Innovation and the Digital Economy".[97]

Overview of the US Digital Policy

The Hegemony of the USA in the Digital Economy

Since 1996, the US Congress and the US Federal Governance were confident that the Internet could lead to a major increase in US productivity of american content industries establishing the US as the uncontested leader in e-commerce, especially in content industries (film, music, etc.) (Wunsch-Vincent 2003:108).

The USA have indeed become the leader in the digital economy for the following reasons:

(a) It was in USA where it all began and is there that the Internet is governed (ICANN)[98]
(b) The USA is a large domestic market with common technical standards which has allowed innovative digital companies in the USA to move quickly across the globe and
(c) The USA has a leading position in technological innovations that has resulted in a critical mass of global digital businesses of large scale since the 1970s: 66 digital companies are listed in the 500 largest companies in the USA (Amazon for books, Blizzard Activision for video games, Google for search machines, VeriSign for security services, e.g. digital certificates, and Symantec for antivirus software) (Marsh 2010).

In addition, the USA benefits from the adoption of the Uniform Commercial Code (UCC) in 1952 by the National Conference of Commissioners on Uniform State Laws (NCCUSL) and the American Law Institute. The UCC is an integrated modern body of law on all transactions, especially on sales of goods across all the US states. Its new version is adopted in 2009.[99] It is not binding but constitutes a recommendation of law that all the states have to adopt with or without amendments (Sukurs 2001:1481 et suiv).

[97]News item: http://www.wto.org/english/news_e/news13_e/pfor_26mar13_e.htm.

[98]See above Chap. 2.2.

[99]See http://www.ali.org/index.cfm?fuseaction=publications.ppage&node_id=86.

The Fundamental Approach of the USA Towards Digital Environment

Because of its local industry demand, the USA has avoided introducing in e-commerce the same trade barriers that have been imposed in the traditional commerce of goods or services (Wunsch-Vincent 2006:103–105). Indeed, the US legislation differs from the European paternalistic ex ante approach and reflects "a hands-off approach" in order to establish a global free trade framework. This approach is established by the US Framework for Global Electronic Commerce, prepared after 18-month consultations of an interagency working group on Electronic Commerce.[100] The Framework is based on the principle that the *private* sector should lead in e-commerce with industry *self-regulation* while governments should avoid undue restrictions on e-commerce. If it is necessary for the governments to intervene, they should support and enforce a predictable, minimalist, consistent and simple legal environment (Wunsch-Vincent 2006:103–105).

In addition, the Framework[101] defines "Free Global e-Commerce" as a tariff-free environment without any new taxes but with free market access and "the broadest possible free flow of information across international borders". This approach has influenced the WTO Work Programme on E-Commerce and the WTO Duty-free Moratorium on Electronic Transmissions (Wunsch-Vincent 2006:103–105).

This framework is based on the 1997 "Presidential Directive on E-Commerce", a very important policy document that outlines the principles of Federal Administration's actions in order to move to the new electronic age of commerce and create the emerging digital marketplace. It also presents a series of policies and establishes an agenda for international discussions and agreements to facilitate the growth of electronic commerce. For electronic commerce to flourish, the private sector must lead. Therefore, the Federal Government has encouraged industry self-regulation wherever appropriate and supported private sector efforts in order to develop technology and practices that facilitate the growth and success of the Internet.[102]

The Internet Tax Freedom Act

The 1998 "Internet Tax Freedom Act" (ITFA)[103] temporarily banned taxes on Internet access and multiple or discriminatory "Internet only" taxes on

---

[100]http://clinton4.nara.gov/WH/New/Commerce/about.html.

[101]See the text in http://clinton4.nara.gov/WH/New/Commerce/about.html.

[102]White House: Memorandum for the Heads of Executive Departments and Agencies. Subject: Electronic Commerce, 1 July 1997 Available at https://www.fas.org/irp/offdocs/pdd-nec-ec.htm.

[103]Internet Tax Freedom Act (ITFA) public Law No 105–277, October 1998.

e-commerce (such as bit taxes, bandwidth taxes and email taxes). The Act empowered the US President to negotiate bilateral and multilateral agreements to remove barriers to global e-commerce through various international organizations, and established the "Advisory Commission on Electronic Commerce". It did not exempt sales made on the Internet from taxation, as these may be taxed as non-Internet sales, or mail order sales. The 2004 "Internet Tax Non-Discrimination Act"[104] prolonged the US Internet tax moratorium for 2 more years, while the "Internet Tax Freedom Act Amendment Act of 2007" extended the prohibitions against multiple and discriminatory taxes on e-commerce until 1 November 2014.[105] The 2015 Trade Facilitation and Trade Enforcement Act extended the moratorium on taxing Internet access permanently (Stupak: 1).

The Digital Millennium Copyright Act (DMCA)

The DMCA[106] implements the 1996 WIPO Treaties[107] amending the US Copyright Act. It also criminalizes the circumvention of technological protection measures employed by copyright owners and prohibits the removal or alteration of copyright management information. It initiated for the first time the non-liability of service providers for copyright infringements by their users.[108]

Communications and Broadband Deployment

The US Federal Communications Commission (FCC) has set the strategic goals relating to broadband deployment.[109] The FCC also issued rules exempting "Internet" services from domestic telecommunication regulation and thus from common carrier obligations (also classification of cable modem services as information service).[110]

[104]47 United States Code Sect. 151.

[105]Public Law 108–435.

[106]Public Law 105–304.

[107]The WIPO Copyright Treaty and the WIPO Performances and Phonograms Treaty.

[108]US Copyright Office: Executive Summary, Digital Millennium Copyright Act, Sect. 104 Report, available at http://www.copyright.gov/reports/studies/dmca/dmca_executive.html.

[109]www.fcc.gov/broadband/.

[110]Title XI of Public Law 105–277.

The Work of the Federal Trade Commission as the Protector of the Consumers and Small Firms in the Era of the Free Internet

The Federal Trade Commission (FTC) aims to protect consumers and competition in the USA and global markets.[111] In this context, the FTC regulates some electronic commerce activities such as the use of commercial e-mails which is reviewed in accordance with the 2003 CAN SPAM Act[112] that established national standards for direct marketing over e-mail, online advertising and consumer privacy. In addition, the FTC ensures that advertising and marketing must be truthful and non-deceptive on the Internet and has enforced the corporate privacy promises on data privacy, including the security of consumers' personal information.[113] As a result, any corporate privacy policy related to e-commerce activity may be subjected to enforcement by the FTC (Wunsch-Vincent 2006:103–105).

The Digital Trade Agenda

In 2000, the USA launched the US Digital Trade Agenda (or "trade policy for the networked economy") in order to anticipate, prevent and remove e-commerce trade barriers and to create a set of international rules for the development of e-commerce trade (Barshefsky 2001).

The US Digital Trade Agenda includes four substrategies:

(i) To make sure that WTO principles and commitments apply to e-commerce and to resolve the classification issues in the most liberal way for digital trade;
(ii) To use the negotiation opportunities to secure improved market access commitments for digital products;
(iii) To create a regulatory trade discipline for e-commerce; and
(iv) To update trade agreements so that new treaties deal with trade-related aspects of intellectual property protection in the digital trade age (Wunsch-Vincent 2003).

The US government and in particular the US Trade Representative obtained the formal delegation of trade negotiation authority from the US Congress with the

[111]Federal Trade Commission Strategic Plan for the Fiscal Years 2009–2014, available at http://www.ftc.gov/sites/default/files/documents/ reports_annual/strategic-plan/spfy09fy14.pdf.

[112]15 U.S.C, ch. 103, Public Law No. 108–187.

[113]http://business.ftc.gov/privacy-and-security.

"Bipartisan Trade Promotion Authority Act of 2002".[114] This Act has established the Trade Promotion Authority as a fast-track authority to conclude trade agreements with a simplified congressional ratification procedure (Wunsch-Vincent 2006:103–105). Both the US Trade Representative and the new fast track authority under the name "Trade Promotion Authority" (TPA) had to pursue a "parallel track" of preferential and multilateral trade negotiations, alongside the ongoing Doha negotiations of the World Trade Organization (WTO). Following the new "principle of competitive liberalization", the USA uses "a concurrent bilateral, regional and multilateral approach to pursue US digital trade objectives" (Wunsch-Vincent 2003).

The Digital Government Strategy

In 2012 the US Federal Government after taking into account the rapidly changing mobile landscape and the fact that almost 50 % of American adults were smartphone owners, has adopted the Digital Government Strategy.[115]

The Digital Government Strategy complements several initiatives such as: the Executive Order 13571 (Streamlining Service Delivery and Improving Customer Service),[116] the National Strategy for Trusted Identities in Cyberspace (NSTIC)[117] and the 25-Point Implementation Plan to Reform Federal Information Technology Management (IT Reform).[118]

Based on the present mix of cloud computing, ever-smarter mobile devices and collaboration tools, the Federal Government aims to deliver and receive digital information and services anytime, anywhere and on any device safely, securely and with fewer resources.[119]

The strategy is built upon four overarching principles:

- An "Information-Centric" approach that will make a shift from managing "documents" to managing discrete pieces of open data and content which can be tagged, shared, secured, mashed up and presented in the way that is most useful for the consumer of that information.

---

[114]Public Law 107–210 under Internet: www.tpa.gov/pl107_210.pdf (20 June 2004); House of Representatives (2001); and Senate Committee on Finance (2002), as mentioned by Wunsch-Vincent, S., op.cit.

[115]Digital Government: Building a twenty-first century Platform to better serve the American People, 23 May 2012, p. 2.

[116]See http://www.whitehouse.gov/the-press-office/2011/04/27/executive-order-streamlining-service-delivery-and-improving-customer-ser as mentioned by Digital Government Strategy, op.cit., p. 3.

[117]See http://www.whitehouse.gov/sites/default/files/rss_viewer/NSTICstrategy_041511.pdf as mentioned by Digital Government Strategy, op.cit., p. 3.

[118]See http://www.cio.gov/documents/25-Point-Implementation-Plan-to-Reform-Federal%20IT.pdf as mentioned by Digital Government Strategy, op.cit., p. 3.

[119]Op.cit., p. 1.

- A "Shared Platform" approach that will aim at working together, both within and across agencies, reducing costs, streamlining development, applying consistent standards and ensuring consistency in how we create and deliver information. The Intra-Agency Governance will be secured by the establishment of a Digital Services Innovation Center to improve the government's delivery of digital services.
- A "Customer-Centric" approach that will influence the way of creation, management and presentation of data through websites, mobile applications, raw data sets and other modes of delivery, and will allow customers to shape, share and consume information, whenever and however they want it. This approach will improve Priority Customer-Facing Services for Mobile Use.
- A platform of "Security and Privacy" that will ensure a coordinated approach on the safe and secure delivery and use of digital services in order to protect information and privacy.

The Digital Government Strategy presupposes the secure architecture of the governmental systems based on common standards, interoperability and "openness from conception". It aims to produce better content and data, through multiple channels in a program and device-agnostic way so that the system will work regardless of the user's device. It will help issue government-wide open data, content and Web API policy, especially by making existing High-Value Data and Content Available through Web APIs.[120]

The US Technology Agenda (the Obama–Biden Plan)

The US Technology Agenda will ensure the full and free exchange of information through an Open Internet (network neutrality) and diverse Media Outlets. It aims at establishing technology and modern communications infrastructure (broadband and wireless spectrum) to create a more transparent, accountable and connected democracy under the supervision of the Chief Technology Officer (CTO). It will improve competitiveness, protect intellectual property both abroad and in the USA (civic discourse), reform the patent system and solve problems such as clean energy, healthcare costs and public safety.[121] It will deploy the Next-Generation Broadband by reforming the Universal Service Fund, giving new tax and loan incentives. It will double federal funding for basic research over 10 years and provide new research grants to the most outstanding early career researchers in the USA. It also plans to make the Research and Development tax credit permanent so that firms can rely on it when making decisions to invest in domestic R&D over multi-year timeframes.

[120]See http://www.archives.gov/digitalstrategy/milestones.html.

[121]See http://change.gov/agenda/technology_agenda/.

Data Management Policy for Research

The US approach on data management for research lacks a coherent federal policy or a forum for a concerted effort. There are many federal, state and private agencies in control of funding as well as many sources for research funding, and each of these funding entities has its own application process, criteria and assessment (Mullins 2008).[122]

In addition, technology offers new digital opportunities, including open access to scholarly publications, access to digital data sets, creation of standards for data set management and national repositories for scanned images while libraries like the Library of Congress are managing massive amounts of data (Mullins 2008). However, no uniform policy is spelled out (Erway 2013).

International Cooperation Between EU and USA in Digital Trade

The EU and US Agreement on Trade-Related Information and Communication Technology Principles

In 2011 the European Commission and the US government, under the framework of the Transatlantic Economic Council (TEC),[123] agreed on a set of ten fundamental principles for trade in information and communication technology (ICT) services. These principles will be incorporated in their trade agreements with other countries, in order to raise the profile of ICT services in bilateral agreements as well as in the WTO, support the global development of ICT networks and services, allow service providers to compete for contracts with local incumbents on an equal footing and open up ICT markets worldwide to the benefit of all businesses and consumers.

The EU-US Trade Principles for Information and Communication Technology (ICT) Services[124] address the following issues: transparency in legislation and regulation; open access to networks and applications; the free flow of information

---

[122]Mullins names the Federal Departments of Agriculture, Defense, and Energy, federal agencies such as the National Science Foundation (NSF), the National Institutes of Health (NIH), and the Institute for Museum and Library Services (IMLS).

[123]The Transatlantic Economic Council is a political body to oversee and accelerate government-to-government cooperation with the aim of advancing economic integration between the EU and the US. Since 2007 the Transatlantic Economic Council has brought together those Members of the European Commission and US Cabinet Members who carry the political responsibility for the policy areas covered by the Framework. See http://ec.europa.eu/enterprise/policies/international/cooperating-governments/usa/transatlantic-economic-council.

[124]http://www.ustr.gov/webfm_send/2780.

across borders; foreign investment in ICT sectors; facilitation of cross-border supply of services; efficiency of spectrum allocation; independence of regulatory authorities; procedure for granting of operating licenses; interconnection between suppliers of basic public telecommunication services; and international cooperation.[125]

The EU-US Trade Principles for ICT Services have been agreed on a best endeavour basis and do not affect the rights of the EU or the USA to maintain their respective policy approaches regarding the protection of intellectual property, privacy and personal data and the enhancement of cultural diversity. The principles will be reviewed every 2 years and are without prejudice to World Trade Organization (WTO) rights and obligations and exceptions under the General Agreement on Trade in Services (GATS).[126]

Therefore, the principles are non-binding but if widely adopted they will support the global development of ICT services that are critical to e-commerce, Internet search and advertising, thus increasing productivity, decreasing the cost of doing business and promoting the growth of cutting-edge industries.[127]

The Negotiations for the Treaty on Trade and Investment (TTIP) for Jobs and Growth

After President Obama announced the beginning of the TTIP negotiations in January 2013, the USA and EU summarized their joint objectives in a joint report of the High-Level Working Group for Jobs and Growth in February 2013. After consultations with a wide range of 350 stakeholders, negotiations began in July 2013 in 24 groups of negotiators, in areas like: Market access for industrial and agricultural products, as well as the rules of origin for those products; technical regulations; the sanitary and phytosanitary regulations primarily in the area of food safety, regulatory coherence and particular sectors; investment and services including in the areas of *telecommunications, electronic commerce and data flows* (*respecting each other's legislation*), *cross-border services and financial services*; government procurement, intellectual property, labour, environment, state-owned enterprises small- and medium-size enterprises, localization barriers to trade,

[125]United States-European Union Trade Principles For Information and Communication Technology Services, http://www.ustr.gov/about-us/press-office/press-releases/2011/April/united-states-european-union-trade-principles-inform.

[126]EU and US agree trade-related information and communication technology principles to be promoted world-wide, IP/11/402, Brussels (4 April 2011).

[127]United States-European Union Trade Principles For Information and Communication Technology Services, http://www.ustr.gov/about-us/press-office/press-releases/2011/April/united-states-european-union-trade-principles-inform.

competition, raw materials and energy, and legal and institutional issues such as dispute settlement.

This TTIP agreement, if signed, will increase growth, jobs and international competitiveness, by including benefits for the small and medium enterprises (SMEs), a basic concern throughout the agreement, but also in a specific chapter within SME-related issues. This is based on the fact that the USA and the EU already spend $4 trillion in foreign direct investment in each other's economies while there is a $3 billion per day in trade in goods and services.[128] Only with international cooperation between the USA and the EU there will be synergies and sustainable growth for both USA and EU. The TTIP negotiations, if completed, will contribute immensely in order to connect the two continents and link together "Technology—Society—International Cooperation—Sustainability".

Conclusions About the Digital Policies of the EU and USA and their Significance for Society and Sustainability

The Internet plays a crucial role as an *economic force and its influence is likely to increase*. In 2010, the Internet economy amounted to $2.3 trillion or 4.1 per cent of GDP across the G20 countries. This number is expected to grow to $4.2 trillion or 5.3 per cent of GDP by 2016. By then, the Internet economy will employ 32 million additional people in G20. More than two billion people are online today, and that number is thought to climb to three billion by 2016.[129]

In addition, in January 2014 in the World Economic Forum[130] in Davos, Switzerland, besides the discussion on Internet governance, the "Davos Declaration" was adopted, in which the industry (Microsoft, Samsung, Telefonica), voluntary groups, education providers and others have pledged to potentially create 250,000 extra courses, 100,000 extra traineeships and thousands of extra jobs.[131]

[128]Department of State, Washington, D.C., Transcript: Chief Negotiators, Dan Mullaney and Ignacio Garcia Bercero Hold a Press Conference Following the Third Round of Transatlantic Trade and Investment Partnership (TTIP) Talks (20 December 2013), available at http://www.ustr.gov/about-us/press-office/press-releases/2013/December/TTIP-Third-Round-Press-Conference-transcript.

[129]Available at http://internet-economy.org/2013/05/launch/#.UvgoLvl_teU.

[130]The WEF is constituted by 1000 of the world's top corporations, global enterprises usually with more than US$ 5 billion in turnover as an independent international organization committed to improving the state of the world by engaging business, political, academic and other leaders of society to shape global, regional and industry agendas. Incorporated as a not-for-profit foundation in 1971, and headquartered in Geneva, Switzerland, the Forum is tied to no political, partisan or national interests. See http://www.weforum.org/our-mission and http://www.weforum.org/our-members.

[131]European Commission: European Commission joins forces with companies to deliver over 250,000 extra training courses and thousands of new digital jobs, IP/14/40 (24 January 2014), available at http://europa.eu/rapid/press-release_IP-14-40_en.htm.

The USA have understood the importance of the digital economy very early. The USA are the undisputed leader in the digital economy thanks to their leading role in technology and innovation. The country has nurtured the digital economy thanks to its "hands-off" policy, articulated since 1998. It has continued to do so with its DCMA, the Digital Trade Agenda, the Moratorium on e-commerce taxation and its Technology Agenda.

The EU has also taken measures but it has mostly focused on the e-consumer and its protection, rather than the Internet governance and business sustainability. Though the EU has adopted Directives on e-signature, on certain aspects on e-commerce and on copyright, its approach was more interventionist in the digital economy than the American one. In addition, the EU has to face fragmented national policies, which even after harmonization, were not uniform, since Directives allowed some flexibility and the adoption of stricter measures if a member state so wished. It was only in 2010, ten years after the USA, that the EU launched its integrated Digital Agenda and its Digital Single Market policy.

The EU Digital Single Market (DSM) will contribute in increasing employment since SMEs could grow to scale in the new digital economy and increase their global competitiveness using the DSM as a platform and open up specialist digital areas across Europe (Marsh 2010). Moreover, the European business at least on telecommunications[132] won't be competitive unless it is connected.[133]

According to our opinion, in case that the EU digital policies do not meet the disagreement of certain member states, the impetus gained may help the EU overcome the economic crisis and achieve sustainable development and growth for the benefit of the European society.

References

Anagnostopoulou D (2011) Law of international transactions, the European dimension. Ant. Sakkoulas

Barshefsky Ch (2001) Trade policy for a networked world. 80 Foreign Aff. 134

Boss AH, Kilian W (eds) (2008) The United Nations convention on the use of electronic communications in international contracts: an in-depth guide and sourcebook. Kluwer Law International, The Netherlands

Erway R (2013) Starting the conversation: university-wide research data management policy, OCLC research report. http://oclc.org/research/publications/library/2013/2013-08r.html. 25 March 2013

Faria JAE (2006) The UN convention on the use of electronic communications in international contracts—an introductory note. 55 International Comparative Law Quarterly 689

[132]Op.cit.

[133]European Commissioner Neelie Croes: Ensuring Jobs for the Future—my blog from Davos, 28.1.2014, available at http://ec.europa.eu/commission_2010-2014/kroes/en/content/ensuring-jobs-future-my-blog-davos. See also http://ec.europa.eu/digital-agenda/en/connected-continent-single-telecom-market-growth-jobs.

Fleischer H (2012) Optionales europäisches privatrecht ("28. Modell"), 76 Rabels Zeitschrift für ausländisches und internationales Privatrecht 235

Grammatikaki-Alexiou A (2001): E-commerce, private international law, international uniform law and Community regulatory efforts – a comparative review, Hellenic Review of European Law 2001, Special Issue, In Memoriam Krateros Ioannou, p. 135 (in Greek)

Igglezakis I (2009) Law on electronic commerce. Sakkoulas Publishers (in Greek)

Karadimitriou K (2008) The e-signature as a means for the security of transactions in e-commerce. Sakkoulas Publishers (in Greek)

Marsh R (2010) Benchmarking Europe and the United States, Verso Economics. http://www.epc.eu/dsm/5/Market_scale.pdf. Accessed 25 March 2013

Mullins JL (2008) A national digital data policy for the United States: to be or not to be. In 2008 IATUL proceedings, paper 2. http://docs.lib.purdue.edu/iatul/2008/papers/2. Accessed 25 March 2015

OECD (2002) Measuring the information economy 2002, Annex 4, Paris

Samuelson P (2013) The US Digital Agenda at the World Intellectual Property Organisation. http://people.ischool.berkeley.edu/~pam/courses/cyberlaw97/docs/wipo.pdf Accessed 25 March 2013

Stupak, J The Internet Tax Freedom Act: In Brief, April 13, 2016 Congressional Research Service 7-5700, R43772. https://www.fas.org/sgp/crs/misc/R43772.pdf. 28 October 2016

Suκurs Ch. (2001) Harmonizing the battle of the forms: a comparison of the United States, Canada, and the United Nations convention on contracts of the international sale of goods. 31 Vanderbilt J Transnatl Law, pp. 1481–1515

Tountopoulos V, Hatzopoulos V (2001) Electronic addresses in the internet, the problem of domain names. Nomiki Bibliothiki, Athens (in Greek)

Tziva E (2007) E-commerce for medicine. Sakkoulas Publishers, Athens-Thessaloniki (in Greek)

UNCITRAL (2008) Promoting confidence in electronic commerce: legal issues on international use of electronic authentication and signature methods. http://www.uncitral.org/pdf/english/texts/electcom/08-55698_Ebook.pdf. 25 March 2013

Wunsch-Vincent S (2003) The digital trade agenda of the US: parallel tracks of bilateral, regional and multilateral liberalization. 58 Aussenwirtschaft, issue 1 (March) 7

Wunsch-Vincent S (2006) The WTO, the internet and trade in digital products: EC-US perspectives. Hart Publishing, Oxford

Part II
Some Applications

Chapter 5
Social Justice Through Aadhaar: An e-Policy Initiative

Anurag Kumar Srivastava and Sangeeta Sharma

Abstract The application of Information and Communication Technology (ICT) in various domains of governance is more than two decade old in India. India's orientation towards welfare driven polity based on socialist pattern of society has been further strengthened by the technological interventions in the form of e-Governance for policy formulation and policy implementation. India's development vision for all has not translated into reality as per the expected lines as the benefits of development have not been equitably shared by all. The distributive injustice has further aggravated the problem of poverty and inequality which has manifested itself in various forms including unemployment, hunger, malnutrition and poverty. Even after 69 years of India's independence fruits of planned development have not reached at the grass root level. It has become evident that a high rate of growth is not a substitute for genuine policy implementation that seeks to ensure equitable distribution of the gains of development. Therefore, there is a need for a public policy through an e-platform for ensuring equitable, fair, transparent, accessible, and participative approach towards development by realising the ethos of social justice. The technology itself can be a biggest leveller which can foster transparency, equity and enhance people's integration into developmental processes which is vital for strengthening the social justice. The Unique Identity of India UID-Aadhaar is an initiative taken by Government of India (G.O.I) under the overall administrative control and management of Planning Commission in the year 2009. Aadhaar has been entrusted with the massive task of ensuring genuine identity of all throughout the country, in order to establish the

A.K. Srivastava (✉)
Department of Public Administration and Policy, School of Liberal Studies, Pandit Deendayal Petroleum University, Gandhinagar, Gujarat 382 007, India
e-mail: anurag.srivastava@sls.pdpu.ac.in; anusri900@gmail.com

S. Sharma
Department of Public Administration, University of Rajasthan, JLN Marg, Jaipur, Rajasthan 302004, India
e-mail: sangeetajpr2004@yahoo.co.in

L.W. Zacher (ed.), *Technology, Society and Sustainability*,
DOI 10.1007/978-3-319-47164-8_5

rightful claims of citizens for accessing the benefits of various governmental services and programmes. The ICT integration in Aadhaar promotes creation of a digital identity which will lead India towards the next level of transition in the sphere of governance. Aadhaar has gained acceptability in administrative functioning and gradually gaining popularity among the masses. It will be interesting to explore the utility and significance of Aadhaar in rolling out public services. The seeding Aadhaar card with various welfare schemes and subsidy programmes has been made to realise the ethos of social justice and equitable development of all through technology. The adoption of Aadhaar for delivering public services/ schemes to the people is a symbiotic process in which both people and government as stakeholders are mutually benefitted. The government gets benefitted in terms of curbing leakages in target/beneficiary identification and validation of people and for the people to avail the government services in a hassle free, transparent, quick, and simple manner through Aadhaar. Aadhaar has become the world's biggest biometric id programme with around 1 billion cards which have been rolled out so far, this clearly reflects substantial investment of policy inputs, capital, administrative will and peoples' participation in making Aadhaar a driving force in creating developmental opportunities, promoting equality and social justice for all.

Keywords Social justice · ICT · e-Governance · UID · Aadhaar · Supreme Court of India

Introduction

Concept of Social Justice

Plato (380 BC) said that justice was achieved when each person received goods they deserved based on their prescribed position in the social order (Grube 1974). Aristotle (384–322 BC) said that justice was a principle that ensured social order by regulating the distribution of benefits (Ross 1980). However, in Aristotle's view, equality and justice applied only to individuals who occupied the same stratum of the hierarchical social order. According to these ideas of social justice, unequals in the social hierarchy are to be treated unequally. These ideas did not challenge the social structures of society, but worked within them. Rousseau (1712–1778) and others who followed him in the "age of revolution" shaped the formation of modern institutions in the west with the view that the pursuit and realisation of social justice was linked to the preservation of individual liberty or freedom, achievement of equality (of rights, opportunities and outcomes) and establishment of common bonds of all humanity (Rousseau 1754).

Justice has been manifested in the Preamble of the Indian Constitution, which talks about three kinds of Justice that is social, economic and political. In fact insightful observation reflects the logical sequencing mentioned in the preamble itself. The precedence has been given to social justice because no economic and political justice can be attained and sustained if the social justice is not materialised into reality. The Directive Principles of State Policy of Indian Constitution also indicate and promote the equitable society in various articles by resorting to create just and equitable society. It talks about social justice in terms of equity of opportunities and development. The social justice discourse further got momentum in the wake of uneven distribution of wealth coupled with fractured development and rising income inequalities. The technology has played a very important role in making human life better in general and Information Communication Technology in particular. The technological expansion and advancements are taking place at an unprecedented pace in the twenty-first century. No aspect of human life, be it personal or professional, community, social, commercial, financial, political, has remained untouched by the technological interventions. The application of technology in the sphere of governance is quite widely practised across the globe. ICT has been utilised in the form of e-Governance in the sphere of government to provide electronic delivery of services to citizens, enhancing participation of the citizens in governmental activities and processes by promoting the concept of e-Democracy. The social justice is at the heart of ethical politics and core of sovereignty. It embraces equality, equal access to resources, well-informed citizenry, fair distribution of public wealth and compliances with law. As the web connectivity is expanding, the interactive plateau for sharing the information regarding violation of human rights, deprivations, marginalisation, socio-economic injustice and other kinds of exploitations is getting echoed, thereby increasing the intensity of delivering social justice. In brief, some of the essential components of social justice are:-

a. Equal access to opportunities and rights
b. Fair system of law and due process
c. Ability to take up opportunities and exercise rights
d. Protection of vulnerable and disadvantaged people individual responsibility
e. Individual capability (emphasises the personal characteristics that enable people to take advantage of opportunities)
f. Recognition of human value and well-being

The role of Aadhaar card is significant in realising the various components of social justice as illustrated above especially in regard to equal access to opportunities and rights.

It is pertinent to mention that this policy initiative will prove to be an effective instrument of social justice and for bringing empowerment. Needless to say that empowerment is the prerequisite for attainment of social equality, a convincing precondition of social justice. The theoretical precept may convey the rationalistic point of view, but the realistic aspect is to distribute the fruits of development, empowerment and progress to everyone. Since Aadhaar scheme will form the key index of inclusiveness in coming era, the discrepancies could be easily surfaced. Once the identification of discrepancies is done, the action would follow, thus perpetuating the ethos of social justice through an e-policy initiative.

Aadhaar an e-Policy Initiative Through E-Governance

The World Bank has defined e-government as "government-owned or operated systems of information and communications technologies (ICTs) that transform relations with citizens, the private sector and/or other government agencies so as to promote citizen empowerment, improve service delivery, strengthen accountability, increase transparency, or improve government efficiency" (Panzardi et al. 2002).

As per the National e-Governance Plan (NeGP) the ultimate objective of NeGP is to bring public services closer home to citizens, as articulated in the Vision Statement of NeGP (National e-Governance Plan 2006).

> Make all Government services accessible to the common man in his locality, through common service delivery outlets, and ensure efficiency, transparency, and reliability of such services at affordable costs to realize the basic needs of the common man (MEITY 2014).

Government of India formally launched its National e-Governance Plan (NeGP), in 2006. It comprises of 27 Mission Mode Projects encompassing 10 Central MMPs, 10 State MMPs and 7 Integrated MMPs. Thus, the initiatives in the arena of e-Governance have been of marathon magnitude. The e-interventions have resulted into the improvisation of quality of services, the cascading effect of which can be seen in the form of delivering social justice to its people.

There lies a sound logic and reasoning behind the efforts of government both Centre and States to stress for e-Governance as it provides ways to the cost cutting or curbing the rising administrative expenses, increased pace of service delivery along with reducing delays, bringing more transparency in the governmental operations. It was in this context that the Government of India took the step towards launching a scheme which was meant to correctly and perfectly target the intended beneficiaries of various governmental schemes and programmes to increase the effectiveness of various government-sponsored programmes and reduce the leakages of funds. This was also done in order to reduce the burgeoning subsidy burden of the Government especially in the social sector. This entire idea was materialised in the form of launching "Aadhaar" the brand name of unique identification number (UID).

Background of Unique Identification Number (UID)

The journey for the Aadhaar started by the creation of Unique Identity Authority of India (UIDAI) on 28 January 2009 as attached office under the Planning Commission of India. The project was initiated by the Planning Commission as an initiative that would provide the identification for each and every resident across the country and would be used as the basis for efficient delivery of welfare services to the people. The idea was also to use unique identification as a tool for effective monitoring of various programmes and schemes of government.

Although this concept is not new, it was first utilised in 2006 when unique ID for below poverty line (BPL) families was approved by the Department of Information Technology, Ministry of Communications and Information Technology, Government of India. An Empowered Group of Ministers (EGOM) was constituted with the approval of Prime Minister on 4 December 2006 to collate the two schemes that is the National Population Register under the Citizenship Act 1955 and unique identification number project of Department of Information and Technology. The mandate of UID is to issue every resident a unique identification number linked to the biometric information. Which can be used to identify by linking demographic attributes with the biometric attributes like finger prints and Iris patterns.

The Mission and Uniqueness

The role that the Authority *envisions* is to issue a unique identification number (UIDAI) that can have online verification and authentication in a cost-effective manner, which is capable enough to eliminate duplicate and fake identities. Unique identity of an individual which cannot be duplicated by others can be established by linking demographic with biometric. This is what has been achieved by rolling out Aadhaar and hence that corrupt practices are curbed and eliminated, as indicated in Fig. 5.1.

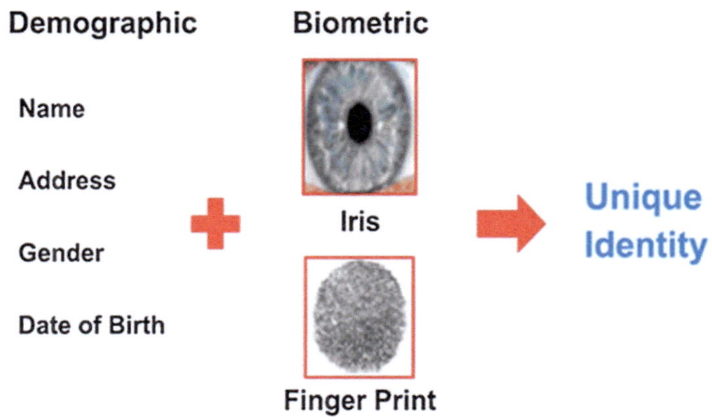

Fig. 5.1 Clubbing demographic and biometrics. *Source* UIDAI

UID Number = Demographic Attributes + Biometric Attributes

Online Verifiability

The online verifiability is significant in facilitating the following dimensions of governance which in turn will guarantee the impartial, impressive and integrity-based system. The effect of this would be in terms of evolving a trust-based system. The major points can be summarised as follows:

1. Can be accessed anywhere in India with online verifiability.
2. Removal of duplicate and fake identities.
3. Authentication of the individual as the same and unique person anywhere and at any time.
4. Greater scalability of services with online authentication.
5. Reduced beneficiary exploitation due to reduced dependency on manual processes.
6. Increased efficiency to service delivery and reduced cost of beneficiary identification.
7. Almost paperless that is no requirement for physical copies of identity documents
8. Electronic audit trail can be created which will allow service agencies to audit their service delivery process more effectively.

Aadhaar: The Foundation and Operating Model

The UID has been referred as Aadhaar. The name Aadhaar itself means the foundation or support in Hindi language. The features of Aadhaar include

1. Guarantee of uniqueness
2. Online identity verification
3. Utilisation for accessing services and resources any time anywhere in the country
4. Provide the identity infrastructure for the financial inclusion

The Aadhaar Operating Model has been mentioned in Fig. 5.2.

Applications of Aadhaar

Aadhaar can be leveraged effectively across industry domains for improving service delivery to the people irrespective of service being delivered by government,

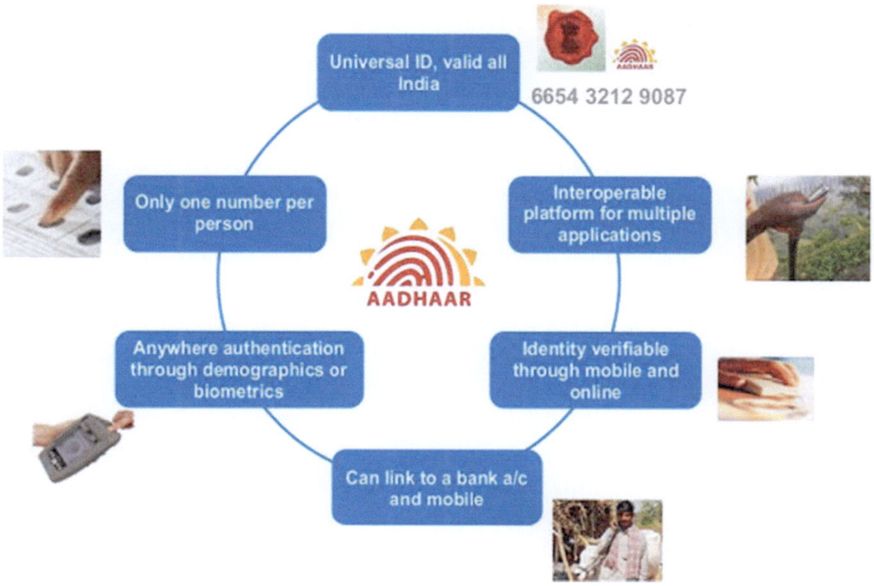

Fig. 5.2 Aadhaar Operating Model. *Source* UIDAI

public sector or private sector. Aadhaar is an IT-enabled identity solution which needs to be leveraged appropriately by service agencies along with required business reengineering and computerisation of their services through e-Governance and ICT initiatives.

1. Government Welfare Programs
2. LPG Distribution and Subsidy Management
3. Financial Inclusion and Electronic Payments
4. Telecom
5. Internet and E-Commerce
6. Aadhaar as Unifier
7. For criminal investigations.

However the ruling of Honourable Supreme Court of India dated 11th of August 2015 regarding Aadhaar not being mandatory is noteworthy. The court ruled that Aadhaar card will remain optional for availing various welfare schemes of the government. However, Aadhaar card will continue to be required for PDS and LPG distribution system. The court has allowed the use of details of Aadhaar card holders for criminal investigations. For detailed discussion pertaining to the social justice this paper has identified two major areas.

1. Government Welfare Programs for, e.g. Targeted Public Distribution System
2. Liquid Petroleum Gas (LPG) Distribution and Subsidy Management

Aadhaar in Targeted Public Distribution System (TPDS)

The welfare programs in India target millions of beneficiaries who depend on these programs, and these programs are often criticised for not having full impact due to delays, leakages, duplications and other delivery-related inefficiencies. Majority of welfare programs use physical identity documents and rely on manual processes for beneficiary identification. These practices result in lack of identity documents, existence of duplicate and fake beneficiaries preventing poor and marginalised people from accessing the benefits of TPDS programme (Planning Commission 2005). Use of Aadhaar to link beneficiaries with their Aadhaar numbers in beneficiary identification and approval process can address these issues. Applicant provides his/her Aadhaar number as part of the application which is saved in welfare programme database along with other application data. In addition, applicant authenticates his/her Aadhaar number through any chosen method (possibly fingerprint biometric) to confirm. The applicant name, address and date of birth provided in the application may also be verified against Aadhaar record with UIDAI through demographic authentication. As part of beneficiary approval step, the applicants Aadhaar number is matched against existing beneficiaries to ensure that the applicant is not an existing beneficiary.

Aadhaar Utilisation for Liquid Petroleum Gas (LPG) Distribution and Subsidy Management

Through Aadhaar-integrated delivery of all welfare programs, it will also allow the verification of beneficiary across different programs in case resident is not allowed to avail certain benefits simultaneously. Many welfare programs require the verification of beneficiary presence in different contexts. Some of the circumstances in which Aadhaar can be utilised are as follows: for subsidy-based welfare programs like LPG, it is important to ascertain whether the person receiving the benefit is indeed the intended beneficiary. Biometric-based Aadhaar authentication can be leveraged to achieve this verification for all scenarios as it confirms the presence and existence of the person. Based on the programme requirement, Aadhaar authentication can be implemented at the time of service delivery (e.g. TDPS).

Social Dynamics

It would be pertinent to explore the impact of this e-intervention on the social dynamics which will be undergoing a massive reorientation in terms of outreach, mobilisation and participation through this e-initiative. This programme will transform the way in which government benefits are delivered to our people who have been marginalised due to complex social system that prevails in India. This requires processing and digitising data, enrolling in Aadhaar, opening bank accounts and seeding these accounts. Hence, the focus is on reengineering the governmental processes. This schemes as pointed out earlier will make a new headway in terms of efficiency, transparency, propriety, in beneficiary identification and changing the entire social scenario. This is instrumental in being an e-policy initiative especially in India where self-created social constructs, corrupt practices; administrative apathy, arbitrariness and inefficiency inhibit the processes of development.

Algorithm

The above analysis is based on certain theoretical precepts but which have been applied in limited regions the following statements can be drawn leading to algorithm which can give some direction for implementing this policy. A simple algorithm can be developed by identifying the categories of beneficiaries and authentication process by Aadhaar server. The algorithm diagram is shown in Fig. 5.3.

Step 1: Start process for C-1 Consumers PDS
Step 2: Enter Aadhaar input information
Step 3: Consumer authentication check via C.I.D.R
Step 4: If authentication successful, then service delivered
Step 5: Else service rejected for consumer
Step 6: End of process
Step 7: Start process for C-2 Consumers LPG subsidy
Step 8: Repeat Step 2 to Step 6.

The diagram explains via simple algorithm that various kinds of entitlement holders, be it PDS consumers or LPG subsidy government schemes, can access their different entitlements by using Aadhaar at the respective points. The C-1 stands for Consumers of PDS, and C-2 stands for Consumers of LPG subsidy. From there onwards the Aadhaar-based information is scrutinised and processed so that the duplicacy and discrepancy can be prevented through Central Identities Data Repository (C.I.D.R). Authentication leads to two kinds of responses one that is yes and the other is no. In case of yes, the delivery of expected service is achieved, and in case of no, the authentication provides negative result the process is redirected towards the start. Hence, it explains in a simple way that if manual manipulations are prevented and discrimination can be avoided through technological progression not only the benefits of various governmental schemes would accrue to the stakeholders in a time bound and reliable manner but also the self-constructed complexities retarding the inclusive development would be abolished up to a great extent, thereby ensuring social justice to its citizens.

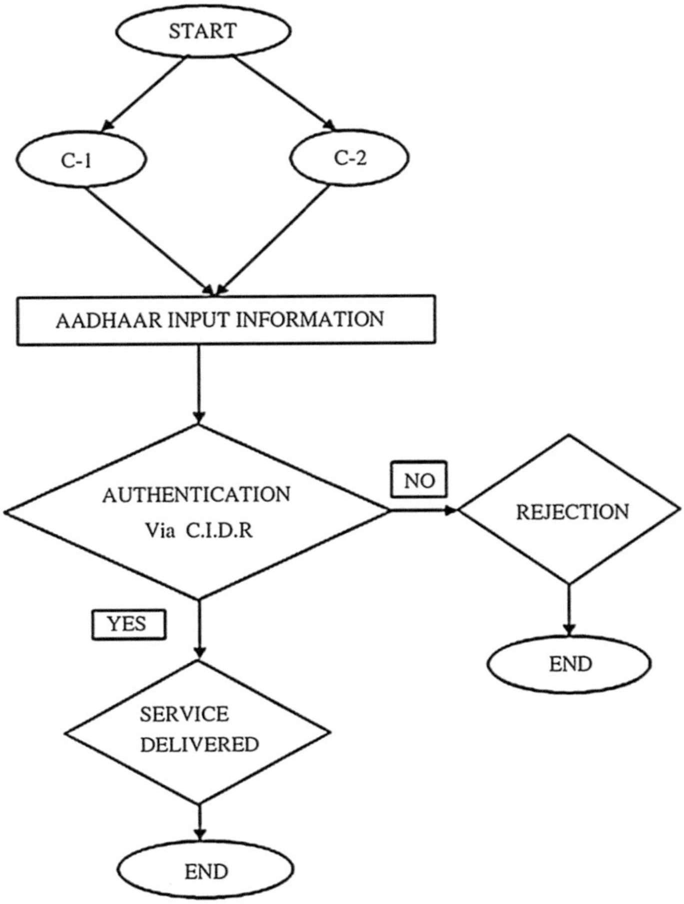

Fig. 5.3 Aadhaar algorithm developed by Anurag

Theoretical Framework and Development of the Hypotheses

The hypothesis was constructed to explore the people's perceptions in regard to Aadhaar. For the purpose of this study, the directional hypothesis has been used. This is stated as below:

H Application of Aadhaar card has enhanced the level of efficiency, transparency and accountability in delivery of public services.

For extracting the responses from the respondents, a questionnaire was prepared containing 10 questions contained in set-A. An interview schedule and mailed questionnaire were utilised to extract responses from the people aged between 18 and 70 years. The respondents were selected through Stratified Random Sampling and their educational and professional background was also

Table 5.1 The binary questions

Question		Yes	No
1	Do you know about Aadhaar Card Scheme?	84	16
2	Do you feel that Aadhaar infringes upon your privacy? (intentionally feeding wrong information)?	70	30
3	Do you find that Aadhaar seeding helps in targeting beneficiary?	78	22
4	Do you feel that Aadhaar should be made mandatory for all government-sponsored schemes?	77	23
5	Do you feel that Aadhaar scheme is a burden on public exchequer	67	33
6	Do you feel that Aadhaar generates greater transparency into the governmental system?	82	18
7	Do you feel that Aadhaar is a game changer in the delivery of social services schemes?	50	50
8	Do you feel that decentralised rolling of Aadhaar cards through private players is a compromise on your privacy?	72	28
9	Do you find that Aadhaar helps in Know Your Customer?	71	29
10	Do you feel that Aadhaar has lowered the corruption levels?	60	40

Source: research survey instituted by Anurag

considered as criteria in identifying the sample in order to make it more representative. The selection of the respondents was done on the basis of stratified random sampling in the city of Ahmedabad (Gujarat). Ahmadabad has over 6.5 million population and is the seventh largest city of India. The total sample size for extracting the responses was 120 (20 respondents each from the six zones of Ahmedabad city). However the actual sample size that could be obtained from sample survey was 100 instead of 120 as proposed. The following are binary questions with the responses of the people as given in Table 5.1 and represented in Fig. 5.4.

Discussion and Analysis

- To the first question the response was 84 saying yes and 16 saying negatively. The difference between the responses appears very sharp, so it can be assumed that people are aware of Aadhaar Card Scheme.
- To the second question the response has been 70 in yes and 30 in negative. This reflects a clear-cut division that Aadhaar infringes upon people's privacy.
- To the third question the responses are 78 in positive and 22 in negative. Aadhaar seeding helps in targeting beneficiary. The reasons could be that Aadhaar information corresponds to the beneficiary identification. And that has been felt by the common people.

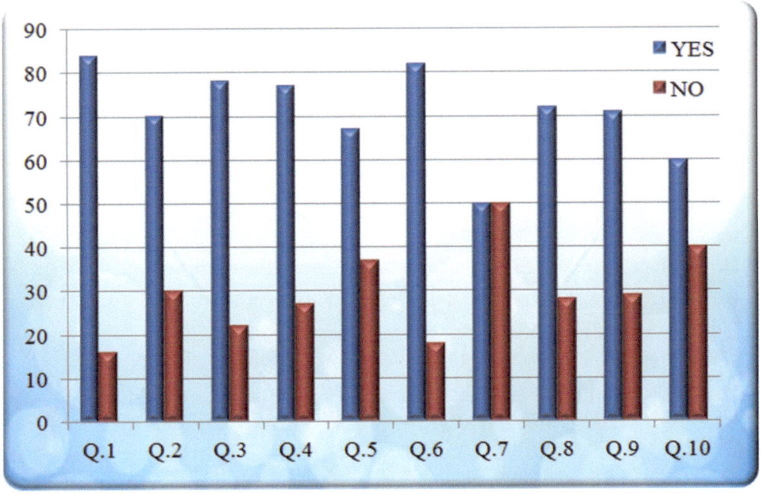

Fig. 5.4 Bar graph representation of binary questions. *Source* research survey instituted by Anurag

- To the fourth question the responses are largely positive where the 77 have responded affirmatively and 23 negatively. This reflects that Aadhaar scheme appears to be a burden on public exchequer as per common perception.
- To the fifth question which addresses the question of accuracy shows that 67 are going for affirmative response and 33 are denying this accuracy. This reflects that Aadhaar promotes transparency into the governmental system by minimising the paper work, avoiding duplications and complicated identity documentation.
- To the sixth question the responses are one sided where 82 responses are registered in the positive category. This suggests that people at large have been given the facility to register their grievances online. The convenience in registering complains could be attributed to e-Governance actually which has facilitated the registration process.
- The seventh question response figure is almost as 50 saying yes and 50 saying no. The perfect 50–50 suggests that opinion of the people is divided. Aadhaar is a game changer in the delivery of social service schemes. Logical explanation inherent in this is that despite Aadhaar being a must for availing benefits of social security schemes it has not been integrated, and popularised effectively.
- To the eighth question it has one-sided response where 72 responded in yes category and 28 in negative that suggest that people trust more to the government agencies than private players in maintaining their privacy in regard to Aadhaar card roll out.
- To the ninth question again the contradictory responses registered where 71 saying yes and 29 saying negative. It is reflected that Aadhaar helps in Know Your

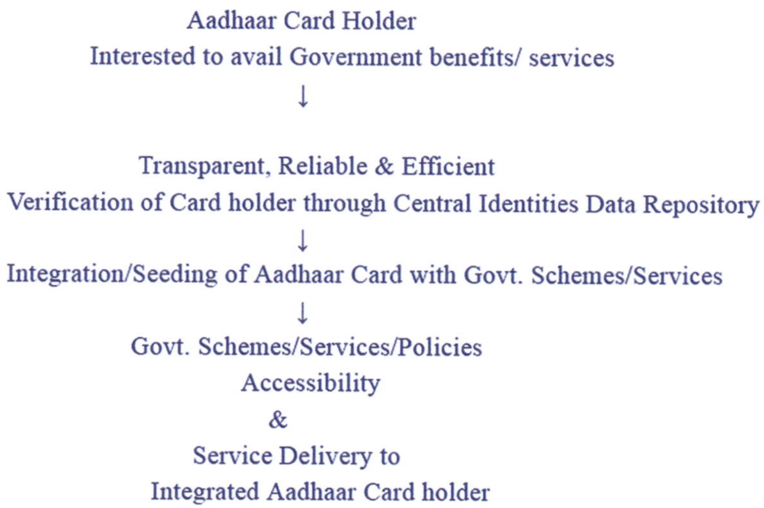

Fig. 5.5 Flow chart of aadhaar enabled service delivery

Customer. The improvement in customer identification has been a major factor contributing towards this overwhelmingly affirmative response.

- The tenth question deals with the fact that Aadhaar has lowered the corruption levels the responses are 60 in positive and 40 in negative. This has been substantiated by earlier responses which have also suggested that Aadhaar-based technological intervention has brought greater transparency, reduced arbitrariness and corruption (Fig. 5.5).

Conclusion

The responses of binary questions authenticate the hypothesis that application of Aadhaar card has brought greater efficiency, transparency and participation in delivering public services. India being a developing country and making strides for faster and more inclusive development has to rely more and more on the technological inputs like ICT to improve the governance process and strengthening the administrative good practices. In a developing country like India where the backwardness and inequality prevail and social justice has yet to materialise in the concrete terms, therefore reliance on government supported and sponsored schemes and programmes becomes much more crucial. In order to ensure that governmental interventions are quantitatively and qualitatively available to the people the ICT has been leveraged as a potent tool. Corruption, delay, nepotism, favouritism are some of the common problems plaguing the implementation of various social welfare schemes in India. Beneficiary identification has been a major problem for providing entitlements to various schemes and programmes. Issues like ghost beneficiaries, fraudulent or fake beneficiaries are quite common while disbursing the fruits of social sector schemes aimed at bringing inclusiveness and promoting social justice.

Another issue pertains to the claims made for such benefits by the genuine and deserving people who should enjoy the social safety net find it extremely difficult to establish their identity because of tedious, time consuming and corrupt process of obtaining such documentation. Aadhaar can with all its features and advantages solve this complex problem in a very transparent and convenient way by eliminating the unnecessary human intervention and arbitrariness. India can usher into a new era of digital governance by making the best use of limited capital and skilled workforce for promoting the socio-economic justice and curbing the delay, corruption and arbitrariness in the administrative functioning has appeared in China Daily stating. China has improved the way it deals with corruption and is increasingly using technology to combat graft, but experts say preventing the technology from being abused must also be taken into consideration. The Chinese Academy of Social Sciences published a report on China's rule of law, stating that the country's disciplinary authorities at every level have developed and applied technology to their anti-corruption work, such as online approval platforms and bribery record systems (China Daily 2013). Having one of the largest technical support pool and IT prowess India is poised to take giant strides in realising the goals of social justice as enshrined in the Constitution of India. Any change in the existing practices of governance has element of apprehension and confusion which is true far Aadhaar also but the possibility of misuse appears less as supported by Aadhaar study. As per the study "Role of Biometric Technology in Aadhaar Enrollment" conducted by UIDAI in 2012 regarding the *Reliability of Biometrics* methods the data revealed that the failure to enrol at 0 %, biometric failure 0.14 %, false rejection 0.057 %, false acceptance 0.0352 % (UIDAI 2012).

The governmental intervention in the social sector is needed in order to alleviate poverty, bringing inclusiveness and leading to empowerment of the people specially of impoverished and the deprived. Aadhaar can be a significant tool in the hands of Governmental agencies for distributing the fruits of growth and development by eliminating corruption, arbitrariness, delays, wastages, leakages and fraudulent practices. Aadhaar-based technology mechanisms helps in identifying the target beneficiary for availing various benefits and economic rights extended by government. Aadhaar enhances the accuracy of factual data by avoiding duplicity, overlapping and through a centralised coordinated set of commands/protocols and operative mechanisms that assist in achieving procedural fairness in the government service delivery beneficiaries. The responses pertaining to the questionnaire also substantiates and justify the hypothesis of the sample survey of this study.

The Supreme Court of India was concerned about the infringement of people's privacy as the court opined that no personal information of Aadhaar card holder should be shared or disclosed and the right to privacy is one of the basic inherent tenets of right to life and personal liberty belonging to Article 21 of the

Indian Constitution (Rajagopal Krishnadas 2015). Therefore, the individual's privacy should be respected and honoured while implementing Aadhaar Card based schemes and programmes.

Amid the concerns raised by Supreme Court of India for making Aadhaar card optional and not mandatory in availing the benefits of government sponsored programmes, which resulted in government taking the legislative route by piloting "The Aadhaar (Targeted Delivery of Financial and Other Subsidies, Benefits and Services) Bill, 2016". With the passage of this bill the attempt has been made to address the concerns of judiciary and allaying the fears of the people in regard to privacy infringement.

Aadhaar is not a citizenship card but it really helps citizens to assert socio-economic rights. In India the underprivileged sections have largely been at the receiving end in terms of accessing government welfare schemes and programmes and exercising their rights as dignified citizens of India. Aadhaar can act as fulcrum for leveraging the opportunities extended by government to the citizens through fair, transparent, rapid, and accountable operative mechanisms routed through technological integration. This would further promote Aadhaar as an instrument of socio-economic justice and empowering people by leveraging e-governance platform.

References

China Daily (2013) Technology used to combat graft http://www.china.org.cn/china/NPC_CPPCC_2013/2013-02/26/content_28060904.html Retrieved on Feb 26

Grube GMA (trans 1974) The Republic Plato. Indianapolis, Indiana: Hackett Publishing Company http://probonocentre.org.au/wp-content/uploads/2015/09/Occ_1_What-is-Social-Justice_FINAL.pdf

Ministry of Electronics and Information Technology, Government of India (2014) National e-Governance Plan. Retrieved from http://nisg.org/files/documents/A05140001.pdf

Panzardi R, Calcopietro C, Ivanovic EF (2002) New-economy sector study electronic government and governance: lessons for Argentina. New-economy sector study: electronic government and governance-lessons from Argentina Retrieved from http://documents.worldbank.org/curated/en/527061468769894044/pdf/266390WP0E1Gov1gentina1Final1Report.pdf

Planning Commission, Government of India (2005) Performance evaluation of targeted public distribution system. Technical report, programme evaluation organisation, Retrieved from http://planningcommission.nic.in/reports/peoreport/peo/peo_tpds.pdf

Rajagopal Krishnadas (2015 August 11) Aadhaar shall remain optional: supreme court. The Hindu Retrieved from http://www.thehindu.com/news/national/aadhaar-shall-remain-optional-supreme-court/article7525976.ece

Ross D, Ackrill J, Urmson JO (trans.1980) (revised) Aristotle. The Nicomachean Ethics, Oxford University Press, Oxford

Rousseau Jean-Jacques (1754) Discourse on the origins of inequality, as cited in National Pro Bono Resource Centre (2011) Occasional Paper: what is Social Justice, University of New South Wales p 5 Retrieved from http://probonocentre.org.au/wp-content/uploads/2015/09/Occ_1_What-is-Social-Justice_FINAL.pdf

Unique Identification Authority of India (2012) Authentication overview Retrieved from https://uidai.gov.in/beta/authentication/authentication.html

Chapter 6
Audiovisuality of the Third Degree: Reality or Fantasy?

Tadeusz Miczka

Abstract The starting point for my reflections concerning contemporary cultural paradigms is category of mimesis indicating one of the most important qualities of audiovisuality which is copying—thanks to image and sound techniques—reality. While characterizing transformation of mimesis, I enumerate three categories of audiovisuality: first step audiovisuality, which is copying based on film conventions oscillating between reproduction and creation, second step audiovisuality, which is copying based on television conventions oscillating between direct broadcast and staging reality, and third step audiovisuality, which is copying based on multimedial virtualisations oscillating between different forms of immersion and signallized simulation of artificial "realities". First and second step audiovisualities create numerous relations and easily interpenetrate. Their relations are shaped mainly by conventions of various games which increasingly replace previous social rules, especially ethical and communicational. Referring to the latest philosophical conceptions of culture and to philosophy of media, I try to prove two theses. Firstly, third step audiovisuality reshapes first and second step audiovisuality makes them more dynamic and replaces them to the same extend to which centres of contemporary power may discipline consumers/prosumers—users of new media by the means of so-called pleasant disorientation. Secondly, development of third step audiovisuality is directly connected with tendencies to move various aspects and forms of life of an individual and a group from real space to virtual space.

Keywords Mimesis · Reproduction/creation · Television conventions · Multimedial virtualizations

---

A slightly modified version of the text has been already published in German: *Audiovisualität der dritten Stufe—Wirklichkeit oder Phantastik?* In: Gerhard Banse, Annely Rothkegel (Hg.), *Neue Medien: Interdependenzen von Technik, Kultur und Kommunikation*, Trafo Verlagsgruppe Dr. Wolfgang Weist, Berlin: 2014, S. 45–52.

T. Miczka (✉)
University of Silesia, Katowice, Poland
e-mail: tmiczka@interia.pl

L.W. Zacher (ed.), *Technology, Society and Sustainability*,
DOI 10.1007/978-3-319-47164-8_6

When 86 years ago cinema "started talking" and audiovisuality appeared, the character of man's orientation in the world started to change quite radically. Moving picture, for the first time in human history, reflected simultaneously our sense of space, time and movement. What is more, the picture used also sound, and a few years later colour and dimension. Almost immediately it became the dominating manner of our orientation, because it met all the basic criteria of imitation. This imitation, "representation" of things, allowed individuals to deepen cognition, gaining, gathering and processing of information and made more active participation in social life possible. It also allowed to fulfil various interests and individual and group needs, which meant better but never full orientation in always "liquid" reality.

Specific character of this imitation is indicated by the notion of "feeling of reality", which is very popular in the theory of film. On the level of meaning, it refers to dual nature of cinema: reproduction and creation. Each kind of art, to some extent, refers to receivers' sense of reality, but in the cinema, no matter how fantastic the screen plot is, audience become its eye witnesses and as if its participants. They acquire emotional faith in reality of shown people, things, events and phenomena. Jurij Łotman—cinema semiotician—characterized nature of film mimetism in the following way: "the feeling of reality and the feeling of similarity between film and life, that film art cannot exist without, is not something essential and do not derive form direct experience. The feeling of reality is an element of complex artistic entity and it is enrooted thanks to numerous relations in group artistic and cultural experience" (Łotman 1983, p. 68).

Cinematic audiovisuality, thanks to which the aforementioned cognitive change took place, may be described from the present perspective as audiovisuality of the first degree. It offered mimetism which was not based on "faithful reflection" but on two kinds of convention: an extremely believable technical reproduction, and less but still believable filmed action. This division between two kinds of film audiovisuality is also visible in division into two kinds of cinema: the first one coming from Lumiere brothers and the other coming from Melies, documentaries and feature films, chronicles and phantasmagorias, cinema of facts and cinema of fiction, "the truth of the screen" and "dream factory" (review of theories concerning these conventions in: Woźnicka 1983). Yet, since the very beginning of the sound, cinema reproduction has mingled with creation, and all the boundaries between them have been efficiently blurred by numerous kinds of conventions such as genre, authorship, avant-garde and new cultural phenomena, mainly its mass character and technical inventions. Since the very beginning of cinema development, mainly demonstrative and narrative techniques have been dominating. These were techniques responsible for disorientation of the audience thanks to presenting particular objects and events and ways of showing them. Still it is worth mentioning that "both: art house cinema and even more genre cinema use disorientation as only local, technical event, so psychologically in some number of cases it is not ≫disorientating<< at all. It was plot postmodernism in its innovative current that finally moved disorientation from the level of technique to the level of strategy and simultaneously brought destruction to the meaning of the film text" (Zalewski 1998, p. 40).

Among all the numerous technical inventions, which appeared in the middle of the twentieth century, television had the biggest influence upon audiovisuality of the first degree, which is now called, after Francesco Casetti and Roger Odin, palaeo-television (Casetti and Odin 1990, pp. 9–26). Increasing popularity of "small screen" not only altered reception of film message and poetics of cinema, but first of all it started process of transformation and then replacing audiovisuality of the first degree with television audiovisuality—audiovisuality of the second degree. The characteristic feature of this new version of mimetism became direct transmission, then less direct and more mixed with directing of the reality presented on the screen. This process became more radical when, as it is described by Casetti and Odin, palaeo-television started to change into neo-television and it quitted traditional TV schedules, genres, forms of narration and presentation, which means it started consciously and openly build its existence upon strategies of active disorientation. The target model of neo-television is 24-h infotainment co-created by individual receivers. That is why the appearance of internet television is perceived as the turning point of development of audiovisuality of the second degree.

New media started new phase, for the time being the last one, in history of audiovisuality. They oblige their users to participate in mass culture more actively (more intensively), redefine the notion of "mimetism" and accept almost unlimited mobility and communicative openness. Multimedia create audiovisuality of the third degree, which transforms and absorbs audiovisuality of the first and the second degree, becoming the dominating manner of orientation of contemporary man in the world. Thanks to interactivity, visualization and convergence, it facilitates all realities of history and the present and all possible forms of creation. Thanks to using "soft technology", which means small, easy to use, multifunctional electronical gadgets, users of the new media can cross borders created by the previous media and reach the next, higher level of communicational disorientation, which complicates structures of hypertexts and multiplies their incontinuity.

The biggest number of emotions and arguments, and probably also consequences in cultural practice, is evoked by crossing this border, which can be referred to as manipulating of the distance between real world and virtual world. Up to now, cinema audience and television viewers felt distance towards perceived world, except for pathologic conditions they could definitely find difference between reality and fiction, materiality and representation, fact and simulation. Multimedia allow their users to preserve the distance. It can be also reduced, extended or shortened, for example, by introducing to screen space their representative, an avatar, their activities or imaginations. They also allow for immersion into virtual world.

Time–space fulfilling definite conditions of simulation, telepresence, interactivity, immersion and so forth presents the problem of mimetism in definitely new light; it is a challenge for all fields of philosophy (cp. i.a. Ostrowicki 2005) and branches of economy, culture, education, politics and the law (cp. i.a. Zacher 2013). As far as philosophy is concerned, first of all it forces ontologist to rethink such notions as matter and idea, reality and fiction, physical aspect and intentionality, and make them more flexible. It forces epistemologist to reconsider the notion of space, time and causality and problems of sense perception, representation, imitation, inclusion, "conscious being", "telematic being" and "pataphysical being". Ethicists, aestheticians and logicians are not indifferent to such space. The first ones have to reconsider the notion of functioning of all kinds of values, subject identity and responsibility. The second ones come back to problems of creation and reception and think over the structure of piece of art in media reality. Logicians consider how the ontological, logical and pragmatic truth can be understood in situations conditioned by audiovisuality of the third degree. The truth which in "culture of participation" do not meet the criteria described by Gianfranco Bettetini, who treated it as "result of social contract, which makes sender and receiver imply certain faith in existence of the truth, contract leading to illusion of the truth, because in this type of ≫conversation<< the notion of ≫effectiveness<< becomes more important than the truth" (Bettetini 1985, p. 94). Of course in this case, we refer to communicational effectiveness in a broad sense, including egoism of consumers changing into prosumers. According to some scientists, it is not extremely dangerous for human cognition. For example, philosopher Józef Sójka emphasizes that "questioning a particular conception of the truth has to refer to existing funds of thoughts, it just moves the emphasis from epistemology to ethics, from metaphysics to rhetoric, from the notion of certainty to the notion of hope, form the notion of objectivity to the notion of solidarity" (Sójka 1992, p. 47). Other scientists notice the problem of "the alternatively pragmatic truth" and prove that today "as a matter of fact there is no deeply enrooted will of the truth in us anymore. [...] There are no measures of reliability. [...] We are pleased with appearances. [...] When the truth is suppose to be replaced by something, the easiest solution is to believe one's own desires. That is where insistency and insolence of demands expressed in public come from. [...] Marketing, in other words ability to seduce, decides about everything" (Filipowicz 2007, pp. 12–13).

In my speech, I am not going to develop or solve any of those philosophical and nonphilosophical problems, because since the beginning of the 1990s of the twentieth century thousands of dissertations have been written on them. They usually result in asking new difficult questions. Fortunately, arguments concerning this current topic take place and reveal next aspects, layers and phases of mimetism. As long as invasion of multimedia takes place both in culture and everyday life, there will be attempts of redefining or even changing of the notion of "mimetism". On the basis of researchers' findings, and in my opinion reflections of multimedia users are similar, we may assume that rejecting of statements, which place reality and fiction, mimetism and abstraction on extremely opposite positions, was right. Its finality and decisiveness do not match the nature of contemporary audiovisuality, its functions in interpersonal communication and other cultural practice. Opposition of notions is not the right methodological way in considering these issues, which I have already explained in detail in other place (Miczka 2009, pp. 231–237).

Charles Jonscher writes: "The ideal, which can be conjured up by science and which cannot be improved by technology yet, is complete sense immersion [...]", but "[...] Stream of bits [...] is a mathematical construction, creation of human imagination, which cannot exist in physical world. The laws of nature do not allow it. Nature is analogue. [...] Digital stream of bits may be approximation, but never accurate reconstruction" (Jonscher 2001, pp. 210–212). Speaking of traditional audiovisualities being replaced by audiovisuality of the third degree, we always speak about approximations, incomplete states, intensified, similar and escaping complete defining. So approximations should be recognized by users of multimedia as artificial versions of real places and worlds.

The issue is not quite simple. Jean Baudrillard was only partly right when he claimed: "We do not need science fiction any more: immediately, here and now, thanks to media and informatics, thanks to circuits and networks, we have the accelerator of particles, which definitely broke referential circle of things" (Baudrillard 1999, p. 9). "The real dimension [...] of virtual machinery is not information, cognition or meeting, but the wish of disappearance. All that is offered by new technologies is a certain kind of pictures in which we can immerse with the possibilities of changing then. How can we think, if we can get into video picture and do with it whatever we want?" (Baudrillard 2001, p. 47). When we read contemporary science fiction prose, it is hard to agree with obvious Baudrillard's diagnosis referring to gloomy future of the literary genre. It is true that thanks to new media the authority of referential function in interpersonal communication is declining, but in my opinion Baudrillard's prediction of total catastrophe of information, cognition, meeting and human thinking was exaggerated. He assumed, just like Virilio (1990), that people are immensely fascinated by possibility of disappearance offered by multimedia.

The notion of "disappearance" is particularly important in scientific approach towards the problem of mimetism. According to Baudrillard and Virilio, today multimedia presentations do not refer to any subject reality, because they exist only on the screen. "Their source is – I quote a scientist summarizing their ideas—a calculated model, algorithm, non-picture information. Ennoblement of numerology and entering the epoch of algorithms means that man has lost contact with reality. Objects disappear and in front of viewer's eyes there is a ≫procession of simulacra≪, delirious world. The viewer himself is facing the ≫metaphysical horror≪ because there is no mirror to reflect in and make sure he exists. Screens of contemporary media do not reflect anything" (Krawczak 2006, p. 182). Even if we do not reject all the detailed argumentation of Baudrillard and Virilio, it is still difficult to agree with their final conclusion concerning complete loss of contact with reality by a man. It is true that thanks to expansion of simulation games, blurring of borders between notions, phenomena and events and relativization of values, functions proving authority of reality lose their meaning in communication. However, as Gerhard Banse rightly concludes his argument on this subject: "adversities of real world cannot be overcome thanks to ≫cunning mind≪ or thanks to replacing them with virtual reality" (Banse 2009, p 48). And this opinion, undermining opinions of Baudrillard and Virilio concerning "wish of disappearance" as well, is confirmed by cultural practices.

Users of new media do not want to "disappear" most of all, which is proved by increasing popularity of such games as Second Life. They prefer those forms of audiovisuality of the third degree which allow them, using their avatars, to test situations remote form their real life and allow them to feel the distance to pictures, based on simulation, which are perfect “approximations” of reality.

Of course, the number of users of new media who take up activities risky for their identity, mind or even physical aspect increases. The number of information used by us, which has already been abundant, increases as well. Conceptions of normality are more flexible in “culture of participation”. Still all this do not prove that foreseen in black pessimistic scenarios, written not long ago by Baudrillard and Virilio and now by their apprentices and disciples, are coming true. On the basis of observation of cultural practice and contemporary state of studies on using the media, I am closer to the opinion of Wolfgang Welsch who says that contemporary man has “a broad sense of plurality competence”. Both chances and dangers for users depend on this competence because it is “protection against dangers of alienation and splitting of identity, it also allows for openness to other forms of life and frames of meaning” (Welsch 1998, p. 33).

So what is the answer to the question asked in the topic of my speech: “what is more favoured by expansion of audiovisuality of the third degree—imitating of reality or creating fantasy?” The answer is: it favours cognition and shaping of both reality and fantasy. There is no doubt that connections change and the borders between reality and pictures wear off, but we finally solve life problem in real life not on the screen. We can add that it happens more often thanks to “support” of screen reality. There are of course intriguing questions: Why connections between mimetism and the truth and experience of the real are broken more often and what are the consequences of the process? The answers take new dissertations including considering at least three fundamental problems: the first one refers to needs, possibilities and justification of moving different forms and dimensions of life of individuals and life of societies from real space to audiovisual space. The second refers to relation—acceptance or rejection of so-called pleasant disorientation by multimedia users—psychological and technical possibilities of conscious “misleading” of others and ourselves, which makes communication more attractive. The third problem needs considering and tracks mentioned by Jacques Ranciere, especially in such books as Na brzegach politycznego (1990), Przeznaczenie obrazów (2003) i Widz wyzwolony (2008), should be continued. It is connected with using specific form of audiovisuality of the third degree by contemporary centres of power to create laws regulating real and virtual life of individuals and communities.

References

Banse G (2009) Rzeczywistość wirtualna“ i jej odniesienie do „rzeczywistości realnej. In: Kiepas A, Sułkowska M, Wołek M (eds) Człowiek a światy wirtualne. Katowice, pp 42–49

Baudrillard J (1999) Das Jahr 2000 findet nich statt. Berlin

Baudrillard J (2001) Rozmowy przed końcem. Warszawa

Bettetini G (1985) L'occhio in vendita. Per una logica e un'etica della comunicazione audiovisiva, Venezia
Casetti F, Odin R (1990) De la paléo a la néo-télévision. Approche sémio-pragmatique. Communications 51:9–26
Filipowicz S (2007) Prawda stała się niepotrzebna. Gazeta Wyborcza 21–22(07):12–13
Jonscher Ch (2001) Życie okablowane. Kim jesteśmy w epoce przekazu cyfrowego? Warszawa
Krawczak E (2006) Cyberprzestrzeń—kreacja realności czy estetyka znikania? In: Sokołowski M (ed) *Definiowanie McLuhana.* Media a perspektywy rozwoju rzeczywistości wirtualnej, Olsztyn, pp 179–184
Łotman J (1983) Semiotyka filmu. Warszawa
Miczka T (2009) Bifurkácia či transverzalita? O vstahoch mezi realitou a fikciou. In: Karul R, Porubjak M (eds) Realita a fikcia. Bratislava, pp 231–237
Ostrowicki M (ed) (2005) Estetyka wirtualności. Kraków
Ranciére J (1990) Aux bords du politique. Paris
Ranciére J (2003) Le destin des images. Paris
Ranciére J (2008) Le spectateur émancipé. Paris
Sójka J (1992) Destrukcja pojęcia prawdy w filozofii i naukach społecznych. In: Kubicki R, Sójka J, Zeidler P (eds) Problem destrukcji pojęcia prawdy. Poznań, pp 8–57
Virilio P (1990) L'inertie polaire. Paris
Welsch W (1998) Stając się sobą. In: Zeidler-Janiszewska A (ed) Problemy ponowoczesnej pluralizacji kultury. Wokół koncepcji Wolfganga Welscha. Część 1. Poznań, pp 11–34
Woźnicka Z (1983) Problem kreacji i reprodukcji w filmie. Warszawa
Zacher L (2013) Wirtualizacja. Problemy, wyzwania, skutki. Warszawa
Zalewski A (1998) *Strategiczma dezorientacja. Perypetie rozumu w fabularnym filmie postmodernistycznym.* Warszawa. Translated by Bogdan Moczko

Chapter 7
Digital Preservation of Cultural Heritage

Nonja Peters, Dora Marinova, Marijke van Faassen and Glen Stasiuk

Abstract This focus of this chapter is the state of the art of digitisation of cultural heritage in Australian archives and libraries from a comparative perspective. Globalisation, mobility and the new techniques that spin off from the digital age bring about new possibilities that stimulate and enhance our capacity to ask new questions about how we perceive ourselves and how we want to preserve our history. It also seeks to make this archival documentation accessible to scholars and community members alike looking for their own family's history in its societal context—within and across the national borders that hold their records. As migration in all its forms can be seen as a metaphor for the journey of the self and the collective, migrant heritage can also serve as a way to prioritise digitisation projects in cultural heritage institutions. However, more global collaboration and partnerships are needed to achieve this "virtual reconnect" the cross-national scattered nature of migrant histories and heritage held in archives around the world.

Keywords Digitisation · Digital age · Migrants · Migration · Heritage · Tran snational · Partnerships · Globalisation · Mobility · Virtual

N. Peters (✉) · D. Marinova
Curtin University, GPO Box U1987, Perth WA6845, Australia
e-mail: N.Peters@curtin.edu.au

D. Marinova
e-mail: D.Marinova@curtin.edu.au

M. van Faassen
Huygens Instituut voor Nederlandse Geschiedenis (Huygens ING), Amsterdam, The Netherlands
e-mail: marijke.vanfaasen@huygens.knaw.nl

G. Stasiuk
Murdoch University, Murdoch, Australia
e-mail: g.stasiuk@murdoch.edu.au

L.W. Zacher (ed.), *Technology, Society and Sustainability*,
DOI 10.1007/978-3-319-47164-8_7

Introduction

Communication technologies have been undergoing swift radical change and this in turn is revolutionising research related to preservation of cultural heritage, as we have understood it in the past. Debates that envision the coupling of the human brain with computers were first introduced into the public scene by one of the fathers of cyberspace—Joseph Carl Robnett Licklider of the Massachusetts Institute of Technology (MIT) in 1965 (Waldrop 2001). Years later at a conference in the Netherlands in 1982, Licklider admitted that although he had foretold many of the technical trends now developed, he had not concerned himself with the more fundamental choice between evolution and goal-oriented change that would accompany computerisation. This is particularly true for libraries, galleries and museums that are trying to collect and preserve cultural heritage. For example would libraries use information technologies mainly to improve the functions and procedures they were using already in the 1960s or would they try to develop new functions and procedures? Together with somewhat radical institutional arrangements such as consortia and associations, would they embark on real goal-oriented change in the organisation of the collections, by finding what is needed and making it available to viewers?[1]

At the same 1982 conference, John Michon of Groningen University contended that the only way to connect a library with a mind was to provide "inputs maximally compatible with representations already held by the user" (Michon 1984, p. 151). In addition, both Licklider and Michon agreed that librarians should commit to the development of knowledge bases rather than provide access to individual documents.[2]

In 1988, at the 11th International Congress on Archives in Paris, Eric Ketelaar, Professor of Archival Studies University of Amsterdam, commented on the dilemma relating to archives often being physically at a distance from their users; and how inevitably, the Internet would obviate the user having to visit holding institutions—in person (Ketelaar 1988). Concurrently, he believed a shift would take place from interest in the place where information is held to how the information could best be retrieved! Ketelaar (1988) also conceptualised a transposition from André Malraux's "musée imaginaire"—museum without walls—to "archives without walls". He visualised this transposition as a removal of the boundaries between archives, libraries and museums since they were based to a large extent on the physical properties of the information objects preserved in them. Indeed he saw the future mission and vision of archives (or information services) without boundaries—as providing access to their information from the home or business computer (Ketelaar 1988).

[1] Also cited by Ketelaar 2003, p. 1

[2] ibid

However, to accept this as the future would also entail recognising that the needs of such an innovative entrepreneurial service would be continually evolving. By way of example, Ketelaar (1988) refers to Bert Looper, Director of the Historisch Centrum Overijssel. To ensure that archives function as a living component of society, Looper, described as an archival entrepreneur, advocated not only a conceptual switch in archivists' thinking from archives to information but also a paradigm shift as our postmodern society moves along from a goods-producing, through to a services-performing and experience-generating economy. In this transition, the new information society is shaping not only the way goods and services are delivered but is also influencing human culture, including how we perceive ourselves and how we preserve our history.

This chapter looks at the digitalisation of cultural heritage as a new way to preserve important historic information. We refer to the Australian experience as an example of innovative ways of implementing digitalisation under budgetary and time constraints.

Connecting Memories

The Internet provides access to various sorts of information and offers exciting new possibilities that people quickly take up. According to Ketelaar (2003:10), the linearity of a paper document has been replaced with interactive relationships, "there is no 'original' any more, only different representations… The information will be presented to every user, at every moment, in a different form and with a different content". He asserted that digitalisation of archives entails not just preserving and providing digital documents but also a techno-cultural challenge to connect the memories in the archives with the memories in people's minds.

Margaret Hedstrom (2002) argued that making the memory metaphor work for scholars and users of archives would require not only a more refined sense of what memory means in different contexts but also sensitivity to the differences between individual and social memory. Such sensitivity is a particularly important issue for countries which have become the new home of migrants, displaced persons, prisoners of war and refugees. In them, the emphasis on national remembrance or social memory has shifted to individual memory and the expression of one's own experience (Raben 1999). Digitalisation further blurs these distinctions and offers individuals the opportunity to connect their personal stories with memories in the social realm.

The current interest in family and personal history is the most significant expression of digitalised connection between social and individual memories. In the globalised world with fading national identities and mass migration, it is increasingly improbable for museums to cope with housing collections that reflect their nation's ethnic diversity. Moreover, the nature of population movements, including for work, family reunion and search for asylum, often makes it impossible or impractical to establish permanent relationships with a particular country or place (NSW Heritage Office and NSW Migration Heritage Centre 1999). The Internet has therefore become the new shared space where individuals can construct, transmit and reflect on the historical knowledge that shapes their own and family's history and identity.

Niche industries emerged and thrive everywhere in response to the need to find out and preserve ethnic cultural heritage. For example, as well as the popular immigration museum exhibitions, Welcome Walls (e.g. in Fremantle, Western Australia) and history books about migrations, we now also have on Australian TV programmes, such as: "Who do you think you are?", "Find my Family" and "Can we help?" with the aim to unravel personal stories and connect people with their historical heritage. There are also innovative consultancies, such as Ancestry (ancestry.com.au), as well as a lot of free advice on the Internet, such as Instructables (http://www.instructables.com/id/Record-Your-Familys-Oral-History-before-it-dies-/), all motivated by people's desire to link with their ancestors but also leave memories for their descendants. The company 'Family Record: For your Future Generations' uses a digitalised documentary style to record the lives of prominent Australians (Macken 2010). The great football star Ron Barassi explains: "You can't show that to anyone but I want copies for my family… Because this is the representation of Ron Barassi that I want my family to see…" (Macken 2010, p. 44). Peter Ivany, the former chief executive of the Hoyts Cinemas, explains: "It wasn't just about capturing family stories but about understanding them better… My parents went though the Holocaust and to understand what the people did at certain times, you have to know what they are thinking. I know their stories, how my father spent 11 months in a concentration camp and came out weighing 32 kilos and then came out here to Australia at the age of 24, but without knowing what he was thinking at the time, it's hard to understand his drive to give his children opportunities, how he went about inventing a life. If my parents had done this, it would have put it in context why they did things and why they seemed so tough at times" (Macken 2010, p. 44).

Digitalisation of Archives

A more meaningful understanding of the person and the influences that impacted and motivated them is achieved by locating their experiences within the socio-cultural, economic and political context of the times. This is best accomplished by reference to research and the documentation and artifacts in national, state and local archives, libraries and museums. Australia is host to an estimated 180 ethnic groups including the dominant British. Should a member of any one of these immigrant groups wish to research and write a comprehensive family history, this requires their access to archival documentation in both Australia and their country of origin. The question is: How readily they can electronically access birth, death and immigration records, shipping lists, passport requests, health clearances, Alien registration, citizenship papers and school or business records in Australia? How close are we to the archives without walls?

Europe has made substantial progress with the digitalising of archives with many initiatives co-funded by the European Union and participant countries. An excellent example is the Archives Portal Europe (portal http://www.archivesportaleurope.net/) where a search for Bulgaria, one of the newest European Union members, generated close to 6000 entries. Other wonderful digital heritage preservation sites include Europeana (http://www.europeana.eu/portal/en); Minerva (www.minervamagazine.co.uk), the Planets project which terminated in 2010 (www.planets-project.eu), its successor the Open Planets Foundation (www.openplanetsfoundation.org/) and the Rijksmuseum collection (https://www.rijksmuseum.nl/en/search).

By comparison, in Australia the digitisation of the records of ethnic groups and of ethnic newspapers, has been slow on the uptake, probably because of the lack of an overarching body. Despite the fact that archives are extremely important because of the multicultural and diverse ethnic heritage, in the scheme of things very little appears to have been digitised.

Digitising Australian Archives

The key institutions, which hold archives in Australia, are the National Archives and National Library, and the various state archival institutions and libraries. National and state libraries also often hold images and other documents relevant to family historians. Whether an immigrant's records are held in state or national archives is directly related to their time of arrival. Until postwar immigration commenced in 1945, which occasioned the founding of the Commonwealth Department of Immigration, arrival and departure records were held in the various state archives. Afterwards, they are kept in the national institutions.

To overcome the budget restraints imposed by the Commonwealth Government, and perhaps the Global Financial Crisis which in fact did not affect the country as much as other parts of the world, the National Archives of Australia (NAA) has been searching for innovative ways to digitalise its rich collection spread over 249 km of shelving. They include business partnerships and digitalisation on demand.

Partnerships

Since 2008, NAA started negotiations and established a relationship with Ancestry.com.au. In 2012, the first official partnership around a specific archival collection was announced. The aim of this project is to create an index and digitise the records of millions of passengers who arrived in Western Australia between

1897 and 1963. It covers most people arriving in Australia by ship (at Fremantle and other Western Australian ports) and aircraft (at Perth Airport), even if they continued their travel onto other ports. According to the project's press release: "Passenger arrival records are an excellent source of genealogical information. They usually include information such as name, nationality, race, age, sex, place of embarkation, occupation, name of the ship, and date and place of arrival" (NAA 2012, n.p.). As this information is made available through NAA, many online users will be able to reconnect social and individual memories creating the picture of Australian life.

Digitalisation on Demand

The NAA collection is located in repositories in all capital cities, and all Australians have a right of access to these documents in accordance with the Archives Act 1983. As each record is unique and many are quite fragile, they cannot be moved from one city to another. Traditionally, this meant that researchers must visit the Archives where the records are located, pay someone else to conduct research on their behalf or pay for photocopies (if possible). Since 2001, NAA started making copies of selected records available on its website, but in 2007 it launched its national digitalisation service as *digitisation on demand* (NAA 2013).

The NAA considered a number of strategies to promote greater awareness of both its collection and public's access to it. From the outset, it was apparent that producing digital images and making them available on the website was one of the most effective ways to do this. Through digitisation on demand, records in the Archives collection are made accessible through images loaded onto the Archives website, which can be consulted via the Internet at any time without cost. Originally the digitisation request was also free; more recently, however, NAA introduced a price starting from A$32.60 with a turnaround time of approximately 30 days. Researchers are no longer constrained by their physical distance from the Archives offices. Further, digital images are available not only to the person who requested them, but to anyone who has access to the Internet.

Digitisation on demand was an immediate success with very high demand of around 2000 requests each month (Ling 2013). The service is very popular with family historians, teachers and students.

The NAA continues its proactive digitisation programme in which each of the states and territory offices select particular whole series to digitise. Some of them are specifically migration records such as Alien registrations, migrant selection documents and passports. It takes more than a year for some of the larger take to complete, and statistics are hard to find as to how many records have been digitised. The importance of digitisation programmes for remote communities, academics and individuals is enormous.

Digitalisation is also a strategic direction for the National Library of Australia (NLA) "so that Australians can engage online with their past, as well as with their

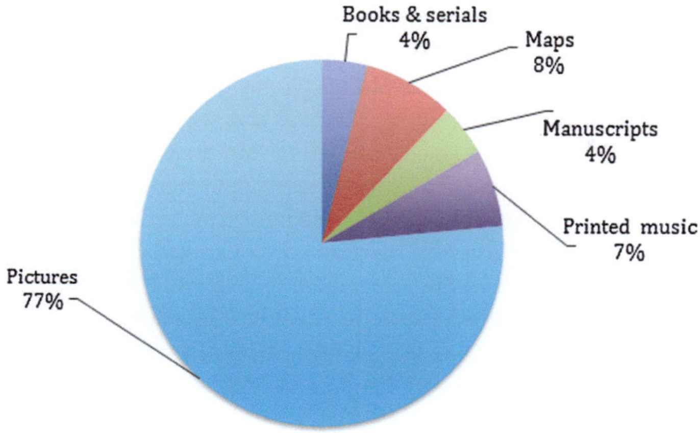

Source of data: NLA 2013.

Fig. 7.1 Digitised items at National Library of Australia, as of June 2013

present" (NLA 2012, n.p.). Collections that have already been digitised include Australian newspapers (e.g. The Sydney Morning Herald with the help of the Vincent Fairfax Family Foundation). The main selection criteria for digitalisation include: cultural and historical significance of the documents; uniqueness and rarity of the material; copyright status; high demand; risk because of physical condition or impending format obsolescence, such as sound and audiovisual recordings (NLA 2012). As of 2013, there were a total of 201,909 digitised collection items, with the bulk of them (71 %) being pictures (see Fig. 7.1).

Digitalisation of museum collections is another important area of connecting memories and stories. Some have started to develop databases relevant to particular projects but a lot more systematic work to digitise documents, photographs and artefacts needs to be done. Further on, a number of heritage websites contain some interesting stories and are proving to be important virtual institutions despite not collecting information in any systematic manner nor having a plan for the site's longevity and technology obsolescence.

Conclusion

Migration in all its forms is a metaphor for the journey of the self and the collective, and increasingly, to preserve this cultural heritage will require partnerships between communities, the business sector, government, national archives, national libraries and museums, genealogical organisations and other cultural institutions. Because of positive digitalisation practices by the Europe Commission, migrants from member countries have a far greater chance of accessing their records there than in other parts of the world. Australia, the country most exposed to the tyranny of distance, is gradually making progress.

Digitalisation is revolutionising the way in which cultural institutions preserve their collections as well as make them more widely known and accessible. More work, however, needs to be completed to fully transform these collections, provide the most needed connections between personal and social stories, and create the history of Australia. Postscript since completing this chapter, the National Library of Australia's (NLA) TROVE - a revolutionary new free search engine (http://trove.nla.gov.au/) about Australia and Australians, has become world renowned. TROVE users can access a wealth of resources - derived from more than 1000 libraries around Australia as well as personal, educational institutions and some international collections with relevance to Australia. Its contents include pictures, unpublished manuscripts, books, oral histories, music, videos, research papers, diaries, letters, maps, archived websites and Australian newspapers from 1803 to 1954. A wonderful resource, it has as yet, however, few ethnic newspapers digitised. On the other hand, the National Archives of Australia (NAA) has over the last couple of years, dramatically improved on the digitisation of its migrant records. See Destination Australia' (https://www.destinationaustralia.gov.au/site/).

References

Hedstrom M (2002) Archives, memory, and interfaces with the past. Arch Sci 2:21–43

Ketelaar E (1988) Exploitation of new archival materials. Report for the 3rd plenary session of the XIth international congress on archives, Paris. Reprinted in Archivum 35(1989):189–199

Ketelaar E (2003) Being digital. Arch Manuscr 31(2):8–22

Ling T (2013) Digitisation on demand. Data issue 6. www.archive.dcita.gov.au/2004/10/issue_6/data_issue_6_-_digitisation_on_demand+The+Archives+currently+receives+about+2000+requests+each+month+digitisation+on+demand&cd=1&hl=en&ct=clnk&gl=au&client=safari. Cited 20 Dec 2013

Macken D (2010) This is your life. AFR Mag 42–44

Michon JA (1984) How to connect a library with a mind. In: Reedijk C, Henry CK, Koops WRH (eds) Large libraries and new technological developments. Saur, Munich, pp 137–152

National Archives of Australia (NAA) (2012) Archival passenger records to go online. http://www.naa.gov.au/about-us/media/media-releases/2012/31.aspx. Accessed 20 Dec 2013

National Archives of Australia (NAA) (2013) The National Archives digitisation service. Fact sheet 249. http://www.naa.gov.au/collection/fact-sheets/fs249.aspx. Accessed 20 Dec 2013

National Library of Australia (NLA) (2012) Collection digitisation policy. http://www.nla.gov.au/policy-and-planning/collection-digitisation-policy. Accessed 20 Dec 2013

National Library of Australia (NLA) (2013) Digitisation of library collections, http://www.nla.gov.au/digitisation. Accessed 20 Dec 2013

New South Wales (NSW) Heritage Office and NSW Migration Heritage Centre (1999) Rich rewards: cultural diversity and heritage practice. NSW Migration Heritage Centre, Sydney. http://www.migrationheritage.nsw.gov.au/mhc-reports/rich/index.html. Accessed 20 Dec 2013

Raben R (1999) Representing the Japanese occupation of Indonesia: personal testimonies and public images in Indonesia, Japan and the Netherlands. Waanders Publishers, Zwolle

Waldrop MM (2001) The dream machine: J.C.R. Licklider and the revolution that made computing personal. Viking Penguin, New York

Part III
Impacts on Societies

Chapter 8
Peril and Promise of Internet Technology for Future Social Order

Alina Betlej

Abstract The author's objective is to assess threats and promises offered to humanity by the already *theoretically personified* Internet technology. Without any doubts we may say that the technology is one of the crucial key areas of human activity. Explorations of its possible evolution and ways in which its different development paths will lead us are still very important. The question of the way in which the *digital status quo* of Internet social order functions and of the principles governing it is more than justified. The future of social civilization connected with the development of modern day *ubiquitous technology* is not obvious. The crucial question: why so many aspects of the understanding of technology changed throughout the twentieth century cannot still be answered. The fact that people misunderstand proceedings of the future has already brought many very serious implications for humankind's condition. To understand a peril and a promise of Internet technology for future social order we should ask three basic questions. What did we believe in throughout the past century? How have we described the technology evolution? How have we tried to understand a process of future order emergence? Searching for answers to the above questions should, in a sense, facilitate identifying new possibilities how to redefine already functioning, often popular visions of possible implications further development of the Internet technology may bring to social life. Visualizing existing notions of future techno-social transformations may bring a myriad of new interesting variants of answers and pressing questions in the article.

Keywords Order of digital chaos · Technological transformations · Internet technology · Principles of network development · Self-organization on internet · Cyberutopias · Cyberunderground

A. Betlej (✉)
The John Paul II Catholic University of Lublin, Lublin, Poland
e-mail: ala@betlej.com

L.W. Zacher (ed.), *Technology, Society and Sustainability*,
DOI 10.1007/978-3-319-47164-8_8

Order of Chaos

> Chaos in not really chaos. Chaos, plainly speaking, is order pretending to be mess. This is a system in which chance and necessity, complexity and simplicity coexist and penetrate one another. Chaos has only recently been discovered by scholars. This is surprising to the extent that it is neither an abstract mathematical construction nor an intangible elementary particle which may be observed only in the interior of stars or in powerful accelerators, but the most natural and most frequently encountered form of reality (Peters Edgar 1996).

In social sciences there prevails a common conviction that a society deprived of an institutional background of formal structures responsible for keeping and protecting social order would be sentenced to chaos and annihilation. Chaos is most frequently described as a state of confusion, disturbance, darkness, disorder, a specific kind of structural mess. A state of chaos signifies lack of order. In colloquial speech chaos is described as a gradable phenomenon. Social processes are characterized by various degrees of chaos, preceding the emergence of ordered structures. Chaos is an inherent element of social order, a ubiquitous and common phenomenon.

In social sciences, however, an expression that appears more often is *social order* which is synonymous with a well-functioning social system, cooperation, consensus, social agreement. The emergence of social order enables the existence and development of a society and its structures. However, proponents of the theory of chaos maintain that it is determined linear processes, and hence those associated with a common way of understanding social order, that are exceptional (Tempczyk 2002).

Therefore, researchers conducting in-depth analyses of modern day Internet society have made the subject of their study the process of specific *cybercosmogony* endeavouring to discover the *nature* of a qualitatively new, technologically induced social order—the Internet era. Researchers of the Internet are trying to find both creators of digital order, aiming at legitimizing it, and its opponents, creative destructors, digital non-conformists of the ideology of Internet society. *Digital order* is based on networks—ontological products of digital social order and its order-creating conditions (White et al. 1976).

What is interesting, in the beliefs of primitive peoples we can find numerous similar descriptions of the emergence of the order of the universe. Examples of this are a belief in a somehow anthropomorphized creative power, very often hidden until the moment of creation, and the existence of a not quite defined *original matter* from which the world was created.

The author's objective is to assess threats and promises offered to humanity by the already *theoretically personified* Internet technology. Internet is defined in many ways: as an example of a powerful technology, a tool for collective action, a society, cyberspace, an information network. Without any doubts we may say that the technology is one of the crucial key areas of human activity. Explorations of its possible evolution and ways in which its different development paths will lead us are still very important. Future starts now. Past prophecies are useless in a great confrontation with an approaching network order. Development stages of the technology, generally, turn into specific, nonlinear revolution phases. Categories of past, present and future coexist in heterogeneous sets of daily semantic dictionaries. A nonuniform flow of innovations impacts in a different way various fields of social order analyses.

The question of the way in which the *digital status quo* of Internet social order functions and of the principles governing it is more than justified. The future of social civilization connected with the development of modern day *ubiquitous technology* is not obvious.

Metaphysical deliberations on being, chaos, potentiality and order will not be the subject of this article; however, creating a vision of the future of formation of Internet society, *a truly technological society*, a society more and more *virtualized* in the near future, requires an in-depth axiological reflection on the general condition of humanity and social effects of its future *technological encounters*.

The crucial question: why so many aspects of the understanding of technology changed throughout the twentieth century cannot still be answered. The fact that people misunderstand proceedings of the future has already brought many very serious implications for humankind's condition. To understand a peril and a promise of Internet technology for future social order we should ask three basic questions. What did we believe in throughout the past century? How have we described the technology evolution? How have we tried to understand a process of future order emergence?

Searching for answers to the above questions should, in a sense, facilitate identifying new possibilities how to redefine already functioning, often popular visions of possible implications further development of the Internet technology may bring to social life. Visualizing existing notions of future techno-social transformations may bring a myriad of new interesting variants of answers and pressing questions. This will not however fill in the semantic gap of the present time should the notions of the past not be dealt with at first.

Technological Transformations

> People who are writing about technology love a theory that says they are the most important people in society. How flattering to be told that 'you are the future', rather than train drivers or hospital workers
>
> (Barbrook 2007).

Throughout recent centuries numerous essentials technological transformations have taken place. Electricity, the telephone, the car, the computer, the mobile phone and a plethora of breakthroughs have contributed to the process of structuration of new social systems—network-based communities. In terms of value judgement great transformations have been addressed in all sorts of ways. The practical aspect of possible future application of technologies to bring about socially desirable effects was of foremost importance.

The attempts to shape the future were motivated by the faith in introducing the socially noble ideas of sustainable development, the knowledge society, counteracting social exclusion, fighting poverty, enhancing innovation (Barbrook 2007). Visionaries of the future monitored the emergence of new social rituals connected with the application of digital tools. The process of designing further technological breakthroughs was based mainly on examining social needs.

The greatest visionaries of the past decade believed in the power of computer. The practical implications and the theoretical consequences of the scenarios grounded on the belief in the unquestionable power of machines were the ambition of creating a hybrid of a man and a machine. A computer integrated with human DNA was to epitomize a grand future, a further stage in the evolution of the humankind offering new possibilities of development of humans and machines (Warwick 1998). The notions of artificial intelligence, smart software, virtual reality, markup language, hypertext markup language, microprocessors and many other high-performance tools quickly mushroomed in scholarly disputes of the twentieth century.

The faith in the power of computerization, information technology, social networks (both virtual and real) was based on the belief in the evolutionary cognitive potential of the humankind, especially the belief in the potential of the Internet technology. Exerting real influence on one's lifestyle, making independent decisions about one's place of residence or job was to be a dream come true due to the development of the Internet and the potential of a computer as a tool of the extension of the human body.

Deliberations over chaos, its origin and consequences as well as on the principles underlying contemporary social order may contribute to a serious interdisciplinary discussion on the essence of social order. Technological advancement has often been associated with the condition of chaos preceding the condition of sui generis techno-social equilibrium, sustainable development, i.e. a positive scenario in which it is the human being that creates and manages technology.

The twenty-first century saw scientists from the American Cooperative Association for Internet Data Analysis at the University of California in San Diego reach the conclusion that the Internet reminds of the human mind, or in micro-dimension even the universe as it is based on the same principles (http://zmianynaziemi.pl/wiadomosc/naukowcy-odkryli-podobienstwa-w-rozwoju-wszechswiata-internetu-ludzkiego-mozgu). At present scientists ponder whether there exist common principles of network development, irrespective of their size and type.

The most serious threats facing the development of the Internet technology are connected with the fear of losing control over structures responsible for maintaining social order. New questions emerge concerning future implications of *acts of creative destruction* on the Internet. Further advancement of the Internet technology may lead to very radical transformations of social order. Disruption of the longue durée structures, axiological micro-revolutions, the increased importance of technological principles in social life, disintegration of social ties, collapse of the previously recognized authorities and hierarchies are but a few consequences of the Internet impacting network-based communities. Digital chaos may wreck havoc and disrupt commonly recognized global narration of sustainable technological development.

Digital chaos is also positively defined as a process that may lead to the emergence of a qualitatively new order often viewed in terms of value judgement as better, more modern, more effective, more future oriented. The interaction between internet technologies and social processes, as it is stressed in this paper, is not unidirectional. Social processes may also affect information and communication technologies, the way they are designed, commercialized, on the direction of development of Computational Social Science.

Assessing positives and negatives of the future development of the Internet entails methodological confusion. The leading notion, the semantic key to understanding the ongoing socio-economic network micro-transformations, seems to be the notion of self-organization. End-of-evolution prophesies, the information catastrophe or network anarchy visions remain very popular.

If however small and big network schemes of different types remain strikingly similar in their functioning, the development of network structures regulated by the similar principles seems equally plausible, though the very principles have not been yet unravelled by scientists. Although nature still eludes human understanding, further research into the Internet structures may prove useful in understanding the regulating rules of social order, with the preceding condition of chaos in the Internet network.

Further important discourse of the future was based on a popular information metaphor. The attention was focused on the practical aspects of measuring the speed, the cost and the volume of information (Toffler 1980). Technological advancement was understood in terms of enhancing the possibilities of information digitalization while minimizing the cost of digital production, its turnover and improving particular parameters. Information efficacy was to become the metaphor for the future. The Information Technology sector instantly earned the reputation of being a highly competitive one of the twentieth century.

Another meaningful transition affected the information paradigm, which transformed into the network paradigm. The resultant revolutionary semantic product of the network technology influenced the way value judgements are made about technological development (Tapscott and Wiliams 2010). Not only they have computers with Internet access affected the décor of the living room of an average user, but they also have replaced the TV sets. The most noticeable change has been acceleration of the speed of the flow of information in networks, i.e. network explosion. A new symbolic culture of digital users has thus emerged. On theoretical level, network cooperation has been viewed as a positive feature, a tool facilitating preservation of democratic values.

Not all visions portrayed bright future. Some feared the era of technological order in which technological principles were to replace the commonly recognized civilization code. A lot of authors while passing a verdict on the Internet referred to its origin as a tool used to wage a war and not a means of providing peace.

Nowadays the belief in technology as a means of extending the human body seems valid. The Internet technology does invade the human body, not only in remotely steered spaces. Contemporary society has been turning into an interactive net of relations reenacted in virtual reality. IT will become a technological base for an alternative social order. New social rituals will emerge; the biological human body will be digitally redesigned.

Self-Organization

> The principles of spontaneous order is embodied in the free market system- a system that does not yet exist in a pure form… The free market allows complex institutions to develop, encourages innovation, rewards individual initiative, cultivates personal responsibility, fosters diversity, and decentralizes power. Market economies spur the technological and social progress essential to the Extropian philosophy… Expert knowledge is best harnessed and transmitted through the superbly efficient mediation of the free market's price signals-signals that embody more information than any person or organization could ever gather (Terranova 2001).

One of the greatest promises given to humanity by the creators and engineers of the Internet is a growing faith in the possibility of discovering the laws governing social phenomena. The Internet is more and more often treated as a specific network laboratory of power, social movements, revolutionary social changes. The reason why scientists became interested in discoveries of biological and exact sciences again was the success of the theory of self-organization (Hejl 1984; Imanda 2008; Levy 1977; Mishra and Zwierlein 1994; Riedl 1984). Its creators and promoters referred to the concepts of order and organization while explaining complex phenomena of the world of nature: order, or a certain model, is the final stage of a dynamic process of changes, while organization is an example of an intentional process.

Researchers of self-organization are interested in spontaneous processes of social order creation (Strogatz 2003). The offered vision of order breaks with the popular premises of the possibility of external control over transformation of chaos into a certain type of organization. The theory of self-organization is based on four principles of synergetics: fluctuation as a source of order, preference for creative individuals, acceptance for chaos, non-recognition of a control centre.

In future, comprehending the process of self-organization can open the door to so far unexplained laws of the nature, defying rational assessment of the scientific world, to discovering one of the greatest mysteries of living systems. The still unresolved questions of the *workings* of evolution, of the order of nature, of human consciousness and the possibilities of its development and practical use, of artificial intelligence and of complicated hybrid socio-technical systems are currently being reformulated and once more raised during serious scientific discussions. The adopted assumption of the common occurrence of self-organization processes at all levels of the world's complexity, from physical systems to cultural ones, is not without an impact on the increasing attractiveness of the discourse.

Will research conducted on self-organization on the Web allow us to understand the process of life and to find out the civilization code of the future? Can we compare ways of organization of nature, its tools of self-reflection, self-reference, self-production, self-maintenance and finally self-organization, which guarantee nature stability, survival and development processes, to the description and characterization of socio-technical systems?

The answer to the question posed is not unambiguous; it depends very much on the context. Transformation of Internet technology at the turn of the twenty-first century led to an increase in interest in potential threats to social order coming from cyberspace. A fear of new, so far unknown categories of threats cyberviolence, cybercrime, cybersurveillance, cyberterrorism, cyberwar gained strength along with attempts to control cyberspace by state authorities and market institutions. *Technology of freedom* is today more and more frequently identified with control, birth of a new quality of cyberauthoritarianism, marketing cybertotalitarianism, symbolic information violence. We are witnesses to another revolutionary change of meaning.

In the twentieth century Internet technology was described in the categories of a new social tissue, a structure creating bonds, a social binder, a technical base increasing a theoretical possibility of peacefully uniting man and a machine into one irreducible whole – a system deprived of a familiar cultural and biological context of reference. In the twenty-first century the Internet is undergoing sharp criticism.

The historical context in which the technology was created is referred to while attempting to answer the question of the degree to which the Web structure affects the formal and informal shape of communication structure, power and diffusion of knowledge in a networked society, in which the technical infrastructure is becoming the dominant link of communication processes (Castells 2000, 2001; Gladwell 2000). There is an enormous capacity for controlling a society which is technicalized and dependent on the Internet to such an enormous extent.

Thus key socio-economic information channels assume a completely new meaning. A dematerialized world of cybersymbols is to an ever greater degree affecting the material dimension of human existence. Internet technology brings with it a serious threat to the democratic principles according to which a networked system should function. There looms a threat of the onset of mega-cyberpanopticon. Modern day Internet users have long become members of a cyberpanoptic system of network control (Krzysztofek 2011).

However, the analysis of the way the Internet is organized, of its network structure, gives certain hopes for a change of direction of its development. It seems that in future Internet technology should promote self-organization in a natural way. Characteristics of the technology itself potentially encourage both this type of actions, as well as the formation of decentralized structures, deprived of a control centre (Benkler 2006; Zacher 2006, 2012).

Cyberchaos, cyberanarchy, destabilization of social order can take on completely new civilization meanings. Widespread glorification of chaos as a process is, however, not very probable in the future. Admittedly, the Internet facilitates a spontaneous formation of large structures and reduces the costs of coordinating activities; nevertheless, the issue of lack of exterior control over the activities remains debatable (Shirky 2008). Is Smart Mob really a wise choice? Was the so-called Egyptian Spring an unplanned cyberproject?

A decentralized structure of the Internet based on the principles of self-organization is very seriously entangled in the market context. Great changes are spreading like viruses in cyberspace. An "epidemic of co-operation" permeating all aspects of social and economic life is a great promise to a future social order. In future, the development of Internet technology will have a significant impact on the approach to problems of networkness. The Internet opens up new ways of exploring the issues of production and distribution of knowledge in an economy and society, and of social self-organization processes. The same hopes, depending on the context of an analysis, can bring many dangers which will ruin the well-known social order.

Cyberutopias

> Dear Patient (first name, last name)! You are presently located in our experimental state hospital. The measures taken to save your life were drastic, extremely drastic (circle one). Our finest surgeons, availing themselves of the very latest achievements of modern medicine, performed one, two, three, four, five, six, seven, eight, nine, ten operations (circle one) on you. They were forced, acting wholly in your interest, to replace certain parts of your organism with parts obtained from other persons, in strict accordance with Federal Law (Rev. Stat. Comm. 1-989/0-001/89/1). The notice you are now reading was thoughtfully prepared in order to help you make the best possible adjustment to these new if somewhat unexpected circumstances in your life, which, we hasten to remind you, we have saved. Although it was found necessary to remove your arms, legs, spine, skull, lungs, stomach, kidneys, liver, other (circle one or more), rest assured that these mortal remains were disposed of in a manner fully in keeping with the dictates of your religion; they were, with the proper ritual, interred, embalmed, mummified, buried at sea, cremated with the ashes scattered in the wind—preserved in an urn—thrown in the garbage (circle one). The new form in which you will henceforth lead a happy and healthy existence may possibly occasion you some surprise, but we promise that in time you will become, as indeed all our dear patients do, quite accustomed to it. We have supplemented your organism with the very best, the best, perfectly functional, adequate, the only available (circle one) organs at our disposal, and they are fully guaranteed to last a year, six months, three months, three weeks, six days (circle one). Of course you must realize that …
>
> Stanisław Lem, from: The Futurological Congress.

Visions of the future based on a strong belief that it is really possible to achieve the postulates of sustainable social development thanks to a growing symbiosis between people and machines are currently undergoing a surprising transformation. Numerous prognoses are still referring to ideas of evolutionism popular in the past. Cyber-Darwinism takes on the role of a new ideology. Once again, people announce ideas of the end of evolution and the onset of an era of perfect superhuman beings, liberated from their biological oppression (compare with: "The body needs to be repositioned from the psycho realm of the biological to the cyber zone of the interface and extension-from genetic containment to electronic extrusion. Strategies towards the post-human are more about erasure, rather than affirmation- an obsession no longer with self but an analysis of structure", STELARC 2001, p. 560). Will digital technology really substitute an imperfect human body in the future?

The question is not likely to arouse much controversy. The issues of belief in the might and speed of calculating power of Internet carriers are much more problematic. In its digital version the most interesting of social rituals, gathering information, also seems to be naïve. Well-known present day scenarios of the future are based on utopian assumptions. The onset of a cyberorder will not lead to creation of a digital society of equality. The Internet will imply a process of forming new, so far unknown social divisions.

The process of transformation of digital utopias was taking place in the background of IT development and changes. The phrase that entered for good the dictionary of Internet terms was the name Web 1.0. Web 1.0 was a one-way model in which the leading role was played by a service creator, who was its founder and editor. The structure assumed a passive participation of remaining users whose role was reduced to reading and receiving information, rendering it impossible to interfere with the text. Web 1.0 showed no signs of interaction, just like other media before it—books, magazines and television, up until a certain point.

A breakthrough in the development of the Internet was the implementation of Web 2.0 model, which took most creative privileges away from service programmers in order to increase users' influence on the way of managing the content. The main idea behind the second generation of the Internet was interactivity and communication. Web 2.0 was an attempt at solving one of the greatest problems concerning the Internet: an overload of content, which is not ordered and properly catalogued. A boom of networking structures draw attention to an interesting feature of collective actions on the Web. Internet structure promoted processes of information and knowledge exchange and sharing.

The future of Internet technology is associated with a new model of semantic network Web 3.0 (O'Reilly 2004). It is assumed that there is a possibility of creating a digital meta-base of human knowledge by connecting all single databases, all computers, users, digital resources. Users and Webs will start co-operating with one another, without the need for keyboards, without the need for computers, telephones thanks to the use of, e.g. brain implants. Computer programmes will find necessary data and commands in a semantic way (combined according to meaning and not without a context). This solution is to deal with the greatest modern problem of a flood of information and an ineffective use of all resources known to humanity. A centralized digital meta-base of human knowledge will enable every participant—man, a programme, a machine—access to all information at any time. At present, most information on the Web can be read only by people.

In practice, this model assumes centralization of so far decentralized structures, conversion of 2.0 system into a model of an independent meta-database, specific mental collectivization of information, digital meta-totalitarianism, a capacity for full control of all users. No one knows on what specific lines the 3.0 Web would function. For the time being, it is a vision putting which into practice, for technical reasons, will take humanity a long time. However, combining this type of vision with the idea of sustainable development seems in itself to be dangerous.

Another scenario of development based on a belief in the power of digital chaos appears to be more probable.

Exit-Node: Cyberunderground

> We are Anonymous. We are Legion. We do not forgive. We do not forget. Expect us (The Anonymous Slogan)

The Onion Router emerges as a competitor of total clear Internet, the cyber-Panopticon enabling identification of each user through IP and sending segmented advertising messages (http://www.onion-router.net/Publications.html; Singel 2006). This virtual computer network created in 2004 has become fertile soil breeding the cyberunderground (http://torstatus.blutmagie.de/). The Onion Router implements the second-generation onion routing thanks to which the net prevents analysing Internet traffic thus providing almost anonymous access to the Internet resources (Perkins 2012).

Onion routing relies on multi-layered encryption of messages which makes it possible for the users to escape content filtering, censorship and other limitations. This comes as a response to the increasing censorship and surveillance of the Internet users. Onion routing has been developing in parallel to the classical network. Though in practice it is not that quick as its classical counterpart and less popularized, onion routing attracts interest. The belief in the speed of the Internet is a thing of the past.

The future may bring a more striking division amongst the Internet users into those who have been consciously or unconsciously subjected to surveillance and those who have escaped into the cyberunderground (Betlej 2012b). The axiological aspect will also change. The right to privacy, to remain anonymous and enjoy the freedom of communication, will reappear in a new context in the public debate[1] (http://english.ruvr.ru/2012_10_22/One-mans-criminal-is-another-mans-freedom-fighter-Anonymous-EXCLUSIVE-interview-part-2/).

The cyberunderground may disrupt the future social order. "Shadow Internet" may breed new types of crime (Sienkiewicz 2012). The Internet black market may become a serious threat to the traditional agendas. One might presume this will

[1]Compare with: "One man's criminal is another man's freedom fighter"—Anonymous, EXCLUSIVE interview, part 2:

Robles: "What is the underlying philosophy of Anonymous?"

Anonymous: Freedom. And the defence of the Internet because it is the greatest tool of liberation in the history of humanity.

Robles: "What would you say to people who say you are criminals?"

Anonymous: What would I say to them? I would shrug and decline the semantic argument. One man's criminal is another man's freedom fighter. Whichever label you choose to attach to us says far more about you than it does Anonymous±.

be the path of development of the Internet in the future. It is extremely difficult to assess the potential threats and opportunities of the IT for the social order and stability as previous forecasts have proved outdated (Betlej 2012a). It does not seem valid to assume that the Internet will become a means of freedom. These are personal aims and intentions of users that will decisively shape the final judgement on the Internet as a communication tool.

References

Barbrook R (2007) Imaginary futures: from thinking machines to the global village. Pluto Press, London

Benkler Y (2006) The wealth of networks: how social production transforms markets and freedom. Yale University Press, New Haven

Betlej A (2012a) Wirtualizacja wspólnoty w społeczeństwie sieci. W poszukiwaniu kodu cybercywilizacji (Virtualisation of the Community in Network Society. In searching for cybercivilization code). In: Zacher LW (ed) Wirtualizacja: problemy, wyzwania, skutki (Virtualization: problems, challenges, implications). Kozminski University, Poltext Publishing House, Warsaw

Betlej A (2012b) Cyberterroryzm versus haktywizm, czyli o wolności wiedzy w sieci. In: Partycki S (ed) Kryzys finansowy-przebieg i skutki społeczno-gospodarcze w Europie Środkowej i Wschodniej. KUL University Publishing House, Lublin

Castells M (2001) The internet galaxy, reflections on the internet, business and society. Oxford University Press, Oxford

Castells M (2000) The rise of the network society, the information age: economy, society and culture, vol I. Blackwell, Cambridge

Gladwell M (2000) The tipping point: how little things can make a big difference. Little Brown, Boston

Hejl PM (1984) Towards a theory of social systems: self-organization and self-maintenance, self-reference and syn-reference. In: Ulrich H, Probst GJB (eds) Self-organization and management of social systems. Insights, promises, doubts and questions. Springer, Berlin

Imanda T (2008) Self-organization and society. Springer, Tokyo

Krzysztofek K (2011) @lgorytmiczne społeczeństwo? Transformacje 3–4:41–42

Levy P (1977) Collective intelligence: mankind's emerging world in cyberspace. Perseus Books, Cambridge

Mishra RK, Zwierlein E (eds) (1994) On self-organization. An interdisciplinary search for a unifying principle. Springer, Berlin

O'Reilly T (2004) Technology and tools for change. http://archive.oreilly.com/pub/a/oreilly/tim/articles/toolsofchange_0804.html

Perkins T (2012) The TOR network and the "Dark Web". http://www.youtube.com/watch?v=9WtbnKRsyuA

Peters Edgar E (1996) Chaos and order in the capital markets: a new view of cycles, prices, and market volatility, 2nd edn. Wiley, New York

Riedl R (1984) Self-organization: some theoretical cross-connections. In: Ulrich H, Probst GJB (eds) Self-organization and management of social systems. Insights, promises, doubts and questions. Springer, Berlin

Shirky C (2008) Here comes everybody. The power of organizing without organizations. Penguin Book, London

Sienkiewicz P (2012) Wirtualizacja cyberprzestrzeni i jej bezpieczeństwo. In: Zacher LW (ed) Wirtualizacja, problemy, wyzwania, skutki. Poltext Press, Warsaw

Singel R (2006) The Onion Router (TOR) is Leaky (Leeky). http://www.wired.com/threatlevel/2006/10/the_onion_route/
STELARC (2001) From psycho- body to cyber-systems. images as post-human entities. In: Bell D, Kennedy B (eds) The cybercultures reader. Routledge, London
Strogatz S (2003) Sync—The emerging science of spontaneous order. Hyperion Books, New York
Tapscott D, Wiliams AD (2010) Wikinomics: how mass collaboration changes everything
Tempczyk M (2002) Teoria chaosu dla odważnych (Theory of Chaos for Braves). PWN, Warsaw
Terranova T (2001) Post-human unbounded. In: Bell D, Kennedy B (eds) The cybercultures reader. Routledge, London
Toffler A (1980) Future shock. The third wave powershift. Bantam Books, New York
Warwick K (1998) What happens when a man is merged with a computer? http://www.kevinwarwick.com/project-cyborg-1-0/
White H, Boorman SA, Breiger RL (1976) Social structure from multiple networks: block models of roles and positions. Am J Sociol 81:730–780
Zacher LW (2006) Gry o przyszłe światy, Prognosis Committee "Poland 2000+". PAN Scientific Publishing House, Warsaw
Zacher LW (2012) Człowiek utechniczniony i zwirtualizowany w hybrydowym świecie. In: Zacher LW (ed) Wirtualizacja, problemy, wyzwania, skutki. Poltext Publishing House, Warsaw

Chapter 9
The Citizen in the Cyberspace: Should There be Any Limits to the Freedom of Speech in the Internet?

Joanna Miksa

Abstract In my article I propose to analyse the internet as means of communication that shapes the public sphere in a similar way as the press and the television or the radio do. More precisely I focus on the issue of the limits of speech that are applicable in any context in which a social agent expresses himself. Yet for its technological characteristics the internet confronts its users with new challenges and obliges them to define the concept of responsibility for their actions in a new way. An example of such a challenge is websites that make a propaganda of the so-called 'positive paedophilia', one of them being a website called 'The Little Prince' which I use a case study in the text. After the analysis of the case of the website in question I put forward the arguments for the appliance of the concept of the public journalism in the context of the Internet.

Keywords Public journalism · Public sphere · Censorship · Pornography · Paedophilia

Introduction

In this paper I would like to focus on the impact that the internet has on us as far as our condition of citizens is concerned and on the challenges that the use of internet brings about for us as participants of discussions that take place in the public sphere. I will claim that for many reasons the internet should be seen as a new element of some more traditional phenomenon which is the public sphere. The public sphere is a concept that started to be elaborated theoretically in the eighteenth century. Its coming into being is closely linked to development of

---

J. Miksa (✉)
Ethics Department, University of Lodz, 16/18 Kopcinskiego Street, 90-232 Lodz, Poland
e-mail: jmalaxer@wp.pl

L.W. Zacher (ed.), *Technology, Society and Sustainability*,
DOI 10.1007/978-3-319-47164-8_9

democracy, a constituent part of which is the freedom of speech and the quest after transparency. One of the most important theoreticians of that concept is Jürgen Habermas, who is a strong defender of the belief that a well-functioning public sphere is a necessary condition of a well-functioning democracy. The quality of the public sphere depends on the readiness of the citizens to make pressure on the state and its institutions in order to assure that they work in a transparent manner. The German philosopher says that probably only in that way the public may make it sure that the people's rights shall be respected (Habermas 1991: 379–383). Although during the two last centuries the mass media have played a crucial role in making the society at large an autonomous political actor, at times it happened as well to pose quite serious a threat to its freedom. Even though it may be claimed that the internet is a phenomenon that is definitely more complex than the press or the television, still there is one important feature which may serve here as a common denominator to all of them: it is the need for education how to use it responsibly.

There is a widespread opinion that internet cannot and at the same time should not be controlled. First of all, it is claimed that the net cannot be controlled because the task is technically far too complicated to be carried out. Probably the most advanced undertaking of that kind is the sum of practices adopted by the Chinese government ever since the internet was born. Even though it is highly successful in making life of the internet users' on the Chinese territory as complicated as it can be, still, the task of controlling the net in China is far from being completed. There is also a legal aspect. The internet sites can be run from any place in the world, which makes it relatively complicated to decide which government should be responsible for prosecuting those responsible for putting some controversial content on the websites. Internet simply seems to be as out of control as the weather is.

Secondly, there seems to be a widespread opinion that the internet should not be controlled and that the unlimited publishing of any possible content can only be beneficial. The systems of law which are limited to some territory only—even if some of those systems already are international as it happens in European Union for example—reflect at least to some point the system of values and historical, contingent tradition of a given community. The internet, which is free from such limitations, provides its users with information and opinions different form their historically and culturally limited environment and hence set them free from such limitations. By gaining access to such a broad perspective from which local events may be reflected upon the internet users may gain a critical distance which is absolutely crucial for everyone to became an autonomous, active participant of debates on the political and social issues. Everything that happened in 2011 in Arabic countries seems to support such an opinion. Taking this into consideration it can be said that internet enlarges the notion of the public sphere.

Yet I want to claim that the internet can and sometimes probably should be controlled. The only agent that is entitled to do so is the public or the society which may only do so though an open and transparent debate carried out in the public sphere. In this article I shall point to such a widely recognised as abhorrent kinds of material that may be found in the internet as the websites promoting paedophilia. I shall specifically concentrate on the so-called 'good paedophilia', which is promoted by the so-called 'child lovers'. In order to show how the internet can be controlled by the public I would like to examine a case of closing down a controversial website that occurred in Poland as a result of a pressure exercised by public opinion, or more precisely by one of leading Polish newspapers Gazeta Wyborcza. I am referring to the page called 'Little Prince', which was active from December 2004 till May 2006 and was closed down after a series of articles published by Gazeta Wyborcza. By achieving that Gazeta Wyborcza acted not only as a detached observer whose role is to inform the public while avoiding at any cost to take a stance, but became an actor. The newspaper was invited to speak out on the problem by people involved in fighting paedophilia in the net and who at the time were working in various non-governmental organisations. That concept of journalists' work, which consists of an active collaboration between community members and journalists in order to solve a problem a community suffers from, was called *public journalism* by an American academic Rosen (1999). His idea was that the journalists cannot limit themselves to convey the news only, and they should recognise that they are an intrinsic part of a society or group which they are supplying with information. The origin of that concept was that the media that limit themselves to observe facts and write about them do not fulfil their role and would eventually loose their readers. The idea of public journalism was thought as well as a response to the process of turning journalism into sheer gain-oriented business like any other. Obviously, Rosen proposed his interpretation of what the journalist profession should consist of in the USA where the state by the virtue of its liberal principles intervenes in the citizens' affairs less than it in Europe, including Poland. Still the question what is the aim of being a journalist is or should be is remains open to discussion. Rosen's proposal of public journalism comes noteworthy in the internet context. Those are the media that have the power to invite the otherwise scattered public to ask fundamental question on how the internet should be used and make the people aware of the dangers that loom in the cyberspace. Those are the media as well that may enforce some actual decisions to be in made in order to make it possible for the society to manage the new technologies. In that way I propose to interpret Gazeta Wyborcza's intervention in the 'Little Prince' website case as an example of public journalism.

The outline of this paper is as follows: firstly I shall present an analysis of an actual case, namely the Gazeta Wyborcza's intervention that lead to making the Little Prince site inaccessible to the internet users. Secondly I shall discuss theoretical issues concerning the censorship of the internet websites in the context of 'good paedophilia' debate.

Before I start my argument I shall elucidate the rather awkward term 'good paedophilia'. The definition of that term is twofold, depending on whom you ask. Objectively from the scientific point of view advocated by psychologists and psychiatrists such a thing as 'good paedophilia' does not exist and the term coined by its advocates is misleading. The supporters of 'good paedophilia' insist on being called 'child lovers'. They claim that there is such a thing as love between an adult and a child that is even more intensive than relationship between parents and their child and which may—although does not have to—involve voluntary sexual relations. There are legal organisations which promote such a point view in the USA.[1]

The pioneer of that movement was a Dutch psychologist Frits Bernard.[2] Bernard died in 2006 as an octogenarian; however, the high point of his activity as 'good paedophilia' defender was the 60s and the 70s. He offered a research that was meant to prove that paedophilia is not an illness or a deviation but one of the many options to live one's sexuality. Probably the most controversial moment in Frits Bernard's life was a 1-h-long interview he gave in 1987 on NBC (National Broadcasting Channel). The host of the programme was Phil Donahaue, and Frits Bernard was supported by young men at the time aged 23 who confessed to having had sexual relations with an adult as child. Even if already in the 80 the public was turning its back on the idea of paedophilia being just one of the options open to any sane sexually active person—no matter how small the number of its sympathisers might have been even in the 60s and 70s—Bernard continues to be an important point of reference for the majority of supporters of campaigns aiming at decreasing of the age of consent.

Case Study: The Little Prince Website

In many ways the Polish case of a man called 'Misiaczek' can be seen as a local mutation of the wider movement of those who think that the age of consent should be decreased.[3] The person who was running the Little Prince website was anonymous, he was known by his nickname Misiaczek (Teddy Bear). The web was operated from outside Poland, more precisely by means of a server located in the Russia and other in the USA, which was one of the most important reasons for which it was so difficult to close it down. Apart from that it was claimed by the person responsible for running the site that the Polish law was not broken by its existence. The police was amply aware of the existence of the site as it was denounced by a couple of organisations specialising in protecting children's rights,

[1]An example of such an organisation is B4UAct, which operates in the USA as well as the more famous North American Man/Boy Love Association which the poet Allan Ginsberg adhered to, in Holland there was an organisation called Vereniging Matijn, it was ruled by the Dutch court to be illegal on the 27th of June 2012. There used to exist a similar organisation in United Kingdom (Paedophile Information Exchange, 1974–1984) and in Denmark (Partij voor Naastenliefde, Vrijheid en Diversiteit—Party for Neighbourly Love, Freedom and Diversity, 2006–2010).

[2]Information on Frits Bernard may be found for example on the website dedicated to him: http://www.ipce.info/ipceweb/Library/dutch_movement_text.htm. The site is interesting as it speaks in favour of Bernard's life achievements. The site is a part of a website run by Ipce, a Dutch child lovers' organisation.

[3]All the information on The Little Prince website is taken from Małgorzata Grzebałkowska's article (Grzebalkowska 2006: Mówi, że jest dobrym pedofilem [He says he is a good paedophile], 2006.05.01, Gazeta Wyborcza—Wysokie Obcasy).

such as Kidproject Foundation or Fundacja Dzieci Niczyje (no one's children foundation). The police told the Gazeta Wyborcza's journalist who investigated the case, Magdalena Grzebałkowska, that the site was under constant scrutiny and that the police was just waiting for any mistake of Misiaczek in order to take legal measures against the Little Prince website. It is a highly important aspect of that case as it proves that the public control of the media and the internet is absolutely necessary to make it safe for the society. More can be said—the level of safety that can be reached in the net depends only on the awareness and the willingness of its users. In that aspect it resembles any other tool that is at hand of citizens of any political system: the rules that govern political institutions can function well only if there is a constant effort of the people to apply them in a responsible manner.

The Little Prince website was widely accessible for a year and a half, up to the moment when its existence drew Gazeta Wyborcza's attention it had had 96 000 entries. During that time the state tried to close it down, yet to no avail. Everything changed in May 2006 when Gazeta Wyborcza published a long interview carried out by Magdalena Grzebałkowska with the man who was using the nickname 'Misiaczek' (the Teddy Bear) and described the content of the site in question. In order to avoid accusation of propagating paedophilia by inviting people to visit a paedophilia-friendly website—and probably feeling responsible for the readers—Wyborcza in its article gave the site another name, 'Peter Pan'. As it was mentioned before everything that was written there could be interpreted as legal, which is what the police originally did. First of all, the Teddy Bear never invited his readers to use violence towards children, he condemned extensively a vicious form of paedophilia which according to him consisted of forcing children to have sex with an adult or even proposing sexual intercourse to children. He emphasised that what is at the core of his concept of paedophilia is friendship with children that might culminate in sexual caress and intercourse if and only if the child in question would ask for it. The Teddy Bear proposed a code of a good paedophile in which it was stated that the child should be treated as a friend and an equal and therefore educated of the legal situation of the friendship between an adult and a child. Sex was not according to the Teddy Bear an imprescindible or even necessary part of the relation.

One can obviously ask how come that expressing such opinions could be regarded as legal. As observes a psychologist Maria Beisert, paedophilia can be defined in various manners depending on the perspective that can be adopted. From the legal point of view paedophilia refers to a deed that involves sexual interaction between an adult and a child, but for a psychologist the actual contact with a child is not necessary, an adult whose sexual phantasies include children suffers from paedophilia, which in its turn is a deviation. A person suffering from it should or at least might undergo a therapy even if he or she has limited themselves to phantasies. The legal definition of paedophilia does not make any difference between a consented or not consented sexual contact between an adult and a child (Beisert 2012: 19–47). Even taking those factors into consideration, even though the Polish authorities were not unaware of the existence of the Little Prince website, up to Gazeta Wyborcza's intervention they just did not see how they could legally oblige the Teddy Bear to close the controversial site down. The legal analysis of the Little Prince website case is far beyond my abilities, it is a subject that may separately dealt with by a lawyer. Asked by the Gazeta Wyborcza journalists and earlier on by the people from Kidproject Foundation the authorities would replay that they were just waiting for any mistake that would entitle them to take legal steps against the Little Prince. Within a month after the publication the site was closed down—although extensive fragments of its content can be found in the net as quotations. It may be safely said that from the legal point of view the case was open to different interpretations and a lot depended on the individuals' decision to intervene or not.

The fact that in the Little Prince case finally a relatively successful intervention occurred reflects a series of multifaceted changes that the Polish society has been undergoing in the recent years. The change that took place in 1989 meant among other things a birth or rebirth of public sphere and a society of citizens. Along with decreasing interest in politics understood as parties' rivalry and quest for power an increase of interest in social problems came into being, issues like domestic violence, civic rights for homosexuals, abortion and eventually safety from paedophilia in the net became themes of information campaigns and matters of a sincere public concern. In that sense Polish situation is not different from that characteristic of any other postmodern society. According to a sociologist Anthony Giddens there is in Western societies an important return to moral thinking and turning to axiological arguments in public debates (Giddens 1991: 202–231). If Giddens is right a postmodern citizen would be willing to interpret problematic situations that may occur in public life in moral terms rather than to think that it is all only a matter of private choices that are morally neutral. As normative statements pretend to be true they need to be followed by a rational discussion. In that sense unlike esthetical choices they are not private and demand something more than actual subjective choice of an individual to justify them. I shall now turn to theoretical issues that should be discussed in order to decide if the internet can be censored and if the answer is yes then how, by whom and what are the limits to which it may be acceptable and justified. At this point I want to stress it that any form of censorship is a last resort that a society might turn to only after careful and well-informed analysis.

Freedom of Speech in the Cyberspace

I propose to analyse the issue of censorship in the internet as a special case of freedom of speech. That issue that is traditionally important in Western political tradition needs rethinking after we all came to live in cyberspace, just as earlier generations had to learn how to live in the press, then radio and almost finally, the television era. Obviously, the contexts in which the freedom of speech will be of interest for me are the child lovers' websites. There are a number of relevant issues that are not new but which are related to a larger problem of the freedom of speech as such, no matter what is the means of putting someone's views forwards to the public. Those will not be taken into account in that article, I shall only refer to those that are important in the debate that focuses on the internet. First of them is the question if the censorship of the controversial contents, which may be widely accepted by the public at large when it comes to a number of issues, but which may nevertheless cause disadvantage to future development of the internet. Another problem is by whom and in which way the internet may be controlled? I shall speak against state control and advocate the opinion that it is the informed public that should be trusted with exerting control over the internet within limits prescribed by the law. If it is the state's role to protect individual freedom, including the freedom of speech, but it is the society that is to interpret in the public debate up to which point the freedom of speech is rather beneficial than harmful. My point is that an unlimited freedom of speech may turn out to be as dangerous as the state's efforts to control it. The first alternative leaves some potentially threatened with aggression, for example groups like ethnic or sexual minorities, while the other one may too easily lead to oppression.

The problem whether freedom of speech may be recognised as child lovers' right to express their views freely in the public sphere is a part of a larger discussion concerning the freedom to present publically materials referring to sex or containing pornography. I want to stress firmly at that point that I am not going to discuss the issue of paedophile pornography as it is simply illegal in most of the countries of the world and is recognised as being immoral mainly because of being harmful. I shall only discuss the child lovers' websites' claim to be morally acceptable. The difficulty with those pages is that they do not present paedophile pornography and not all of them incite adults to propose sex to their underage victims, whom they call friends. The author of the Little Prince website claimed that the actual intercourse is not a necessary part of a friendship between a loving adult and a child. Sexuality is a subject that has so far raised a great deal of controversies every time when the freedom of mass media was a matter of a public debate. The '68 generation inspired by philosophers such as the Frankfurt School adherent Herbert Marcuse or later on by Michel Foucault spoke against legal constraints on sexual freedom, with special attention given to the way the media presented the subject. Owing to the generational change in the public sphere that took place in the 60 and the 70 most of the undertakings to censor the media began to be looked upon with certain suspicion in the Western World. Nevertheless, a point was reached at which even those who identified themselves as liberals started to ask the question if less constraints really means more freedom. Those were the feminists like Andrea Dworkin, Catharine MacKinnon, Robin Morgan, Diana Russell among many others who observed that freedom of pornography might raise criticism as far as the image of women is concerned and who therefore advocated limiting the access to pornography in the internet. Another case where some kind of censorship is widely accepted in democratic societies is websites supporting racism or political doctrines including racist views such as Nazism. I shall refer to discussion on the pornography as it sets a context that is in my opinion of use while discussing the censorship of websites advocating 'good paedophilia'.

My point of reference will be a book written by Amy E. White, *Virtually Obscene. The Case for an Uncensored Internet* in which she speaks out against any form of censorship in internet. The arguments she puts forwards in the context of pornography as such are relevant in the case of websites promoting 'good paedophilia'. The author uses the example of pornography in order to explain why efforts to limit the freedom of speech or expression are unjustified. Agreeing with her on that point I will refer to the most important of her arguments in order to show why they cannot be of use in the case of the 'good paedophilia'. I will list White's arguments not in the order she gave them but taking the weight she ascribed to them as a point of reference.

The first argument that was taken into consideration in the American discussion on the censorship of the pornography in the net and which White comments upon reflected mainly on possible consequences of the decisions that could be taken for the freedom of the internet. The feminists authors advocating censorship of pornography that White quotes, Catharine MacKinnon and Andrea Dworkin, claimed that it strengthens the wrong image of a woman in the popular culture as it reduces women to an object that may be used to satisfy men's sexual needs. That strengthens a widely accepted stereotype that is harmful in public sphere as well as in private (White 2006: 89–107). Another point White comments on is the possible harm caused to children who may come across pornographic materials in the net accidentally. Those in favour of censorship warn that contact with pornographic content in early age may deviate psychological development of the young (White 2006: 69–88). To these objections Amy E. White responds in two ways. First of all she quotes psychological research showing that there is not a link between being exposed to pornographic materials in the young age and a risk of becoming a sexual pervert nor is there any proof that men watching pornographic films depreciate women when compared with those who do not watch them. She observes as well that popularity and development of pornography in the twentieth century took place mainly in democratic and women-friendly countries, which proves that pornography does not denigrate women's image in the society. She concludes that the harmful character of pornography in the internet cannot be shown and suggests instead that it is rather a matter of conviction. Secondly White points to the fact that visiting pornographic sites is only voluntary and in the end those are the subjects and their moral choices that should matter when it comes to deciding the fate of pornography in the web. That makes a difference between the internet on one hand and television and the radio on the other, which in many countries are not allowed to present pornographic materials. Owing to voluntary character of using the internet, the pornography issue should therefore not be regulated by the law (White 2006: 69–107).

The second White's argument is that any regulation of the internet's accessibility is potentially dangerous as it may impede the development of the internet as such. She claims that an unlimited freedom of the internet is one of the requirements that must be fulfilled if the internet as a technology is to have a future. White refers to unwanted consequences of the internet filters, for example blocking certain phrases in order to block access to pornographic sites in public libraries results in blocking access to completely innocent or even morally laudable ones, like those containing information on HIV protection or gay and lesbians' rights among many others. She suggests that those fairly benign examples of censorship on a minor scale confront us with the necessity to ask the question what may happen to the internet if we start to believe that the constraints on the internet may be justified (White 2006: 132–147).

The third argument that is important for us in White's reasoning is a certain measure of distrust towards the public opinion. She points to the psychological research that shows that decisions are more often taken under influence of emotion in spite of reasonable arguments rather than owing to them. There obviously is a community of university scholars who are trained to reason according to rationally defined methods, but there is no reason to expect that the public at large is capable of such a feat (White 2006: 56–60). That argument is very important indeed, the discovery of irrational character of decision-making is almost half a century long that initially implausible for many, but well-proved thesis was proposed in the 50. by Leon Festinger and his young at the time assistant Elliot Aronson. Since then a number of experiments carried out by a number of scholars supported that view. Recently it was still further supported by neurological tests. The psychologists concluded that we observe actively and pay more attention to the elements of reality that are consistent with our worldview while tending to ignore anything that would force us to change our opinions (Aronson and Tavris 2007). The trust in society's at large capability to reach an informed and well-thought-out consensus on any matter, so dear to philosophers like Jürgen Habermas and John Rawls, would in the light of those findings be too optimistic. Therefore, it may be claimed, as Amy E. White does, that society consists of people who are irrational and only very few are able to live up to any reasonable standards that be applied in the process of any decision-making.

The arguments proposed by Amy E. White in the context of pornography may be discussed in the case of websites promoting 'good paedophilia'. First of all there is a question of the influence such materials may have on the public. The article from Gazeta Wyborcza denouncing the Little Prince contains an opinion expressed by a worker of a non-government organisation Kidproject that fought against the site in question. That person, whose name is Paweł Śpiewak, says that the danger resulting from sites run by the so-called 'child lovers' consists in the building of a morally neutral or even positive image of paedophilia. The Teddy Bear (Misiaczek) justified his views by referring to child's autonomy and respecting its will as well as by love and care. However, it is difficult to state to which degree the internet user may really be influenced by that kind of messages. We may safely agree that someone who is not a paedophile will not be affected and turned into one after reading a website, but on the other hand someone who actually is a paedophile will suffer from that deviation regardless of having or not having access to the child lovers-run websites. For the time being we may accept as a hypothesis that the existence of sites like The Little Prince does not change in any way the number of paedophilia cases.

Yet I advocate the view that still there is a good moral reason for the society to intervene against the child lovers' freedom on the internet. I fully understand that to enter any site you must take decision to do so and that in most cases you do not come across them accidentally. Still there are options that should not be allowed to be accessible to as it may be claimed that people are not only responsible for what they do but for what they let happen as well. The internet is a part of a public sphere and therefore is subject to rational examination, although as a means that allows its users to express freely their most intimate views it may deceptively produce an impression that by entering the internet we remain in the privacy of our homes or informal discussions. This, however, is misleading, as many reckless users of social websites had sometimes quite painful a chance to observe once they shared too much private information with others. As we cannot imagine toleration without limits, in that same way every society needs to define the limits of freedom. The unfortunate characteristics of moral values are that very often some of them cannot be made part of our factual life simultaneously. Freedom is doubtless crucial for every democracy, yet it is not the only value that needs protection. I shall not speak for safety in that place, which is a famous rival of freedom, especially in arguments that engage liberals and communitarians. Instead I shall claim that every democratic society owes protection to the weakest and there is no reason not to do it on the normative level. We are not always able to foresee all the possible consequences of some social phenomena, for example such as the actual influence on child lovers' websites on children's safety. Still even regardless of the possible consequences there is a good normative reason to step against the freedom of propagating paedophilia as there is a good reason to fight the freedom of expressing racist or pro-Nazi sympathies. Letting such opinions circulate freely in the public sphere is not consistent with the idea of human rights that need protection and which are a foundation of modern democratic societies.

I proposed earlier to take for granted for a while a hypothesis that child lovers-run websites do not cause harm. Yet there are reasons to suspect that the child lovers' websites are potentially dangerous. According to the psychologists and sexologists one of the fundamental traits of a paedophile is immature personality that makes it possible for him or her to interpret children's behaviour as an introduction to start a sexual relationship (Beisert 2012: 90–94). Child lovers' websites only support that kind of thinking. Now as in an overwhelming majority of cases the paedophile is a relative of the child or at least a person of trust, it is often complicated to detect the problem as it is easy for them to force or to coax their victim into silence. At the same time there currently are voices of paedophiles who understand that they suffer from a deviation that have no control over and are ready to seek for help. It appears that information people get from any kind of mass media do have an influence on individuals. The argument that would say that there was a problem of paedophilia before the internet era would be simply demagogic. Even if paedophilia had always been an unfortunate part of human life, our latest experience shows that speaking out loud about the problem and education is probably our most effective weapon against it.

There are as well a lot of sound arguments that demonstrate that pornography has influence on the psychological development of children exposed to it as well as it may affect sexual life of adults consuming it. The problem with the empirical evidence we obtain from experiments and clinical observation is that it is ambiguous, it is very difficult indeed to prove that there is a real causal link between pornography and criminality (Curran and Renzetti 2008: 403–414). Even if pornography was to some extent harmful, still it is not a valid reason to censor the internet, especially if someone, as Amy E. White does, takes liberal definition of freedom as a point of departure in her reasoning. Adult people are free to make any use out of their freedom as long as they do not enslave others. Obviously it is not crystal clear what enslavement may mean in various cultures, but we can reasonably accept that putting pornography on the net can hardly be called an enslavement. We may mention even journalists writing on human trafficking, most important of whom are probably a Canadian Victor Malarek and a Mexican Lydia Cacho. They claim that pornography culture is potentially dangerous as it promotes sex devoid of feelings and that approach is later on put into practice by the prostitutes' clients. Now the problem with the latter is that an alarming number of the sex-business workers are victims of human trafficking. The pornography would, Malarek and Cacho claim, make popular the view that prostitution equals a means of having fun as many others and therefore makes us indifferent to the sex slavery problem (Malarek 2009: 193–207; Cacho 2010: 260–276). However, the kind of use made out by adults out of their freedom is completely irrelevant in the child lovers' websites context. Pornography may be potentially harmful and yet there is no reason to prohibit it as long as everyone involved, especially the spectator and the performers, can be reasonably called able to freely consent to producing or watching such materials. However, it is not the case once it comes to sites or any other bits of information promoting child lovers' point of view as children may not consciously take an informed decision. The eventual consent may only be called a result of a seduction and for that reason cannot be called free. It is for that reason that Western legal systems penalise paedophilia regardless of consent of the child involved.

The second argument against censorship White proposed was that censorship may impede the internet's future as it needs an unreserved freedom of speech in order to function and develop. I shall respond to that that it is an overstatement. The conclusion drawn out of the examples given is exaggerated. It is true that internet filters may bring about unwanted results such as blocking access to important information. But it only shows that censorship of the internet is a very complicated undertaking, not that it should not be done at all. Universal filters may not be adequate tools of controlling the internet but in the case of Little Prince website case what was needed was very precisely defined and punctual intervention. I agree that using too grossly defined a set of norms to censor internet may be dangerous for its growth, but it does not lift from us a responsibility to react in particular cases. The examples Amy E. White lists in order to prove that internet should not be censored at all because owing to the use of filters useful information may be blocked are not enough to prove that the freedom of speech in the cyberspace should be unlimited. The argument is superficial as it only shows that filters are an unsatisfactory means of making the internet as safe as it can be and not that it is impossible to react successfully to unwanted phenomena that may take place in the cyberspace.

Another question to raise is the technical question who should be made responsible for putting unwanted material on a website. From the agents that are involved we may choose the ISPs (Internet Service Providers), the administrator of the website or even the user of the computer on which an illegal material was watched. For example in virtually all European countries it is illegal to view explicitly paedophile materials, such as photos or films. Participating in discussions on an overtly paedophile forum is in many cases tantamount to breaking the law as well. Obviously in the last case it is often difficult to prove if the owner of a computer is at the same time the user, but it does not stop the European countries from prosecuting such behaviours. I shall not propose an answer to such a technical question here, in practice there are many ways in which that problem is resolved within various legal systems. I mention that problem here in order to stress that once again the complexity of the issue cannot be interpreted as an argument for impossibility of exerting some kind of control over the internet. In the case of closing 'The Little Prince' site it turned out to be possible to end or at least greatly reduce the problem. It was possible because there was a group of responsible citizens, non-government foundations workers and journalists, who within the limits prescribed by the law managed to take effective steps in order to block one morally unwanted phenomenon.

Finally the case of closing down the Little Prince website after citizens' intervention shows as well the limits of the argument built the irrational character of our thinking as proved by modern psychology and neurobiology. It is a clear example of a naturalistic fallacy. It is true that actually we have certain characteristics which are a part of our description. Nevertheless, those characteristics do not define normative rules we should live according to. From the fact that we tend to be irrational in our thinking does not follow that we should not make efforts to become rational in public life. The mankind has shown throughout history that we can reasonably expect ourselves to live up to certain normative standards even if we have all the reasons to suspect that those standards will never lose their normative character and will not become part of our actual characteristic instead. Even if we will never become angels or perfectly rational beings it does not mean that for that reason we are free from moral responsibility for our choices. Rather because we actually too often behave irrationally and tend to harm others easily while producing with ease rational justifications for such deeds, we need to constantly analyse our acting from the moral point of view. In the same way the fact that there is a plethora of possible meanings we may rationally ascribe to the term 'morality' cannot be an argument to drop moral thinking altogether. Instead it is an argument to enter into open public discussion. The open public discussion seems to be the way in which problems within democratic societies should be solved, and it is congruent with basic normative standards which define democracy in postmodern societies and states.

Conclusion

The story of closing down the Little Prince website shows that no matter how complicated the media happen to be the rules that apply to the user are fairly similar. The case of people publically advocating abhorrent ideas—including educated ones and doing a research as it was in the case of Frits Bernard—is not peculiar to the internet era, it only puts an old discussion about the limits of the freedom of speech in a new context. From the technical point of view the internet is a means of communication far more complicated than the traditional ones, and it introduces as well serious changes in the relations which link the actors of the public sphere. As the example of Wikileaks shows, owing to the internet the need for transparency can be fulfilled on an unprecendented scale, it can as well promote democratic changes or at least bring about information necessary for the people to be autonomous political subjects. And yet on the other hand it has one characteristic that is common with any other tool, no matter how simple that tool might be—it can be used to promote progress as successfully as it can be used to bring about many unwanted phenomena. A specialist in mass media studies can explain us in detail how the society is being changed by the internet, and it is obvious that its characteristics have an important impact on the society in many dimensions. Yet from the moral point of view the very coming about of all those changes does not make us free from the responsibility for the use we make of the opportunities that the internet offers. Therefore, in order to promote civilisational growth it is important to ensure an education for a responsible use of technology that differs from many other achievements of human thought by its omnipresence. Internet might not be ascribed to democratic societies only, but it is probably one the most democratic technologies due to its being dramatically more accessible than for example the latest techniques of saving human lives that medicine has come up with.

References

Aronson E, Tavris C (2007) Mistakes were made (but not by me): why we justify foolish beliefs, bad decisions and hurtful acts. Pinter & Martin Publishers, London

Beisert M (2012) Pedofilia. Geneza i mechanizm zaburzenia. Gdańskie Wydawnictwo Psychologiczne, Sopot

Curran DJ, Renzetti CM (2008) Kobiety, mężczyźni i społeczeństwo. Państwowe Wydawnictwo Naukowe, Warsaw

Cacho L (2010) Esclavas del Poder. Un viaje al corazón de la trata sexual de mujeres y niñas en el mundo. Random House Mondadori S.A., Barcelona

Giddens A (1991) Modernity and self-identity. self and society in the late modern age. Stanford University Press, Stanford

Grzebałkowska M (2006) Mówi, że jest dobrym pedofilem, 2006.05.01, *Gazeta Wyborcza—Wysokie Obcasy*

Habermas J (1991) The structural transformation of the public sphere: an inquiry into a category of a bourgeois society. MIT Press, Cambridge, MA
Malarek V (2009) The Johns. Sex for sale and the men who buy it. Key Porter Books, Toronto
Rosen J (1999) What are journalists for?. Yale University Press, New Haven
White AE (2006) Virtually obscene. The case for an uncensored internet. McFarland & Company, Inc., Publishers, Jefferson

Chapter 10
Progress in Science and Technology Versus Threats to Civil Liberties—Selected Issues

Ewa Polak

Abstract The purpose of this study was the analysis of the challenges and threats to the contemporary society, human rights, and democracy, because of progress in research and technology and the new scientific and technical achievements (especially in means of communication). The essential question arising in view of the above concerns the long term consequences of the observed processes—which of the opposing processes is dominant today—broadening of the civic liberty, or consolidation of the invisible control? The state has gained new tools of exercising power. The contemporary technology offers almost limitless opportunities, and as such tempts to control, watch, monitor, and intervene in the private lives of individuals, also in democratic countries. The development of computer techniques contributed for cultural changes too, among others to form cultures of spying. This type of activity is coming increasingly into the focus of the local authorities, banks, telecom operators, and other enterprises, managers of the web portals and browsers, advertising corporations, internet site owners, schools, employers, parents versus, e.g. the preschool teachers and other carers of their children, and even the neighbours and spouses. At the work was used a literature on the subject, including well known forecasts and the press reports.

Keywords Community web · Civic rights · Invisible control · Invigilation · Bugging system · Sinopticon consept · Transparent society

Progress in research and technology and the new knowledge gained in effect, i.e. in technical science, sociology, and psychology, induce changes in the opportunities open to individual participants in the social, political, and economic life, as well as in the nature of the specific government–citizen relationship and the

E. Polak (✉)
Department of Global Economy, Faculty of Economics, Institute of Maritime Transport and Seaborne Trade, University of Gdansk, Armii Krajowej 119, 81-824 Sopot, Poland
e-mail: polakewa@wp.pl

L.W. Zacher (ed.), *Technology, Society and Sustainability*,
DOI 10.1007/978-3-319-47164-8_10

interpersonal relationships in general. The occurring changes are moulded specifically by the ever more numerous technological systems: technical systems—e.g. the Internet, satellite telecommunication systems, and stock exchange systems (Zacher 2006, p. 86). The web information society is forming. The state has gained new tools of exercising power, whilst the society has gained new instruments of monitoring the politicians and one another. The essential question arising in view of the above concerns the long term consequences of the observed processes. Will the new scientific and technological achievements (especially in means of communication) add transparency to the social and political life and pertain to increased power of the single human being and further democratisation of the political systems or will the direction be just the opposite with increased variety and effectiveness of the instruments of public control in the hands of the political, economic, and cultural powers? Which of the two processes is dominant today—broadening of the civic liberty, or consolidation of the invisible control? (Krzysztofek 2006, p. 61) Will this in effect bring higher concentration or dispersion of power? Will the humans turn to using the modern technology to control and exert an even stronger impact on one another, or will the technology grow to become a factor making life easier, more safe, and comfortable? On the one hand, access to information is gaining in democracy; modern tools of communication are becoming more common enabling tighter public control over those in power, exerting pressure thereon, and self-organisation to broadly voice and solve own problems. On the other hand, the margin of control, invigilation, manipulation, and subjection of the public is broadening. One could say that the more civic rights and liberties the man has, the more information he must accumulate to exercise his liberty. On the other hand, the more information on individuals is publicly available, the higher potential risk the fact carries for their liberty.

The situation has triggered many forecasts and visions of the future, mostly pessimistic, warning against the temptation of abusing the new knowledge and skills, particularly by those at rule.

The best known prognosis was constructed by G. Orwell in 1984, in his book: '*Year 1984*' (Orwell 2006) The author claims prophetically that the modern technology and instrumental use of psychology create the potential of control and manipulation available to the rulers resembling the power of the omnipotent Big Brother. The tool is claimed to serve total control and intimidation of everyone around, particularly anyone that challenges the reality. Power, of not loved, should be feared. The technical means in our possession make it ever easier to track down, identify, manipulate, penalise severely, and absolutely subject those who violate the existing status quo.

M. Foucault, on the other hand, in his famous work: '*Discipline and Punish. The Birth of the Prison*', compares the contemporary society to a prison designed to the *Panopticon* concept J. Bentham. Panopticon is the design which enables observing the inmates continuously and—more importantly—without them being aware. Panopticon stands for permanent invigilation (the privileged minority observes the subjected majority).

Addressing the *Panopticon* concept, Z. Bauman believes that the phrase more adequate for the contemporary times is *Synopticon*, or invigilation of local space for the global purposes (the majority observes the minority) (Bauman 2000).

In his '*The Transparent Society*', D. Brin hypothesises that although technological progress cannot be stopped, balance could be attained, if everyone had access to the tools of public invigilation and everyone could democratically eavesdrop on everyone else (Brin 1998). Here, one develops the doubt whether the decision-makers would be interested in having the society watch all their steps.

According to N. Klein, a way to lessen the public resistance and keep people in obedience is to give them a shock through inspired social, political, and economic upheavals or natural catastrophes (Klein 2009).

A. Huxley, on the other hand, writes about the great new world where people—ideal employees—have no extravagant needs or expectations, never revolt, but instead are always happy, whilst the rulers do everything to get '*them to like their inescapable social destiny*', (vide: Huxley 2011) by subjecting them to comprehensive psychological and biological conditioning and turning them into peculiar automatons.

The contemporary technology offers almost limitless opportunities, and as such tempts to control, watch, monitor, and intervene in the private lives of individuals, also in countries deemed democratic. Telephone calls, web data transfer, and satellite transmission can be monitored through automatic bugging systems. These employ, e.g. the tool of analysing the words used in telephone calls, SMS messages, or e-mails. The ever perfected tapping devices and increasingly dense network of cameras spying on people in the streets, on motorways, at airports, railway stations, shops, banks, and other service outlets enable various institutions accumulating information on the citizens' lives whilst at the same time exert total control of the society.[1] Electronics has caused that wherever an individual turns, he/she leaves traces of his/her presence. Tracking them is possible through, e.g. mobile phones, the GPS, electronic archives at hospitals, outpatients' clinics, banks, payments made in regulated financial transactions using pay cards or the Internet, the use of loyal client cards, air flights, or apparently anonymous questionnaires. The Google browser has launched the Street View option attached to the navigation maps of major cities and thus enabling web surfers watch the developments there in real time. The tyres used in American cars have a special chip built in. The chip emits a characteristic radio signal, which enables owner identification (Bendyk 2004, p. 176). The Dutch started building an intelligent motorway in 2012, and Poland intended to hold a procurement procedure for the construction of an

[1]There are over 5 million cameras in the UK, monitoring the lives of private individuals in 2013. The New York Civil Liberties Union (NYCLU) has documented nearly 2400 cameras in public places on Manhattan alone; the streets of Warsaw are watched through more than 5 thousand cameras, there are 1800 cameras at schools, and 4615 on public transport. Gdansk has had more than 2 thousand cameras installed, out of which nearly 1 thousand at schools. Unmanned aircraft vehicles, or the so-called drones, were introduced in 2013 to monitor the New York city.

intelligent transport system in 2013 (Zboralski 2013, p. 4). When crossing some borders at airports, the travellers are exposed to the procedure of taking their fingerprints and scanning the whole body before they board the plane. Control can use such elements as the PESEL or NIP numbers *[population register ID and tax ID]* allocated to each citizen, tags in the ear of each cow, the numbers on eggs in shops, the manufacturer's identity markings of each TV set or computer, the permanently mounted GPS tracing the movements of the car owner, the eye readers meant to substitute the door keys, or voice identifiers intended to make life easier and operate the household equipment. The mass media have recently and ever more frequently reported on biometric identification of humans, or even chipping them. The volume of such biochips could be equated to ca. 2000 pages of the A-4 format (Lachowski 2006). They would enable not only locating their holder, but also gaining access to any information he/she accumulates over his/her life, or even (nearing science fiction) controlling his/her behaviour. For instance, one of the discos in Barcelona charges its fees using subcutaneous chips. The American TV has broadcast an advertising spot promoting chips that enable the doctors do remote check ups of their patients. Children in Texas have chips fixed to their ID cards, they are obliged to carry on them all the time, and school pupils in Japan have to wear chip bracelets. School kids in Maryland are issued lunch at school based on their palm scans. The parents, mainly in western countries, can have a current insight into the academic achievements and school attendance of their children through electronic school register, whilst the teachers can use the same tool to monitor the parents' involvement in the children's schooling and education. Cameras are present at schools, in lecturing halls, on the public transport, at stadiums and in the streets, and on private estates. The ideas of chipping hospitalised patients and medication are already being put in practice.[2] It is a common practice to fit the company cars with GPS transmitters. Ever more countries are introducing biometric passports and ID cards (containing a single biometric identifier—the digital representation of the biometric features, e.g. the DNA, blood group, face geometry, fingerprints); some of the developed countries will even not let anyone in without such document. In Italy, fingerprints are taken of some immigrants and national minorities (small children included), e.g. the Gypsies. In the USA, Pentagon is supervising works on the creation of a 'second generation passenger screening' computer system, which will enable instant compilation of the life story and personality profile of each of the millions of passengers travelling by air (Szymborski 2003)

An ever-growing number of states (Poland currently being at the stage of preliminary considerations) are installing intelligent electricity meters, which may make it possible for the users to save on power. Thanks to such meters, the user

[2]Poland was planning to introduce them in 2013. The chips will contain facts about the patient, his health, features, and genetic predispositions. If the patient's data are fed to the WWM (World Wide Med), they will be available anywhere in the world where he happens to take treatment. There are plans to introduce the electronic lineage for medical drugs to guarantee their authenticity.

will only pay for the actually consumer energy, and the supplier will be automatically informed of a power failure and of the energy consumption peak hours. This will enable improving the management of the power transfer grid and in effect reduce the costs and allow for compiling a tailored solution offered to each individual client. On the other hand, the energy consumption data obtained in this way pave the way to gaining access to detailed information on the individual's actions and behaviour, identifying the hours when the power buyers are at home, the number of individuals in a homestead, the type of the devices they have, and sometimes even the title of the daily watched serial. This gives rise to serious reservations as to intrusion on human privacy and the apprehension of potential use of such data by unauthorised individuals or entities, particularly if their protection system is poor (Liczniki 2013).

Some states and companies commission various types of customised spyware. We mention, e.g. scientific, technological, or industrial intelligence. Systems are being created to manipulate mass awareness through community webs. Among many other devices, intelligence service world wide has for several years used a facility called Syborg 36v, which enables passive and active invigilation of the citizens through mobile phones. In the year 2011 in Poland, telecommunication operators spent PLN 74 million to process inquiries for the subscriber data (about 2 million inquiries). In order to meet the demand from the prosecution authorities, the operators had to invest in building a separate technical infrastructure to obtain and accumulate data (Królak 2013).

The American CIA has announced that it is very close to calculating all pieces of information people generate all over the world. Its objective is to store for an unlimited time and analyse all obtained news fragments which gain specific value only once they have been put together with other pieces of information and events in specific points in time (Ogórek 2013).

The European Union, on the other hand, is currently pursuing 45 projects in '*security research*' under the 7th Framework Programme with the funds earmarked for it for the years 2007–2013 adding up to 1.4 billion Euros. The group includes the INDECT project, i.e. the project focused on an intelligent and comprehensive IT system. The project is headed by the Kraków Academy of Mining and Metallurgy. The system is intended for observing and searching for people, as well as for curbing criminality, mainly through anticipation. The scale of the project can be visualised in figures—60 months of research, 15 million Euro budget, and 17 cooperating European partners. INDECT can be used, e.g. for digital face recognition, detection of unnatural behaviour and potential hazards on the monitored area, analyses of the data coming in from cameras, microphones, or even aerial and satellite photographs, creation of prototype devices serving the tracking of moving objects or extracting information from the mobile telephone network. The system is meant to serve continuous and automatic monitoring of such public resources as websites, discussion groups, 'Usenet' groups, file servers, p2p nets, and individual computer systems. It can be programmed to eavesdrop in any desired group of people. The INDECT system is to be installed in automobiles produced by some manufacturers. In order to reduce the costs of the project, the system microprocessors are to be installed in the already existing surveillance cameras of the monitoring systems in cities. With the INDECT functions combined with, e.g. biometric passports, it could become possible to obtain remote information on the passport holder, whatever the nature. Other similar include, e.g. the ADABTS, i.e. 'Automatic Detection of Abnormal Behaviour and Threats in Crowded Spaces', focused, e.g. on analysing the human voice and body language, or HUMABIO oriented on biometric authentication. The common belief is that the main sponsors of the 'security research' policy originate from the large business shark circles (State Watch 2012).

This means that employing technology to obtain information is not restricted to the sphere of politics, but ever more often spreads into economics. Here, examples include not only business intelligence, but also application of the RFID processors (identifying tags read through radio frequencies) in every day life. These tags are used, e.g. by Unilever to follow the life of its products, whilst Tesco and Wal Mart make use of them to check whether the shop shelves are stocked with their products. The chips can read human faces and control the spots displayed so that they match the clients' preferences. They enable a hypermarket shopper to pay for a full trolley of 'goodies' within a blink of an eye. Swiped through the exit gate, it triggers automatic bill calculation, and an American client of the Exxon Mobil filling station pays for the petrol bought by letting his/her car keys come close to the fuel distributor. The same chips are also built into proximity pay cards.

The possibility to locate the holder of a mobile phone or even record the voices and noises in his/her whereabouts contributes to the development of intelligent services. For instance, if the phone holder passes a Chinese restaurant, the telephone will let him know of the *happy hours* offered there at the very moment (Bendyk 2004, p. 173). All this injects convenience into human lives, gives those having modern technology at their disposal a competitive edge in the market, but all along causes that we leave ever more traces of our own activity. These traces allow the firms dealing in the so-called predictive analysis (the *Target* chain best known in this respect) to accumulate huge numbers of customer data, the latter not restricted to the shopping the customers do, but going on to their life situation, preferences, habits, problems, or views. One might find it almost anecdotal to hear a story of a father whole learnt that his daughter, a high-school girl, was pregnant, from the employees of the said chain stores. The staff had followed the shopping habits of the young girl and her words spoken on the Internet, and putting things together deduced that the girl was pregnant and sent her a token for baby clothes (Uchatius 2012). One of the *Target* slogans is: *Just a little further, and we will be sending you coupons for goods you desire even before you realise you actually want them'.*

The press of the threshold of 2012/2013 reports that *Facebook* tracks the pages visited by the Internet surfers, even those that do not have their account on that particular community service. For instance, clicking 'I like it' triggers instant transmission on the user's IP, resolution, the operation system, the web browser, and the visited pages. Based on the accumulated data, *Facebook,* just like other community portals, profiles the user, and this will enable, e.g. the traders to address the user with an appropriate advertising offer (Wchodzisz 2013).

The spheres of politics and business embark on cooperation and sharing information on the behaviours of the consumers, employees, investors, etc. For instance, in March 2013, CIA admitted having signed a 10-year contract with the Amazon, valued at the estimate of USD 600 million, for the services of data processing in the so-called computer cloud (Ogórek 2013).

The employers too gain greater possibilities to invigilate their employees, despite the dispersed company form, with its flat and fluid organisational structure and varied forms of employment and working methods. Over 80 % of the American employers make use of active electronic monitoring of their employees. The ratio grew twice over the years 2000–2005 (Bendyk 2004, p. 176). Employers can easily monitor the electronic correspondence, telephone calls, actual whereabouts when abroad via the GPS, track the activity of their employees on the community portals, or programme the maximum time for the employee to perform a specific operation on the computer. For example, in February 2012, the employees of one of the widest known banks in Poland had to sign a declaration of consent for having their work audio/video recorded. The arising question is who will have access to the interview with the clients and how they will be secured. One of the presidents of the bank stated on TV that the bank would like to be the leader in implementing the Big Data idea, in other words seek information on its clients not only in its own databases but also, e.g. on community portals or internet forums (Samcik 2013).

The economic significance of the global web stems primarily from the time compression it enhances, which translates to the possibility of transmitting information over large distances and creating new social relationships not delimited in space, all resulting in increased productiveness, effectiveness, and innovation in economics (Sienkiewicz 2008, p. 143).

This presentation of exemplary information on the technical aspects of public surveillance is not intended to serve an analysis of any specific scientific or technical solutions (this is not the topic hereof). Instead, it is meant to give and insight into the immense human potential in the area and visualise the multitude and variety of the solutions applied in various spheres of life.

The improvements in the techniques of spying on the society go hand in hand with amendments to the law conducive to limiting the civic liberty, increasing the restrictive edge of the law and informal censorship in an ever-growing number of states, all under the banner of care for public safety.[3] Under the pretext of protecting personal property or counteracting threat ever more powerful databases are compiled, the freedom of expression, meeting, and the right to privacy is being limited whilst the authority of the police, military, special forces, courts, and control agencies is being broadened.

Following suit of the modern devices serving monitoring and invigilation is ever more modern techniques of social engineering, also geared on tighter control

[3]E.g. the Patriot Act and the *National Defense Authorization Act in the USA*, EU Directive 2006/24/EC, 'Anti-Counterfeiting Trade Agreement + ACTA, EC Draft: Clean IT', the American projects: Terrorism Information Awareness and Terrorist Threat Integration Center.

over and subjection of the citizens, consumers, and employees. Public behaviour can be manipulated appropriately, dissatisfaction channelled, and attention distracted from the vital issues. The contemporary man lives in a simulated world where he is subject to a specific social engineering geared on building his artificial, simulated awareness. Means of 'improving' the human being are met with ever more increasing interest, to name, e.g. Prozac, neuro-linguistic programming, or plastic surgery. The modern technology combined with the multitude of the available channels of exerting an impact on the society enables implanting essential cultural changes. It limits the options people take into considerations, removes tricky issues and uncomfortable view-points from the forum of public debate, and induces consent to subjection (Scott 2006, p. 113). States and people are managed through defining and creating aspirations, desires, challenges, and goals to be pursued, and through building and atmosphere of rivalry around the effort aimed at reaching the promoted goals. In other words, we are witnessing imposition of the system of values, aspirations, standards, sense of humour, tastes, lifestyles, preferences, and morality. The tool of moulding attitudes and behaviours consists in creating the so-called authorities, idols, and celebrities setting the example in terms of behaviour, advertising campaigns, and the so-called expert opinions. Censorship is redundant, as the idea is that people should read only specific messages consistent with the interests of those at rule and discard other solutions, even if advantageous for them. The means reached for to that aim include, e.g. the so-called expert statements echoed again and again, appropriate compilation of information, and triggering specific associations. Sharing specific viewpoints gives a sense of belonging to the elite. Words–keys code the subconsciousness triggering pleasant or unpleasant associations in specific situations.

In apprehension of exclusion, marginalisation, and stigmatisation, people hold back their true views, attitudes, and values. Stigmatisation has become a tool of exercising power. The obligatory attitudes are: optimism, positive thinking, inducing positive emotions, and enthusiasm. To succeed, one has to find his way in the imposed convention.

A way to consolidate the political or economic power is to disintegrate the society. The sense of community, civic activity, authentic democracy, the sense of national autonomy, tradition, and the related values are the enemy to the contemporary order and the interests of its promoters.

The new systems of organisation and management in a business enterprise also put an emphasis on keeping the staff in continuous stress, uncertainty, rivalry, euphoria, and enthusiasm for the performed jobs. Specialists devise ever new (preferably cost-free) methods of motivating and inducing self-control on the employees. The bosses' main concern is to delegate duties and responsibilities onto their inferiors to the maximum possible extent whilst simultaneously making them believe that they are becoming self-leaders in this way. Given the profile of the nowadays desirable traits of character and the prerequisites of professional success such expressions are making career as flexibility, mobility, creativity, availability, high self-worth, self-assurance, assertiveness, mental resilience, effectiveness, success, competitiveness, enterprise, innovation, positive emotions, rankings, or mandatory optimism. Professionalism today is identified not as knowledge or experience, but as no emotions. The worst allegation a member of a community may face, especially an employee, is that he/she lacks the skills of working in a team, which exposes the team to the risk of underperforming (Sennett 2006, p. 153). The skill of 'fitting in' is a core advantage of the contemporary employee and citizen.

What are the causes of the presented phenomena? The temptation most certainly comes from the evolving circumstances, i.e. the fast expanding technical potential and availability of increasingly more sophisticated devices and social engineering techniques. The ever-broader access to information and possibility of exerting total control are synonymous to infinite power. Information and communication constitute one of the fundamental resources of the contemporary civilisation. This resource is used as a vehicle to impose the way of thinking, behaving, the nature of social relationships, and the way the reality is comprehended. Public communication may become an instrument of manipulation. The argument raised by authorities of all kinds in favour of expanding the public control system based on modern technical devices is the declared will of ensuring public safety, the will of protecting the citizens from any threats, and protection of the law and order apparatus. Invigilation is justified by painting out the increasing threat of crime, especially terrorism, and the need to wage an effective battle against the phenomena. Perhaps it is also intended to inject stricter discipline among the citizens and subjecting them to the decision-makers; perhaps the idea is to create the impression that each move they make, word they speak, or thought they conceive can be watched and assessed, so it is better they control themselves and their families.

The phenomenon of spying on the society intensifies as the society disintegrates and mutual human trust slumps. It is also a reaction to the growing stratification of the income and the accompanying dissatisfaction among a part of the society. It further accompanies the changing role of the state. Not so long ago, the state was the guarantor of a specific living standard in the society. It played the role of the regulator to the economy, the guardian especially of the weaker social groups, and the protector of the employees. It took responsibility for the education, the housing, the public health, and development of culture. What we witness today is privatisation of social problems and the ever more commonly accepted rule: 'Count on yourself, if you can count'.

The state compensates its dwindling guardianship function with ever deepening penalisation and repressive nature of the law. By multiplying threats and seeking new enemies of the public order, it substantiates its existence. The English language has two words to denote safety: security (safety of the society) and safety proper (physical, bodily safety). With the state governments unable to ensure public security, they are ever more frequently pointing to safety at risk (e.g. from terrorists, hooligans, criminals, the homeless, pushy beggars, irresponsible drivers, oppositionist parties, immigrants, other faiths, intruders from other planets) (Bauman 2006). This is the reaction to the crisis of the legitimate nature of power.

Consequently, invigilation intensifies taking any of the following forms:

- political, economic, and social;
- private, public, and state;
- active and passive;
- open and secret;
- legal and illegal;
- serving the interests of the rulers and those of the society;
- direct and indirect.

Invigilation is primarily practised by state institutions led by, e.g. the special forces and the police formed with that particular aim in mind. They spy on both their own citizens and on foreign entities and states. This type of activity is coming increasingly into the focus of the local authorities, banks, telecom operators, and other enterprises, managers of the web portals and browsers, advertising corporations, internet site owners, schools, employers, parents versus, e.g. the pre-school teachers and other carers of their children, and even the neighbours and spouses. It is enough to key in the word 'tapping' in any website browser to see offers of various devices intended for both private invigilation and protection against it. At the outset of the 21st century, the Americans spent USD 6 million daily on spy gear (Bendyk 2004, p. 176). In Poland, the commonly available so-called spyware costs between PLN 1.2 to 3 thousand. Obviously, the potential and the goals of spying are highly varied among the exemplary entities mentioned.

The consequences of the increasing opportunities to control and invigilate and the growing interest in it among various entities are not straightforward. They can include, e.g.: higher national security, higher effectiveness in tracking down crime, growing sense of social security, higher business efficiency, and competitiveness among the businesses. One should, however, remember that the situation also carries numerous threats to the functioning of states and their citizens, particularly if in hand of degenerated rulers or criminal groups or in the situation of political, social, and economic instability. This is because it is easy to go beyond the borderline between justified control and surveillance on the one hand, and oppression and attempt to slave the society on the other hand. It is easy to overlook the fact that democratic, civic, and social solutions are ever more frequently substituted with instruments of invigilation and manipulation. The dangers arising from wrong cooperation between modern technology and various kinds of political, economic, or public authority include, e.g.:

- transgressing the law-permissible borderlines of surveillance;
- driving at attainment of unrestricted access to all information;
- attempting to obtain information in illicit ways;
- developing a sense of impunity of those at rule;
- intruding on the private lives of the citizens or even eliminating the private space;
- restricting the civic freedom;
- harassing the citizens by electronic means;

- accumulating exhaustive information on individuals, their personalities, lives, viewpoints, careers, problems, and difficulties in protecting them, all in one location;
- endangered safety of the accumulated information from, e.g. hacker or terrorist attacks, or a system failure;
- transforming the accumulated date into keenly sought for merchandise;
- the way paved to abusing the obtained information in disdain of the law (e.g. pressure, blackmail);
- the possibility to use the power held to misinform the public;
- subjugating the individuals to the omnipotent power of the state, or even slaving them—the modern slavery;
- excessive authority of the special forces;
- redistribution of the central and local budget income to the disadvantage of the citizens. For instance, installation of one street camera (the simplest monitoring device) costs PLN 60—100 thousand in Poland. This should be topped with the cost of 24-h work of the camera operators;
- the threat of the emergence of autocracy and totalitarian regime;
- the threat that Orwell's vision of the elite rule of those 'having knowledge', though not necessarily wise, or honest, or righteous might come true; (Zacher 2006, p. 184)
- the resulting cultural changes, e.g.: the growing mutual mistrust among people, mutual hostility, alienation of the state rule, the sense of uncertainty, no courage to share one's views, hampered spontaneity, withdrawal from public life, availability, conformism, passivity, and indifference as the dominating social attitudes, self-censorship, the culture of spying, seeking for being onto people, and denunciation;
- the dwindling social capital in companies in effect of the listed cultural changes;
- market discrimination of certain individuals in effect of their personal data becoming known to employers or insurance companies.

The above named threats are growing in significance in the context of the broadening meaning of the notion of invigilation as used in certain articles on the topic. They refer to actions connected with spying, surveying, controlling, or monitoring (behaviours, utterances, thoughts, and possessions) as passive invigilation, whereas active invigilation tends to be associated rather with the consequences thereof and various ways of using the information obtained. Hence, active invigilation will include, e.g.: exerting pressure, blackmailing, inducing, misleading, tampering with the data, formulating accusations, stirring, persecuting, intimidating, forcing obedience and availability, disorienting, manipulating, isolating the invigilated individuals and social groups.

The outlined tendency to reduce civic rights and the growing position of special forces can be observed, more or less intense, all over the world. Some political scientists claim that the global transformations in technology and in the functioning and quality of the democratic system may lead to autocratic rule (Thurow 1999, p. 335). In the circumstances of social dissatisfaction, no economic security, the sense of helplessness, and the necessity of adjusting to the ever more rapid rate of changes the society develops a longing for a strong arm at rule that would inject order and help solve all problems. On the other hand, the expanding spheres of poverty and the related pathologies, conflicts, social protests, and perplexity of the rulers faced with the global forces may lead governments to developing the desire of stabilising the situation through autocratic rule, and the modern technology may make it easier for them. Furthermore, centres of pan-state power are emerging (e.g. ECB, EC, IMF, WB) attempting, and ever more effectively so, frequently in collaboration, to shape the fate of countries, human behaviours and their political choices, or the decisions of state governments.

References

Bauman Z (2000) Globalizacja. I co z tego wynika, Warsaw

Bauman Z (2006) Praca, konsumpcjonizm i nowi ubodzy. Krakow

Bendyk E (2004) Antymatrix. Człowiek w labiryncie sieci. W.A.B, Warsaw

Brin D (1998) The transparent society. Perseus Books

Huxley A (2011) Brave new world, Warsaw

Klein N (2009) Doktryna szoku. Warsaw

Królak J (2013) Telekomy żądają zapłaty za usługi dla służb. Puls Biznesu 22

Krzysztofek K (2006) Cyberopium dla cyfrowego ludu. In: Kurczewski J (ed) Wielka sieć. Eseje z socjologii Internetu. Trio, Warsaw

Lachowski K (2006) Sprzężenie zwrotne procesów globalnych. Transformacje

Liczniki inwigilacji w naszych domach (2013) http://tech.wp.pl/kat,130038 Accessed 23 Mar 2013

Łukasiewicz A (2012) Wielkie ucho. Uważam Rze 6(53)

Ogórek P (2013) CIA chce gromadzić wszelkie możliwe dane. http://wiadomosci.onet.pl/swiat. Accessed 15 Mar 2013

Orwell G (2006) Year 1984. Warsaw

Press release issued by the Registrar Chamber judgment. Not Final.Uzun v. Germany (application no. 35623/05)

Samcik M (2013) Alior Bank jak Wielki Brat? Chce nagrywać rozmowy pracowników z klientami? Kto będzie oglądał te taśmy? http://wyborcza.biz/finanse/1,105684,13461794,Alior_Bank_jak_Wielki_ Brat__Chce_nagrywac_rozmowy.html#ixzz2OTGMWYln, Accessed 26 Feb 2013

Scott J (2006) Władza. Warsaw

Sennett R (2006) Korozja charakteru. Osobiste konsekwencje pracy w nowym kapitalizmie, Warsaw

Sienkiewicz P (2008) Systemowe aspekty globalnego społeczeństwa informacyjnego, in: Ku przyszłości. Księga jubileuszowa dedykowana prof. L. W. Zacherowi. Warsaw

Sokołowski M (ed) (2005) Oblicza Internetu. Internet a globalne społeczeństwo informacyjne, PWSZ, Elblag

Splichal S, Calabrese A, Sparks C (eds) (1994) Information society and civil society. Contemporary perspectives on the changing world order. Purdue University Press West Lafayette, Indiana

State Watch (2012) Report "NeoConOpticon"

Szymborski K (2003) Obywatel na podglądzie. Polityka 4

Thurow L (1999) Przyszłość kapitalizmu. Jak dzisiejsze siły kapitalizmu kształtują świat jutra, Wrocław

Uchatius W (2012) W sidłach manipulacji. The New York Times, after: Forum 11
Watson PJ, Ch. (2013) Washington Legislative Bureau ACLU—web portal: Infowars
Wchodzisz na te strony? Jesteś na podglądzie! (2013) http://tech.wp.pl Accessed 18 Mar 2013
Zacher LW (2006) Gry o przyszłe światy. PAN KP Poland 2000 Plus, Warsaw
Zboralski Ł (2013) Cybernetyczna jazda. Do Rzeczy

Chapter 11
Impact of ICT on the Law

Małgorzata Skórzewska-Amberg

Abstract We live in an ever expanding information society, in an era where global computer networks, mobile telephony and digital audiovisual transmission are not only covering practically all kind of communication, but are more and more the fundamentals of the functioning of the society. It is therefore worth to consider what influence these technologies have on the law, whose basic task is to ensure the functioning of the state and the society and, above all, the safety of the individual. Leaving the ICT technologies without not only precise, but also adopted to the new communication environment, legislation, creates a danger of infringement of the interests of individuals, especially the right to privacy. It is essential to provide comprehensive—both technical and legal—protection of the ICT systems. Maintaining the balance between freedom and security of the individual is an old and still valid complex challenge in the new technological communication reality. Legal standards are and should continue to be inevitably aimed at preserving both these fundamental values.

Keywords ICT · Freedom of the individual · Security of the individual · Anonymity · Cryptography · Cyberspace · Cyberspace law · Cybercrime · Surveillance

Introduction

The history of computer science dates back to distant times. Already in the year 3000 BC, the inhabitants of Babylonia tried to construct devices whose main task was to facilitate calculations, primarily in commercial transactions. So the abacus appeared.

M. Skórzewska-Amberg (✉)
School of Law, Kozminski University, Jagiellonska 57/59, 03-301 Warsaw, Poland
e-mail: mskorzewska@kozminski.edu.pl

L.W. Zacher (ed.), *Technology, Society and Sustainability*,
DOI 10.1007/978-3-319-47164-8_11

Moving closer to more contemporary times, the seventeenth century saw the invention of early calculating machines by Wilhelm Schickard and Blaise Pascal's Pascaline machine. Most notably, it was during this time that the mathematical foundations were laid for the construction of a machine capable of performing calculations in a binary system. In 1671, Gottfried Wilhelm von Leibniz began to build his calculating machine which used his own mathematical deductions as a basis and was capable of addition, subtraction, multiplication, division and of extracting the root of a number.

In 1833, the English mathematician and inventor Charles Babbage completed the project of an analytical machine (often called the arithmetic mill), which was to perform basic mathematical operations, and was also capable of remembering the data input as well as intermediate and final results of calculations. Punched cards were used to input and output data. Due to practical difficulties, Babbage's machine was never built in his lifetime.[1] Had it been, it would have been the first programmable computer in history.

In 1842, a paper on Babbage's analytical machine was published by Professor Luigi Federico Menabrea. His work was translated into English by Ada Augusta Byron Lady Lovelace, with her own additions and comments i.a. about Bernoulli number calculation methods.[2]

In an article published in 1843, *Scientific Memoirs* (and published under the editorship of Richard Taylor), Lady Lovelace presented bold visions of the use of Babbage's machinery in order to create complex music, graphics and other purposes—both purely scientific and some practical. Although it seemed to be only fantastic dreams, a future was to prove the truth of the conjecture.

In 1979, the programming language developed by the United States Department of Defense was named ADA—in honour of Lord Byron's daughter.

The great digital revolution began in the fourth decade of the twentieth century. One of the first computers was the ENIAC (*Electronic Numerical Indicator And Calculator*)—which weighed 30 tons, occupied 72 m^2, consisted of 17468 tubes[3] and could perform 5000 additions or 300 multiplications per second as well as hold twenty ten-digit numbers in its memory.

In 1946, John von Neumann defined the rules of computer design which are still applied today. EDVAC (*Electronic Discrete Variable Automatic Computer*), built at the turn of 1949 and 1950, was the first internally programmable machine designed on the basis of von Neumann's architecture.

In 1952, the first series-produced computer—IBM 701 appeared. About thirty years later—the IBM PC introduced the personal computer which was to have such a significant impact on our personal and professional lives.

At the beginning of the seventies, universities and companies operating within the contract signed with the agency of the US Department of Defense

[1]In 1991, the Second Differential Machine was built in the Science Museum in London, proving the overall correctness of the structure proposed by Charles Babbage.

[2]This algorithm description is often acclaimed as the first computer programme in the history.

[3]And also close to 6000 commutators, 10000 condensers, 50000 resistors and 1500 transmitters.

(ARPA—*Advanced Research Projects Agency*) created a computer network called ARPANET (Garfinkiel, Spafford 1997:435). This network was to develop over to ensuing years to create the Internet covering most of the free world.

When in the forties experimental computers appeared, nobody believed that within just a few dozen years their successors would be omnipresent, and life without computer systems would be practically unimaginable. Ray Tomlinson of *Bolt Beranek and Newman*, who in 1971 sent the world's first *e-mail*, also probably did not believe that twenty years later it would be the primary way of communication among people.

"Throughout the world, information and communications technologies are generating a new industrial revolution already as significant and far-reaching as those of the past. (…) Technological progress now enables us to process, store, retrieve and communicate information in whatever form it may take—oral, written or visual—unconstrained by distance, time and volume".[4]

At the beginning of the twenty-first century, the information society—defined as a highly developed society in which full access to services and information is ensured through ICT technology,[5] enabling the acquisition, processing, storage and dissemination of audio, visual, textual and numeric information through computer networks—became a reality. Space of communication with the aid of such networks is defined as a virtual space (cyberspace) (Maurer 2011; also http://techsty.art.pl/hipertekst/cyberprzestrzen.htm).

We live in an era of global computer networks, mobile telephony and digital audiovisual media. It is really difficult today to imagine daily life without the assistance provided by ICT. However, these technologies and the processes that they help create, facilitate and drive raise many questions as to what impact they have on the law, whose basic function is, after all, to ensure the functioning of state and society and, above all, the safety of the individual.

Interactions of the ICT and the Law—Selected Examples

Cryptography

The development of cryptographic techniques, including critical public key systems, enables the design of protocols that allow i.a.: distance data transmission—including

[4]Europe and the Global Information Society. Bangemann report recommendations to the European Council, http://ec.europa.eu/archives/ISPO/infosoc/backg/bangeman.html, (15.04.2012).

[5]cf. Bangemann's Report http://ec.europa.eu/archives/ISPO/infosoc/backg/bangeman.html (15.04.2012); Directive 98/34/EC of the European Parliament and of the Council of 22 June 1998 laying down a procedure for the provision of information in the field of technical standards and regulations, OJ L 204, 21/07/1998 s. 0037–0048; amended by Directive 98/48/EC of the European Parliament and of the Council of 20 July 1998 amending Directive 98/34/EC laying down a procedure for the provision of information in the field of technical standards and regulations, OJ L 217, 05/08/1998 s. 0018–0026.

data exchange in banking networks, e-commerce, e-government and primarily—the generation and authentication of digital signatures.

The technical possibilities of communication and data transmission give rise to the need for legal settlements in the spheres they relate to i.a. the need to secure electronic trade, to secure rights and determine the obligations of parties to electronic transactions, to secure and determine the probative value of documents in digital form. Work in this area undertaken in international fora, including: the United Nations,[6] the Organisation for Economic Co-operation and Development[7] and the European Union[8] resulted in the adoption of standards i.a.: in the area of data exchange in ICT networks, electronic commerce and services, as well as the recognition of electronic signatures.

The use of cryptographic techniques is one of the most effective methods protecting data in digital media and computer networks (i.a. data transmission). Problems start when the content intercepted by subjects authorised to intercept data transmission is encrypted. Modern cryptographic techniques enable very effective encryption, where the only way to uncover the content is through uneconomical, brute attack. Therefore, legislators seek to protect the interests of the state, where there is authorised interception of data transmission. Provisions concerning authorised data interception often impose on entities that provide cryptographic support an obligation to decrypt intercepted content (e.g. British *Regulation of Investigatory Powers Act 2000*; article 11-1 of the French law No 91-646 of 10 July 1991 on protection of secrecy of correspondence passed by telecommunications measures[9]; in Poland, § 7 and § 8 of the regulation of Minister of Justice dated 24 June 2003 on technical preparation of the network used for transmission of information (…) to control transfers of information and manner of making, registration, storage, playback, and destruction of records with controlled

[6]ia: UNCITRAL Model Law on Electronic Commerce (http://www.uncitral.org/uncitral/en/uncitral_texts/electronic_commerce/1996Model.html), UNCITRAL Model Law on Electronic Signatures (http://www.uncitral.org/uncitral/en/uncitral_texts/electronic_commer-ce/2001Model_signatures.html).

[7]ia.: OECD Guidelines for Consumer Protection in the Context of Electronic Commerce, 9.12.1999, (http://www.oecd.org/sti/consumerpoli-cy/34023811.pdf).

[8]ia.: Directive 98/84/EC of the European Parliament and of the Council of 20 November 1998 on the legal protection of services based on, or consisting of, conditional access (OJ L 320, 28/11/98, p.0054–0057), Directive 2000/31/EC of the European Parliament and of the Council of 8 June 2000 on certain legal aspects of information society services, in particular electronic commerce, in the Internal Market (Directive on electronic commerce, OJ L 178, 17/07/2000 p. 0001–0016), Directive 1999/93/EC of the European Parliament and of the Council of 13 December 1999 on a Community framework for electronic signatures (OJ L 013, 19/01/2000 p. 0012–0020).

[9]Journal Officiel Numéro 162 du 13 Juillet 1991.

transfers[10] (Journal of Laws No. 110, item 1052); article 19 Swedish Electronic Communication Law[11]). This can be effective only with respect to cryptographic systems incorporating the possibility of breaking the security using a so-called *trapdoor*, i.e. additional information, which enable authorised entities to compromise security in a particular situation. Of course, the owner of a cryptographic key can always decrypt the message. The law allows for the empowered authorities to demand and obtain the necessary tools for cryptographic services, decrypting encrypted content. Even if we assume that the entity offering the cryptographic algorithms and systems (e.g. algorithms available in scientific publications) may not be considered as providing cryptographic support, this does not give rise to a substantial extension of the problem of legal order also to ensure the possibility of declassifying a cipher for those entities. From a legal point of view, the most difficult is a situation in which the person who constructed the cipher itself (created algorithm or transferred an existing one) refuses to decipher data as requested by the entitled services. In accordance with the generally applicable legal principle that no one may be forced to disclose an evidence incriminating him or herself—the law, in these cases, is generally powerless.

Privacy in the Net

Leaving the computer systems and methods of managing such systems without precise legislation creates a danger of infringement of individual interests, notably the right to respect the privacy. It becomes necessary to provide solutions which enable comprehensive—both technical and legal—protection of computer systems.

The European Convention for the Protection of Human Rights and Fundamental from 1950—referring to the Universal Declaration of Human Rights, adopted 10 December 1948 by the UN General Assembly—is the main link in the European system of protection of those rights. Signed by Poland on 26 November 1991 and ratified on 19 January 1993 (Journal of Laws of 1993 No. 61, item 284 and 285), article 8 provides the right to respect private and family life, home and correspondence and stating that interference by a public authority in the exercise of this right is unacceptable—except in cases which are in accordance with the law and deemed necessary in a democratic society because of the interests of national security, public safety or the economic well-being of the country, preventing

[10]Original title: Rozporządzenie Ministra Sprawiedliwości z dnia 24 czerwca 2003 roku w sprawie sposobu technicznego przygotowania sieci służących do przekazywania informacji (...) do kontroli przekazów informacji oraz sposobu dokonywania, rejestracji, przechowywania, odtwarzania i niszczenia zapisów z kontrolowanych przekazów (Dz.U. Nr 110, poz. 1052).

[11]Original title: Lag (2003:389) om elektronisk kommunikation.

counter-insurgency or crime, protection of health and morals or the protection of the rights and freedoms of other persons. The same standard of protection of rights is defined in art. 17 of the International Covenant on Civil and Political Rights of 1966, ratified by Poland (Journal of Laws of 1977 No. 38, item 167). Here, it is stated that nobody will be subjected to arbitrary or unlawful interference with his private life, family, home or correspondence and that everyone has the right to legal protection against such interference.

The transfer of a wide range of activity to the virtual sphere, in conjunction with the low awareness of the risks associated with disclosure, storage and transmission of information in ICT networks, has resulted in the disclosure of information concerning network users by them and on an extremely large scale. This primarily concerns the wide-ranging social networking services, including *Facebook*. It has come into vogue to move most of the people-to-people contacts into the virtual sphere—and it is not just constrained to teenagers. The opinion that anyone who does not have an *Facebook* account almost does not exist has become common. Social networking services users publish their personal data, data about their interests, photographs—including these concerning other people, often without their knowledge and consent.

Access to profiling data (e.g. as posted in social networking services) causes the development of targeted advertising (e.g. behaviourally targeted, online behavioural advertising), which entails new legal and legislative challenges—in particular, the need to create a specific directory of behaviours and to specify the actions that go beyond the accepted framework and what should be treated as a violation of privacy.

ICT helps create many new applications, but does not always take into account the legal implications of privacy protection. *Google Street View* or *Google Earth* can be mentioned here. These very interesting applications help to illustrate the need for a formal legal policing of current activities. There is no doubt that the taking of pictures of public areas, and buildings from a publicly accessible street which may include images of individuals who are part of a group (as against individuals who are on their own) can not be treated as violation of privacy. It is important, however, from what level the pictures are taken, as it may indeed occur that the photographed place is someone's private zone, e.g. private garden separated from the street by a non-transparent fence. Problems encountered during preparation of *Google Street View* can be also mentioned. During these preparations, unauthorised interception of data transmitted with the use of ICT (e.g. Wi-Fi) was made repeatedly. Moreover, for such actions a French court imposed quite high fines on Google Street View.

Cloud computing, which is developing very dynamically, also makes it necessary to rethink the existing legal solutions—including those concerning privacy aspects. Legal issues related to cross-border flow of personal data can be indicated here—existing solutions do not envisage a situation in which data managed in a particular country may be processed in the cloud—i.e. in a place which may be both uncertain and variable. Given that cloud computing provides a lot more of computer power and software tools are cheaper than existing "stationary" solutions, it is expected in the not too distant future that programming and processing outside the cloud may become uneconomic and the law should be prepared for this.

Inextricably linked with the right to privacy is the right to the protection of confidentiality of communication and matters related thereto which is governed by i.a.: directive 2002/58/EC of the European Parliament and of the Council of 12 July 2002 concerning the processing of personal data and the protection of privacy in the electronic communications sector (Directive on privacy and electronic communications, OJ L 201, 31/07/2002, p. 0037–0047) in conjunction with the Directive 2002/21/EC of the European Parliament and of the Council of 7 March 2002 on a common regulatory framework for electronic communications networks and services (Framework Directive, OJ L 108, 24/04/2002 p. 0033–0050).

The EU Member States are obliged to adopt such provisions in national law, which ensures, among others, confidentiality of communications and publicly available electronic communications services. In particular, the listening to, recording, storing or capturing of messages in a different manner—without the consent of the communicating parties—should be prohibited.

Limitation of any manifestations of the right to privacy, including the freedom of communication, is legally permissible in some situations. Article 15.1 of the directive on privacy and electronic communications, which provides that Member States may, by law, restrict the right to privacy, if necessary, appropriate and proportionate within a democratic society in order to ensure national security, defence, public security and prevention, investigation, detection and prosecution of criminal offences or of the unauthorised use of electronic communications systems. Similar authorisation is implemented by the provisions of the Convention on cybercrime rules of procedure (articles 14–21).

In national law, the necessity to secure data transmission from unauthorised access is the subject of regulation through criminal law and criminal procedure law. Substantive law prohibits the interception of data transmission (including breach of secrecy of correspondence and the use of tapping) performed without authorisation (i.a.: article 267 Polish Criminal Ccde; article 201–202b, 206 German Criminal code, article 85, 87 of the law on telecommunications,[12] article 226-1, 226-226-3, 226-15, 432-9 of the French criminal code; British *Regulation of Investigatory Powers Act 2000*). Authorisation to acquaint with the content of transmitted information is commonly included in the rules of procedure and made subject to a number of conditions, of which the most important are: suppression of the possibility of data transmission interception in a situation in which there is a reasonable suspicion of committing the offence; the need to obtain the consent of the competent authorities (usually of the Court) to intercept the information content and the supervision of the competent authorities on the process of information

[12]Original title: Telekommunikationsgesetz (TKG) vom 25.Juli 1996 (BGBl. I, S. 1120), zuletzt geändert durch das Gesetz zur Bekämpfung des Missbrauchs von 0190er-/0900er-Mehrwertdiensterufnummern vom 9. August 2003 BGBl. I S. 1590.

interception (i.a. in German law—article 100a-100d, 218 of the rules of procedure, the provisions of the specific laws concerning the surveillance other than the trial one; French law No 91-646 of 10 July 1991 on the protection of the secrecy of correspondence passed telecommunications measures[13]; British *Regulation of Investigatory Powers Act 2000*; article 218, 218a, 230a, 236 237-242 of the Polish code of criminal procedure, article 159, 160, 161, § 2, 179, 180 of the telecommunications law[14] and the provisions concerning the surveillance other than trial one contained in specific laws).

Anonymity in the Network

The Internet, in its initial phase of development, was primarily a channel for communication and data exchange within the scientific community. For many years, only scholars, universities and research institutions had access to the network (Garfinkiel, Spafford 1997:435; Glister 1995:29; Gołaczyński 2007). Even in 1994, after the opening of the Internet to the world, it was expressed that "the unusual freedom of expression and the availability of information in the Internet suggest that it may launch a new era for the exchange of information between human beings" (Cerf 1995:23). Today, less than twenty years later, the world faces the problem of whether and how the exchange of information should be controlled.

The Internet evolved on the basis of lack of fixed, strict rules of conduct, but this has not meant an absence of rules in general. The scientific community was able to effectively enforce respect for freedom of all its network users, where the excessive freedom of action of some menaced the freedom of others. The development of a global network and its free and open public access has forced a legal intervention. This is to cover the Internet as a public medium used by individuals whose freedom must be protected but whose unlawful action must be punished.

Until recently, anonymity was the strength of the internet, but it may soon become one of the major threats to the legal framework governing modern society. Total freedom of action in the net, including hiding someone's identity, becomes a supreme right of many cyberspace's users. More and more applications are becoming available which enable parties to an Internet transaction to maintain their anonymity. Any attempt to introduce any kind of Internet monitoring causes controversy. Questions about the freedom of expression and protests against censorship appear. Issues about the interference by authorities in the sphere of network users' privacy are also raised.

---

[13]Original title: Loi 91-646 du 10 Juillet 1991 relative au secret des correspondances émises par la voie des telecommunications, Journal Officiel Numéro 162 du 13 Juillet 1991.

[14]Original title: Ustawa z dnia 16 lipca 2004 roku – Prawo telekomunikacyjne (Dz. U. Nr 171, poz. 1800 z późn. zm).

Increasingly a growing number of states are introducing legal measures of order and censure in order to ensure that the law retains the ability to counteract the legal infringements that the Internet facilitates or even allows. It seems unavoidable to introduce solutions which can limit the anonymity of network users, whether it should be clearly noted, that it is not about the disclosure of someone's identity to other users, but the security capabilities of reaching the particular person in the case of law infringement.

The Internet is often accessed with the use of a mobile phone card. Some European countries[15] such as Germany, Italy, Greece, Slovakia and including non-EU member—Switzerland, have introduced compulsory registration of prepaid SIM cards purchasers and presumably users. Others such as Poland allow voluntary registration of purchasers of prepaid telephones and in return offer them equal with subscribers rights with regard to service provider obligations. On the other hand, some countries treat such users completely anonymously. Inconsistency of the law in this area may result in a failure to achieve the demands of some legislators who seek to establish the identity of a prepaid card user. A person seeking anonymity will simply purchase a card or device in a country where there is no legal obligation to register the buyer. Therefore, the obligation of the registration of prepaid cards buyers can at least, partly, limit the problem of anonymity in the network.

It is necessary to remember that most currently used devices with access to the Internet are equipped with a webcam (e.g. laptops, tablets and netbooks), and it is not difficult to imagine a situation in which such a device can be used as a remote-controlled tapping device, acting without the knowledge of its user. This is just one of the many cases in which anonymity of the network user, who remotely runs processes in other devices connected to the network, can prevent legal protection of the individual.

Legislative Problems

Technology development often causes legislative problems resulting from insufficient understanding of the technology language by the language of law. Two examples may be cited under Polish criminal law—both concerning the protection of digital information.

The primary problem which had to be faced was the necessity of another perception of the protected good. Usually law, protecting information, secured the information as such—ICT networks appearance has resulted in the need to protect not only information itself but also the access to such information.

---

[15]Outside Europe regulations obliging to register pre-paid phone users are in the force in i.a. Mexico, Japan, Australia, Thailand, South Africa.

Article 267 of the Polish Penal Code might be a good example. The provision protects unauthorised access to information. The penalty for such a breach of a computer system is on a par with opening of an enclosed letter, connecting to telecommunications network or breaking or omitting electronic, magnetic, software or other specific information security, as well as an unauthorised installing and handling and tapping, visual or any other device, software or technology.

Article 267 Polish Penal Code in its previous wording,[16] apart from the protection of communication secrecy also ensured the protection of messages against unauthorised access. Only the information was protected, hence in cases when access to a system was gained without visibly obtaining any information (e.g. the offender has obtained information in completely different network, another country and in a manner which does not necessarily can link the losing of information in one system with an unauthorised entry to another system), the perpetrator could avoid the responsibility for his action. Doubts also arised whether activities such as bypassing security or taking advantage of software gaps is equivalent to protection breaking.

The change made by the legislator moves protection of information to the access to it, essentially changing the scope of protection, allowing prosecution of gaining access to information without entry into possession of its content.

In turn, article 269b § 1 penalises the manufacture, disposal or the sharing with others any devices or computer software adapted to commit enumerated offences. Punishment also includes those activities in respect of computer passwords, access codes or other data enabling illicit access to information stored in a computer system or an ICT network. The legislator's intention was to indicate the illicitness of action intended to breach the information integrity. However, the lack of a precise definition of what kind of information stored in the computer system or network that should be protected, results in a provision that can penalises any behaviour which accesses any information in the ICT network (e.g. links to Web pages). The simple addition to the provision content may completely change its application in practice—in accordance, moreover, with the likely intention of the legislator. Instead of "access to information stored on a computer system or an ICT network" should be therefore "access to secured information stored on a computer system or an ICT network".

The above examples show how easy it is to skew the legislator's intentions due to the incompatibility of technology and law languages.

Technology is developing much faster than the legislative process. At the same time, various forms of risk arise from the broad use of ICT systems which are evolving rapidly, and with which new and appropriate legal sanctions can not keep pace. Language, which modern technology uses must be translated to the language of the law, and the created law should include the widest possible range of behaviours in cyberspace.

[16]The changes were introduced by the Penal Code Amnebdment Act, original title: ustawa z dnia 24 października 2008 r. o zmianie ustawy – Kodeks karny oraz niektórych innych ustaw (Dz.U. Nr 214 poz. 1344).

Conclusion

The technological revolution has always been and will be a challenge. More closely—not so much the technological development in itself, but the adaptation of social standards, including in particular the regulatory framework.

Legal standards, presumably, should regulate certain social behaviours—on the one hand, it should stabilise the existing social order, on the other—give impetus to change. The primary function of the law is, however, the determination of certain values which should be specifically protected. Freedom and security of the individual belonging definitively to them.

The technological revolution is moving a large part of human activity to the virtual sphere and is making society strongly dependent on ICT systems. For the first time, the technological revolution, whilst giving man a powerful tool has made him vulnerable to a extent not previously encountered. In virtual space, everyone can get access to extremely wide resources of information on any subject, including information about specific people, their interest, their wealth status, businesses, and even weaknesses. Sooner or later any information becomes globally available—sometimes as a result of the actions of the person directly affected, sometimes without his or her knowledge and consent. It is natural for humans to require information about ourselves and our environment. The Internet has given us unprecedented and powerful tool to access wide and readily available information.

Attempts to manage the information chaos cause concerns with regard to possible restriction of freedom. Freedom of the individual is, however, the value of the parent unit, but this freedom ends where it infringes the freedom of another. The role of the law is to safeguard the rights and interests of individuals—it is therefore necessary to define the rules for limiting freedom. Balance between freedom and security becomes the fundamental matter—an old challenge in a new reality.

References

Cerf VG (1995) In preface to Paul Glister. Internet. Przewodnik użytkownika. Wydawnictwa Naukowo-Techniczne, Warsaw

Garfinkiel S, Spafford G (1997) Bezpieczeństwo w Unixie i Internecie, O'Reilly & Associates Inc. Wydawnictwo RM, Warsaw

Glister P (1995) Internet. Przewodnik użytkownika. Wydawnictwa Naukowo-Techniczne, Warsaw

Gołaczyński J (2007) Umowy elektroniczne w prawie prywatnym międzynarodowym. Wolters Kluwer S.A, Warsaw

Maurer T (2011) Cyber Norm Emergence at the United Nations—an analysis of the activities at the UN Regarding Cyber-Security [in:] Science, Technology, and Public Policy Program Explorations in Cyber International Relations Project, Discussion Paper #211-11 Explorations in Cyber International Relations Discussion Paper Series, Harvard Kennedy School, Belfer Center for Science and International Affairs, September 2011

Chapter 12
Legal and Moral Dilemmas of Targeted Killing by Drones

Wojtek Lamentowicz

If we are to ignore the pain of others, how are we to heal the damage?

Former drone operator, Brandon Bryant

Abstract The essay makes a critical review of the legal debate in the USA and in the United Nations on moral and legal issues involved in military use of drones in wars of today. The main goal was to study the main lines of argument ant its relevance to military practice of targeted killing. We found that legal criticism is on the increase but the military practice continues. We stress the moral risk of using the autonomous weapons. In conclusion, we suggest the need for new both domestic and international regulations of any use of drones before they become fully autonomous and beyond control. Timing is crucial. If humans will not control the technology, the technology will control humans. New UN convention on smart weapons and the conditions under which its use should be allowed is a matter of practical necessity as the number of states using it increases so fast.

Keywords Targeted killing · Legality of drones · American debate · United Nations' reports · Limits of law

What is the Military Drone and for What Purposes We Build It?

An unmanned aerial vehicle (UAV), known as a drone, is an aircraft without a human pilot on board. Its flight is controlled either autonomously by computers in the vehicle, or under the remote control of a pilot on the ground or in another vehicle.

W. Lamentowicz (✉)
Academy of Business and Administration in Gdynia, Gdynia, Poland
e-mail: wl.667788@gmail.com

L.W. Zacher (ed.), *Technology, Society and Sustainability*,
DOI 10.1007/978-3-319-47164-8_12

They are deployed predominantly for military applications, but also used in a growing number of civil applications, such as policing, firefighting, and non-military security work, such as surveillance of pipelines or state borders. UAVs are often preferred for missions that are too "dull, dirty, or dangerous" for manned aircraft. Governments, corporations, cops, and criminals want them—peace and privacy and human rights lovers left and right don't. You can run but you cannot hide from the drone—this an obvious threat to human privacy and dignity.

The drone-industrial complex wants 30,000 eyes in the sky looking down on Americans by 2020. In 2015, a new law on civilian uses of drones in the United States shall be adopted. A legal and political debate on this issue will get hot.

A Short History of the UAVs as Combat Weapons

The vision of Nikola Tesla, an American constructor, who imagined a fleet of unmanned aerial combat vehicles in 1915 was the beginning of this story. The first reported use of a "drone" was in 1919, when the inventor of autopilot technology and the gyroscope, Elmer Sperry, sunk a German battleship with a pilotless aircraft.

More were made in the technology rush during World War II; these were used both to train antiaircraft gunners and to fly attack missions. Nazi Germany also produced and used various UAV aircraft during the course of WWII.

Jet engines were applied after World War II. Nevertheless, they were little more than remote-controlled airplanes until the Vietnam War in the 1960.

The birth of US UAVs began in 1959 when United States Air Force (USAF) officers, concerned about losing pilots over hostile territory, began planning for the use of unmanned flights. This plan became intensified when Francis Gary Powers and his secret U-2 were shot down over the Soviet Union in 1960. Within days, the highly classified UAV program was launched under the code name of "Red Wagon." The August 2 and August 4, 1964, clash in the Tonkin Gulf between naval units of the USA and North Vietnamese Navy initiated America's highly classified UAVs into their first combat missions of the Vietnam War. During the Vietnam War, the United States fitted drones with cameras and deployed them for reconnaissance.

The United States used predators drones for the same purpose during the Gulf War of 1990–1991 and the Balkan conflicts of the 1990s. President William J. Clinton used drones in both Bosnia and Kosovo. They flew out of a hangar in Albania that the CIA had rented in exchange for two truckloads of wool blankets. Drones were first used (like balloons) for surveillance.

Targeted Killings by Drones: Where and When? Who Are the Targets and Who Gets Hit?

By 2001, the United States began arming drones with missiles and using them to strike targets during combat in Afghanistan. Predators were successfully armed at a Nevada Air Force base just as George W. Bush was taking office, and drone strikes began in Afghanistan in the weeks after the September 11 attacks. A November 2002 Predator strike in Yemen killed Abu Ali al-Harithi, a mastermind of the 2000 bombing of the USS Cole. It was the first time in decades that the United States had publicly confirmed an assassination outside a declared war zone, and the strike also claimed the first American drone victim, Ahmed Hijazi (Luban 2013). On November 3, 2002, CIA agents in Djibouti fired laser-guided Hellfire missiles from a drone at a passenger vehicle in Yemen, killing all passengers on board, including an American citizen. During the invasion of Iraq that began in March 2003, the United States regularly used reconnaissance and attack drones. That use seems to have ended along with combat operations in 2009. The United States has been using combat drones in Somalia since at least 2006.

Israeli-developed Samson drone successfully triggered Syrian radar systems in the Beqaa Valley. This allowed Israel to destroy Syria's considerable surface-to-air missile arsenal. Israeli drones also proved effective in monitoring Syrian-operated air bases and allowed the Israel Defense Forces (IDF) to track Syrian and Palestine Liberation Organization fighters.

The United States began using attack drones in Pakistan in 2004. President G. W. Bush expanded the drone war in Pakistan in his final year in office. "Signature strikes," targeting as-yet-unidentified suspects on the basis of their behavior alone, entered the American repertoire around this time and the Obama administration expanded it still further. The number of attacks jumped dramatically in 2008 and continued to climb in 2009. There were 51 reported strikes in 2009, more than during the entire Bush administration, in which there were 45 only. In 2010, the United States launched twice as many drone attacks in Pakistan as in 2009. So far 2010 stands as the peak of the drone war, according to the Bureau of Investigative Journalism, with 127 strikes in Pakistan alone.

Suspects of terrorism and civilian noncombatant persons, non-US citizens and US citizens are becoming victims of targeted killings. It is well documented, but the number of victims presented by various reporting sources differs quite a lot. Between 2004 and 2013, CIA drone attacks in Pakistan killed up to 3461 people—up to 891 of them civilians, according to research by the Bureau of Investigative Journalism. The 2010 *Report of the United Nations Commission of Inquiry into the Facts and Circumstances of the Assassination of Former Pakistani Prime Minister Mohtarma Benazir Bhutto* (Bhutto Report) illustrates the extent to which Islamist terror has come to debilitate the Pakistani State and establish northwest Pakistan as a virtually autonomous terrorist controlled territory.

From June 2004 through mid-September 2012, available data indicate that drone strikes killed 2562–3325 people in Pakistan, of whom 474–881 were civilians, including 176 children (Covert War on Terror 2012).

Peter Bergen and Katherine Tiedemann of the New America Foundation in their latest study into the use of drones in Pakistan claimed that overall, during the almost six years and two months covered by the study, the United States carried out 114 drone attacks, resulting in between 830 and 1210 total deaths. A third of those have been civilians; two-thirds have been militants. Pakistanis hate the drone attacks. Only 9 % of Pakistanis approve of their use (The Year of the Drone 2012).

The bottom line, however, seems to be that drones' usefulness is limited: The US drone strikes don't seem to have had any great effect on the Taliban's ability to mount operations in Pakistan or Afghanistan or deter potential recruits, and they no longer have the element of surprise. Still, heavy use of drones is likely to continue, despite strategic concerns about blowback and the possibility that the strikes themselves are illegal—both because they've been successful at hitting certain high value military targets and because it's the only way for the USA to target its enemies inside Pakistan. Drones hover 24 h a day over communities in northwest Pakistan, striking homes, vehicles, and public spaces without warning. Their presence terrorizes men, women, and children, giving rise to anxiety and psychological trauma among civilian communities.

The USA has deployed more than 11,000 military drones, up from fewer than 200 in 2002. John Pike, a defense expert at the think tank GlobalSecurity.org., predicts that the F-35 Lightning II, now under development by Lockheed Martin, might be "the last fighter with an ejector seat, and might get converted into a drone itself."

At least 50 other countries have drones, and some, notably China, Israel, and Iran, have their own manufacturers. Aviation firms—as well as university and government researchers—are designing a flock of next-generation aircraft, ranging in size from robotic moths and hummingbirds to Boeing's Phantom Eye, a hydrogen-fueled behemoth with a 150-foot wingspan that can cruise at 65,000 feet for up to four days (Horgan 2013).

More than a thousand companies, from tiny start-ups like Miser's to major defense contractors, are now in the drone business—and some are trying to steer drones into the civilian world. Predators already help Customs and Border Protection agents' spot smugglers and illegal immigrants sneaking into the US NASA-operated Global Hawks record atmospheric data and peer into hurricanes. Drones have helped scientists gather data on volcanoes in Costa Rica, archeological sites in Russia and Peru, and flooding in North Dakota.

Threat of abuses of drones with the capability of launching missiles and dropping bombs increased and must be managed somehow. Each new development in military weapons technology invites assessment of the relevant international law.

Moral Legitimacy and the Fundamental Strategic Choices

The fundamental strategic choice in the counterterrorism policy planning is as follows.

- Is it better to be less efficient and more legitimate in both moral and legal terms?
- Is it better to be more legitimate and more efficient?

It refers to a military wisdom *might is right*. If this is true, there is no need of moral or legal justifications for the use of military might. If we feel that there shall be some valid and morally relevant reason or ground for the use of military power by the state or non-state actors, the military wisdom *might is right* cannot be recognized by critical human mind.

Legitimacy of war and the war on terror, as many strategy makers call it nowadays, may not focus on combat efficiency alone, but must consider rules and underpinning values both moral and legal. If there is some valid reason to kill the enemy, it is not allowed at any moment and at any situation. Modern military strategy of counterterrorism is based on the permanent search for a proper balance between the military/political efficiency (might prevails over right) and moral legitimacy for the use of military power (right prevails over might). The stronger the state the more its strategy makers tend to make the first choice. It is perhaps the reason why the American military and political establishment is keener to believe in the use of pure power than the ruling elites of the European Union which would never accept the principle of *might is right* (Lamentowicz 2009).

Under Bush II administration after 9/11 legal changes were made in order to say that these targeted killings were not assassinations, they were just a simple extension of combat in a global war on terror. It is obvious that the US top leaders wanted to provide a legal legitimation for their secret police operations overseas.

Legal Debate on Targeted Killings: Criticisms by Some Lawyers and NGOs

The legal debate got started in the United States first and was launched by the critics of the military policy of President Bush against terrorism across the globe. The legal argument of the critics of targeted killings was and still is as follows.

- The way the drone attacks are made is an extrajudicial killing without the due process of law; thus, it is illegal death penalty executed by the administration of the President.
- This killing happens somewhere else than in the battlefield and during the military conflict that is not in accordance with the *ius in bello (The Hague Treaties).*
- Another breach of the *ius in bello* is the fact that many victims are among civilians, as well.
- The killings is organized and implemented under the authority of the CIA or by the CIA and not by the US armed forces. The legal problem is that the CIA has no authority to operate outside the USA as if it was a military (O'Connell 2010).
- Targeted killing by CIA is an act of assassination that is not allowed under US law.
- Current US targeted killings and drone strike practices undermine respect for the rule of law and international legal protections and may set dangerous precedents. International law of self-defense has been damaged and distinctive lines were blurred.

Jane Mayer's outstanding report of October 26, 2009, in New Yorker on the ethics of drone strikes in Pakistan made another strong point (Mayer 2009). There are two separate drone killing programs; first is run by the military as an extension of operations on a specific battlefield and the second is run by the CIA as a covert operation of killing instead of arresting suspects and bringing them to the court of justice. Jane Mayer claimed the CIA drone strikes in Pakistan were probably illegal but quite efficient in military sense, as the Taliban and al-Qaeda have failed to capitalize on anger surrounding the strikes because of their own brutal reactions, which involve killing nearby civilians who they suspect of being CIA informants. Cruelty and suspicions by Taliban fighters are presented by Mayer as a kind of excuse for drone US military operators and their commanders.

Gerald Ford in 1976 banned American intelligence forces from engaging in assassination. Before 9/11 the US was against the targeted killing by Israel because it was perceived in Washington as an illegal assassination. In this article, Mayer rightly points out that the use of predator drones is directly in conflict with this executive order and against American law. The article further argues that the use of drones is against international law: In order to target civilian terror suspects, the terrorist group must be engaging in armed conflict, the use of force must be a military necessity, and the force used must be proportionate to the threat. This is clearly not the case. The CIA leaders and the President perceive the entire world as a battlefield within the framework of the war on terror, perhaps.

Mayer's argument continues "some Predator pilots suffer from combat stress that equals, or exceeds, that of pilots in the battlefield. This suggests that virtual killing, for all its sterile trappings, is a discomfiting form of warfare. Meanwhile, some social critics, such as Mary Dudziak, a professor at the University of Southern California's Gould School of Law, argue that the Predator strategy has a larger political cost. As she puts it, "Drones are a technological step that further isolates the American people from military action, undermining political checks on… endless war." The article concludes that while this operation goes on and the benefits of drones are not without costs, there don't appear to be any good alternatives.

Recent reports from Amnesty International and Human Rights Watch have also challenged the legality of drone strikes. The protests reflect a general unease in many quarters with the increasingly computerized nature of waging war. Looking well beyond today's drones, a coalition of non-governmental organizations—the Campaign to Stop Killer Robots—is lobbying for an international treaty to ban the development and use of "fully autonomous weapons."

Computerized weapons capable of killing people sound like something from a dystopian film. So it is understandable why some, scared of the moral challenges such weapons present, would support a ban as the safest policy (Losing Humanity 2012).

Lethal autonomous robots (LARs), more advanced drones, are the reason for concern of campaign groups such as Human Rights Watch, the International Committee for Robot Arms Control, and the Stop the Killer Robots campaign.

The groups had been granted an audience with the UN General Assembly First Committee on Disarmament and International Security where 13 countries, including Canada, Egypt, the USA, the UK, India, Ireland and S. Korea, were represented.

Recently, the American Civil Liberties Union (ACLU) requested records from the CIA on January 13, 2009. The key questions asked by ACLU were as follows: Who may be targeted? How to limit civilian casualties? How the victims are identified? Can drones be used for military purposes outside Afghanistan and Iraq? Whether drones can be used by the CIA or any other government agency aside of armed forces?

CIA responded to this request in an arrogant way: We can neither confirm nor deny that we have any drone records because of security concerns. The ACLU filed in 2009 a lawsuit demanding the legal justification for the use of drone strikes in countries where the USA is not at war, following the government's refusal to provide any documents relating to their initial request on the subject. In early spring 2013, Federal Court of Appeals panel of three judges ruled unanimously that response by CIA to ACLU request was not sufficient and should be more transparent. Philip Alston, the U.N. Special Rapporteur on Extrajudicial Executions, has previously suggested the CIA operations may break international laws against assassination. He did it in many of his ten reports over 2005–2010 period and in his final report on targeted killing in a most systematic manner of course (Alston 2010).

The critics of drone killings proved that military efficiency of this strategy is doubtful. The number of high-level targets killed as a percentage of total casualties is extremely low—estimated at just 2 %. Furthermore, evidence suggests that US strikes have facilitated recruitment to violent non-state armed groups and motivated further violent attacks. Many people believe that drones have replaced Guantánamo as the recruiting tool in Islamist communities. Both Guantamo interrogations and targeted killings by drones increased an anger and hatred among Islamic communities. It makes easier to find young *shahid* (a fighter for Islam who is ready to die for this cause) when hatred to the West is high.

A detailed legal analysis, *Weapons under International Human Rights Law*, has been drafted by international experts and was published by Cambridge University Press in 2013. Edited by Stuart Casey-Maslen, it covers a range of weapons law issues, including the use of firearms, "less-lethal" weapons, drones, and chemical agents; cyber warfare; the use of weapons in prisons or for riot control; weapons in peace operations and armed conflict; the transfer of weapons; the use of weapons by non-state actors; corporate responsibility for the use of weapons; weapons and economic, social, and cultural rights; and remedies for the unlawful use of weapons. However, there are no new critical ideas on drones in contributions to this textbook collection of papers. The European voices in this debate seem to be hardly raised and heard. The outstanding exception is the group of experts in UK which makes valuable contributions at the Web sites http://www.article36.org and openly critical about war http://dronewars.net.

Harold Koh and John Brennan: First Approach to Legitimacy Reasoning by Obama Administration

On March 17, 2010, State Department Legal Adviser Harold Koh, the senior official responsible for international legal issues, shared his views after his public remarks at an American Bar Association speech (ASIL Keynote Highlight 2010). "I have studied this question," Koh said. "I think that the legal objections that are being put on the table are ones that we are taking into account. I am comfortable with the legal position of the administration, and at an appropriate moment we will set forth that in some detail." In a March 2010 speech, Koh voiced his strong support for the legality of targeted killing by aerial drone strikes in Pakistan, Yemen, and other countries included by the US government as being within the scope of the war on terror. The then State Department's legal adviser said that "US targeting practices, including lethal operations conducted with the use of unmanned aerial vehicles (UAVs)," which the Obama administration has leaned on heavily in its efforts to eliminate al-Qaeda and other terrorist groups in Asia, "comply with all applicable law, including the laws of war," citing the principles of distinction and proportionality. He said that the USA adheres to these standards, and takes great care in the "planning and execution to ensure that only legitimate objectives are targeted, and that collateral damage is kept to a minimum." He said the USA is in "an armed conflict with al-Qaeda, the Taliban, and the associated forces," and therefore has the lawful right to use force to protect its citizens "consistent with its inherent right to self-defense" under international law. Dr. Koh identified three elements that the USA considers when determining whether to authorize a specific targeted drone killing:

- Imminence of the threat;
- Sovereignty of other States involved; and
- Willingness and ability of those States to suppress the threat the target poses.

He also said that the drone strikes against al-Qaeda and its allies were lawful targeted killing, as part of the military action authorized by Congress, and not assassination, which is banned by executive order. Under domestic law, US targeted killings against 9/11-related entities are authorized by the Authorization for Use of Military Force against Terrorists.[1] The speech earned praise from the editorial board of *The Wall Street Journal.*

---

[1]Harold Hongju Koh is the Legal Adviser of the Department of State, the 22nd to serve in that position. He is one of the country's leading experts on public and private international law, national security law, and human rights. He is on leave from Yale Law School, where he is the Martin R. Flug '55 Professor of International Law at Yale Law School. From 2004 to 2009, Koh served as the 15th Dean of Yale Law School. From 1993 to 2009, he was also the Gerard C. and Bernice Latrobe Smith Professor of International Law at Yale Law School. From 1998 to 2001, Koh served as Assistant Secretary of State for Democracy, Human Rights and Labor. He previously served on the Secretary of State's Advisory Committee on Public International Law. A Marshall Scholar, Koh graduated from Harvard, Oxford, and Harvard Law School, and has received eleven honorary degrees and more than 30 awards for his human rights work, including awards from Columbia Law School and the American Bar Association for his lifetime achievements in International law.

One year later, when the criticism of drone targeted killing was mounting, John Brennan, a close adviser to President Obama on counterterrorism and home security, elaborated the basic ideas of Harold Koh. Brennan in his speech at Harvard University on September 16, 2011, made a point that drone strikes are based on the four core principles of counterterrorism strategy of President Obama. "First, our highest priority is—and always will be—the safety and security of the American people. Second, we will use every lawful tool and authority at our disposal. Third, we are pragmatic, not rigid or ideological—making decisions not based on preconceived notions about which action seems "stronger," but based on what will actually enhance the security of this country and the safety of the American people. We address each threat and each circumstance in a way that best serves our national security interests, which includes building partnerships with countries around the world. Fourth—and the principle that guides all our actions, foreign and domestic—we will uphold the core values that define us as Americans, and that includes adhering to the rule of law. And when I say "all our actions," that includes covert actions, which we undertake under the authorities provided to us by Congress. President Obama has directed that all our actions—even when conducted out of public view—remain consistent with our laws and values" (Brennan 2011).

Brennan has not put his message very sharp and said—"President Obama has made it clear—we must reject the false choice between our values and our security. We are constantly working to optimize both"(Brennan 2011).

In his 2011 remarks, John Brennan proposed a definition of the conflict. He said it was a "war with al-Qaida" and did not mention the war on terror as President G.W. Bush used to define it. So the enemy is named and identified by President Barrack Obama in a clear and sharp way. The critics of the war on terror were questioning very blurred definition of the enemy by Bush administration. Brennan admitted that the US authorities and their international allies such as many NATO members disagree on the geographic scope of the conflict. The United States does not view its authority to use military force against al-Qaida as being restricted solely to "hot" battlefields like Afghanistan. They believe to have the authority to "take action against al-Qaida and its associated forces without doing a separate self-defense analysis each time. And as President Obama has stated on numerous occasions, we reserve the right to take unilateral action if or when other governments are unwilling or unable to take the necessary actions themselves" (Brennan 2011).

Brennan and his supreme commander are aware that even some of the US closest allies and partners take a different view of the geographic scope of the conflict, limiting it only to the "hot" battlefields in Afghanistan and they argue that, outside of those restricted areas, the United States can only act in self-defense against al-Qaida when "they are planning, engaging in, or threatening an armed attack against US interests if it amounts to an "imminent" threat."

Brennan argued that understanding of "imminence" when dealing with terrorist groups should be more flexible and broader, because "al-Qaida does not follow a traditional command structure, wear uniforms, carry its arms openly, or mass its troops at the borders of the nations it attacks. Nonetheless, it possesses the demonstrated capability to strike with little notice and cause significant civilian or military casualties."

Obama and his closest assistant in targeted killing operations believed in convergence of views with their allies—"The more our views and our allies' views on these questions converge, without constraining our flexibility, the safer we will be as a country."

Brennan concluded his 2011 speech in a pathetic manner—"As a people, as a nation, we cannot—and we must not—succumb to the temptation to set aside our laws and our values when we face threats to our security, including and especially from groups as depraved as al-Qaida. We're better than that. We're better than them. We're Americans."

In his second presentation of moral and legal legitimation of targeted killing policy in April 2012 at Woodrow Wilson Center, John Brennan stressed legality of drone attacks again by saying—"The Authorization for Use of Military Force—the AUMF—passed by Congress after the September 11 attacks authorizes the president "to use all necessary and appropriate force" against those nations, organizations, and individuals responsible for 9/11. There is nothing in the AUMF that restricts the use of military force against al-Qaida to Afghanistan. As a matter of international law, the United States is in an armed conflict with al-Qaida, the Taliban, and associated forces, in response to the 9/11 attacks, and we may also use force consistent with our inherent right of national self-defense. There is nothing in international law that bans the use of remotely piloted aircraft for this purpose or that prohibits us from using lethal force against our enemies outside of an active battlefield, at least when the country involved consents or is unable or unwilling to take action against the threat it's useful to consider such strikes against the basic principles of the law of war that govern the use of force" (Brennan 2012).

Self-defense argument has been supported by Brennan remarks of 2012 by quoting briefly four basic principles of international law about the use of force:

- The principle of **necessity**—the requirement that the target has definite military value. "We have the authority to target them with lethal force just as we targeted enemy leaders in past conflicts, such as German and Japanese commanders during World War II."
- The principle of **distinction**—the idea that only military objectives may be intentionally targeted and that civilians are protected from being intentionally targeted. Brennan expressed the view that drones had "unprecedented ability to precisely target a military objective while minimizing collateral damage," and that "never before has there been a weapon that allows us to distinguish more effectively between an al-Qaida terrorist and innocent civilians."[2]
- The principle of **proportionality**—the notion that the anticipated collateral damage of an action cannot be excessive in relation to the anticipated

[2]The critics say this damage is unacceptable as too many civilians got killed or wounded.

military advantage. Again the point about precision of strikes has been made by J. Brennan at this junction.
- The principle of **humanity** which requires to use weapons that will not inflict unnecessary suffering.
  Brennan concluded his legal reasoning with the strong thesis: "For all these reasons, I suggest to you that these targeted strikes against al-Qaida terrorists are indeed ethical and just."

Brennan claimed that President Obama did a lot to foster transparency of military and security policies. He recalled –

- A new Executive Order on classified information that, among other things, reestablished the principle that all classified information will ultimately be declassified.
- A Freedom of Information Act Directive mandating that agencies adopt a presumption of disclosure when processing requests for information.
- The legislation that revised the process for reporting sensitive intelligence activities to Congress and created an Inspector General for the Intelligence Community.
- The combined budget of the intelligence community, and reconstituted the Intelligence Oversight Board, an important check on the government's intelligence activities.

Despite all these steps, the 2013 sensations of Edward Snowden about National Security Agency secret operations made it clear that transparency is not strongest feature of Obama style of governance and his trust to his best allies is actually quite low.

Rand Paul, a Republican senator from Kentucky, commenced an old-fashioned country-style talking filibuster of Barack Obama's nominee for CIA director, John Brennan, in protest over the administration's policies on the use of drones. American drones are being used to kill suspected terrorists in entirely too cavalier a fashion, killing at a minimum hundreds of innocent civilians, and the administration's procedures for deciding when to authorize drone strikes are opaque and lack due process. Senator Rand Paul tried to alert public opinion before appointment of John Brennan as a new CIA chief, by talking Senate filibuster, which continued for 13 h. After this filibuster, Brennan was confirmed by a Senate vote of 63–34.

Brennan was sworn into the office of CIA Director on March 8, 2013. President Obama confirmed his trust in professional skills and moral integrity of John Brennan with whom they operated a so called "kill list" for targeted executions by drones.

New Strategy of Obama Administration: The President's Speech on May 2013

President Barack Obama, who vastly expanded US drone strikes against terrorism suspects overseas under the cloak of secrecy, has been openly seeking to influence global guidelines for their use as China and other countries pursue their own drone programs. Obama's new position is not without irony. The White House kept details of drone operations—which remain largely classified—out of public view for years when the US monopoly was airtight. That stance is just now beginning to change, in part under pressure from growing public and Congressional discomfort with the drone program. US lawmakers have demanded to see White House legal justifications for targeting US citizens abroad, and to know whether Obama thinks he has the authority to use drones to kill Americans on US soil.

Remarks by the President at the National Defense University, in Washington, DC, on May 23, 2013, were quite soft. His speech was a kind of pragmatic adjustment to a new situation and a first draft of a new counterterrorism comprehensive strategy. He reminded first of all that USA was "at war for well over a decade" since September 11, 2001. "What is clear is that we quickly drove al-Qaeda out of Afghanistan, but then shifted our focus and began a new war in Iraq." The US President admitted that "in some cases, I believe we compromised our basic values—by using torture to interrogate our enemies, and detaining individuals in a way that ran counter to the rule of law" (Obama 2013).

On the other hand, the President presented his efforts made so far "to change the course" of this war. "We relentlessly targeted al-Qaeda's leadership. We ended the war in Iraq, and brought nearly 150,000 troops home. We pursued a new strategy in Afghanistan, and increased our training of Afghan forces. We unequivocally banned torture, affirmed our commitment to civilian courts, worked to align our policies with the rule of law, and expanded our consultations with Congress" (Obama 2013). After referring to positive results of his strategy (Osama bin Laden is dead, and so are most of his top lieutenants. The core of al-Qaeda in Afghanistan and Pakistan is on the path to defeat. There have been no large-scale attacks on the United States, and our homeland is more secure. Fewer of our troops are in harm's way, and over the next 19 months they will continue to come home. Our alliances are strong, and so is our standing in the world. In sum, we are safer because of our efforts), he turned to some costs of the counterterrorism war such as over a trillion dollars spent on war by American nation over the last decade, 7000 dead service members and much more wounded and sufferings of their families.

To the set of premises for a new strategy, Barrack Obama added a very deep moral concern in a long term view to the future: "From our use of drones to the detention of terrorist suspects, the decisions that we are making now will define the type of nation—and world—that we leave to our children. So America is at a crossroads. We must define the nature and scope of this struggle, or else it will define us. We have to be mindful of James Madison's warning that "No nation could preserve its freedom in the midst of continual warfare."

The components of a new counterterrorism comprehensive strategy are to be the following:

1. Better supervised by the Congress and by the public opinion but a continued, targeted action against terrorists. The goal is to finish the work of defeating al-Qaeda and its associated forces. In Afghanistan—to complete a transition to

Afghan responsibility for that country's security in order to get American troops come home. Beyond Afghanistan, we must define our effort not as a boundless "global war on terror," but rather as a series of persistent, targeted efforts to dismantle specific networks of violent extremists that threaten America—B. Obama defined a war in a new way because the old war on terror was over. Drone attacks he presented as effective,[3] more precise that any other military action, and legal and just. "To say a military tactic is legal, or even effective, is not to say it is wise or moral in every instance. For the same human progress that gives us the technology to strike half a world away also demands the discipline to constrain that power—or risk abusing it. And that's why, over the last four years, my administration has worked vigorously to establish a framework that governs our use of force against terrorists—insisting upon clear guidelines, oversight and accountability that is now codified in Presidential Policy Guidance that I signed yesterday."[4] The need of unmanned strikes will be reduced and collateral damage and civilian casualties, as well.
2. Effective partnerships with many countries in the gathering and sharing of intelligence, the arrest and prosecution of terrorists.
3. Diplomatic engagement in many countries. President asked for gratitude to American diplomats because *America cannot carry out this work if we don't have diplomats serving in some very dangerous places.*
4. Economic assistance in development and modernization of world regions where the conditions of living may foster extremism and aggressive political behavior—"for all the focus on the use of force, force alone cannot make us safe. We cannot use force everywhere that a radical ideology takes root; and in the absence of a strategy that reduces the wellspring of extremism, a perpetual war—through drones or Special Forces or troop deployments—will prove self-defeating, and alter our country in troubling ways." While solving "deep-rooted problems like poverty and sectarian hatred… We must help countries modernize economies, upgrade education, and encourage entrepreneurship—because American leadership has always been elevated by our ability to connect with people's hopes, and not simply their fear… the peaceful realization of individual aspirations will serve as a rebuke to violent extremists…"

While discussing tactical aspects of drone combat the President reminded that "we only target al-Qaeda and its associated forces. America does not take strikes when we have the ability to capture individual terrorists; our preference is always

[3]"In the intelligence gathered at bin Laden's compound, we found that he wrote, "We could lose the reserves to enemy's air strikes. We cannot fight air strikes with explosives." Other communications from al-Qaeda operatives confirm this as well"—*Remarks by the President of May 23, 2013.*

[4]*US Policy Standards and Procedures for the Use of Force in Counterterrorism Operations outside the United States and Areas of Active Hostilities* is a short brief made available on the White House Web site. The full document is classified.

to detain, interrogate, and prosecute. America cannot take strikes wherever we choose; our actions are bound by consultations with partners, and respect for state sovereignty.

America does not take strikes to punish individuals; we act against terrorists who pose a continuing and imminent threat to the American people, and when there are no other governments capable of effectively addressing the threat. And before any strike is taken, there must be near-certainty that no civilians will be killed or injured—the highest standard we can set…

Conventional airpower or missiles are far less precise than drones and are likely to cause more civilian casualties and more local outrage. And invasions of these territories lead us to be viewed as occupying armies, unleash a torrent of unintended consequences, and are difficult to contain, result in large numbers of civilian casualties and ultimately empower those who thrive on violent conflict".

The very new component of domestic security presented by B. Obama is the idea about the cooperation with the Muslim American community which has consistently rejected fundamentalism. Obama said that "these partnerships can only work when we recognize that Muslims are a fundamental part of the American family. In fact, the success of American Muslims and our determination to guard against any encroachments on their civil liberties is the ultimate rebuke to those who say that we're at war with Islam."

The new strategy of targeted killing and the old one is supported by majority of American public opinion, and President Obama must not be afraid that his efforts will be disregarded by his fellow citizens. Although the United States has been killing suspected terrorists with drone strikes in non-battlefield settings for over ten years, public opinion polling of the controversial tactic began only a year and a half ago. Averaged together, the polls demonstrate that 65 % of Americans support the targeted killing of suspected terrorists, and 51 % approve killing US citizens who are suspected of terrorism[5] (Zenko 2013). It is revealing that, according to a February 2013 poll, support for US military drone strikes (75 %) is considerably higher than for those conducted by the CIA (65 %). Finally, it is remarkable how normalized drone strikes have become over the years, to the point that Americans are more comfortable with killing suspected terrorists than torturing them for information.

The context and framing of the question in the polls varies widely. For example, targets are alternately described as "suspected terrorists," "extremists," those "deemed a threat to the United States," and "high-level terrorism suspects."

[5]Polling in this area of low knowledge is of limited value, however. It is instructive of public opinion any more than asking people on the street about the Higgs Boson particle. People are only aware of vague details and have absorbed willfully incorrect memes and images from the media. Explaining the military drone program and the issues surrounding it requires a minimum of a 20 min conversation.

United Nations Investigations and Reporting 2005–2013: Alston, Heyns, Emmerson, UN Secretary General

The source of the pressure upon Obama administration were not just lawyers and NGOs, and there were reports to the United Nations as well. First, it was Philip Alston, professor from the New York University School of Law, appointed in 2004, who presented 10 annual reports to the UN authorities over the years 2005–2010 (Alston 2010). Philip Alston, the UN Special Rapporteur on Extrajudicial Executions, made it clear in his reports that extrajudicial executions might break international prohibitions: In law enforcement operations there is no legal right of using lethal weapons but in case of self-defense of police forces ONLY. Therefore, launching missiles and throwing bombs in such circumstances on human beings is not allowed by the law. CIA operations are a variety of police actions.

Combatants and civilians do not wear uniforms and do not fight and thus are similar to undefended objects. Therefore, they should not be hit by missiles. It's not possible to differentiate precisely between militants and non-militants because militants live among the population and do not wear uniforms, and because government sources have the incentive to claim that only militants were killed, while militants often assert the opposite (Alston 2011).

Alston reporting paved the way to next reports of 2013 by Christof Heyns, by Ben Emmerson QC, and finally by Secretary General Ban Ki Moon.

Christof Heyns, the South African law professor, the next UN special rapporteur on extrajudicial, summary or arbitrary executions, submitted his report to the UN Human Rights Council in Geneva in April 2013 (Heyns 2013). His report called for moratorium on weapons that can kill targets without human involvement. Machines lack morality and mortality and should not have life and death powers over humans. "Lethal autonomous robotics" (LAR)—weapons systems that, once activated, can lock on and kill targets without further involvement of human handlers—are most questionable. Fully autonomous weapons have not yet been developed and exist only in the imaginations of military planners. However, experts in warfare technologies warn that the world's leading military powers are moving so rapidly in this direction that a preemptive ban is essential. In his submission to the UN, Heyns points that a drone technology has already moved a step closer to a fully autonomous state in the form of the X-47B, a super-charged UAV developed by the US Navy. Britain is developing its own next generation of drone, known as Taranis that can be sent to tackle targets at long range and can defend itself from enemy aircraft. Like X-47B, it has two in-built weapons bays, though is currently unarmed.

Heyns challenges the US position, most apparent in the leaked Department of Justice White Paper, of a much broader concept of "imminence" which would mean in effect that no immediate threat is with regard to using lethal force under self-defense rules. Heyns states: "The view that mere past involvement in planning attacks is sufficient to render an individual targetable even where there is no evidence of a specific and immediate attack distorts the requirements established in international human rights law" (Para. 37). Heyns also argues that only a State's highest authority can give permission to another State to use force on its territory and if that permission is withdrawn, such force must cease (see Para. 82–84). This is clearly a reference to arguments within the USA that despite Pakistan Government announcements urging an end to US drone strikes, authority has previously been given or alternatively that secretly, Pakistan continues to give permission for the strikes through the ISI, the Pakistan security service. Heyns also calls follow-up drone strikes, if aimed at the wounded, rescuers and medical personnel—dubbed as "double-tap" strikes by the media—war crimes. Heyns also challenges the uncritical acceptance that drone are more precise than other weapons. There is little if any empirical data in the public domain for such claims. This leads to the main thrust of Heyns' report—the need for greater transparency on the use of drones—not just from the USA but from all States using armed drones. "A lack of appropriate transparency and accountability concerning the deployment of drones undermines the rule of law and may threaten international security."

Article 2(4) of the UN Charter prohibits the threat or use of force by one state against another. Two exceptions to the Article 2(4) prohibition on the use of force are particularly relevant to the question of whether US targeted killings in Pakistan are lawful: (1) when the use of force is carried out with the consent of the host state; and (2) when the use of force is in self-defense in response to an armed attack or an imminent threat, and where the host state is unwilling or unable to take appropriate action.

Heyns did not say the use drones for executions is or must be inherently illegal. He suggests that targeted killing blurs some lines in legal reasoning and might be dangerous. His conclusion is as follows:

"While it is not clear at present how LARs (Lethal Autonomous Robotics) could be capable of satisfying IHL and IHRL requirements in many respects, it is foreseeable that they could comply under certain circumstances, especially if used alongside human soldiers. Even so, there is widespread concern that allowing LARs to kill people may denigrate the value of life itself."

His main recommendation is that "the High Commissioner for Human Rights shall convene, as a matter of priority, a High Level Panel on LARs consisting of experts from different fields such as law, robotics, computer science, military operations, diplomacy, conflict management, ethics and philosophy."

At the press conference on October 25, 2013, he said "drones were not illegal, but raised challenges… In general, drones should follow the law and not the other way round." There was not a need for new treaties, but the application of the existing system more rigorously (Press Conference 2013).

The issue was moving rapidly up the international agenda after China and Russia in October 2013 jointly issued a statement at the UN Human Rights Council, backed by other countries, condemning drone attacks.

A second UN rapporteur, the London-based barrister Ben Emmerson QC, who monitors counter-terrorism since January 2013, has not completed his final report on his inquiry on behalf of the UN into the use of drones in counter-terrorism operations, launched in January 2013. The inquiry was originally to be completed in time for the UN General Assembly in October 25, 2013, but has taken longer than expected and there is therefore only an interim report available. The complete findings are to be presented in 2014. While originally focusing on a sample of 25 'case studies' of drone strikes, this has now been expanded to 33 case studies (Emmerson 2013).

Like Heyns, Ben Emmerson examines the "principal areas of legal controversy" surrounding the use of armed drones, focusing on when an individual may or may not be targeted and whether the US can be said to be acting in self-defense. The report contains a brief "Conclusion and Recommendations" section, which reads as follows:

- "If used in strict compliance with the principles of international humanitarian law, remotely piloted aircraft are capable of reducing the risk of civilian casualties in armed conflict by significantly improving the situational awareness of military commanders.
- Having regard to the duty of States to protect civilians in armed conflict, the Special Rapporteur considers that, in any case in which civilians have been, or appear to have been, killed, the State responsible is under an obligation to conduct a prompt, independent and impartial fact-finding inquiry and to provide a detailed public explanation. This obligation is triggered whenever there is a plausible indication from any source that civilian casualties may have been sustained, including where the facts are unclear or the information is partial or circumstantial. The obligation arises whether the attack was initiated by remotely piloted aircraft or other means, and whether it occurred within or outside an area of active hostilities.
- The Special Rapporteur identifies herein a number of legal questions on which there is currently no clear international consensus. He considers that there is an urgent and imperative need to seek agreement between States on these issues. To that end, he is currently consulting Member States with a view to clarifying their position on these questions. He urges all States to respond as comprehensively as possible.
- In particular, the Special Rapporteur urges the United States to further clarify its position on the legal and factual issues raised herein; to declassify, to the maximum extent possible, information relevant to its lethal extraterritorial counter-terrorism operations; and to release its own data on the level of civilian casualties inflicted through the use of remotely piloted aircraft, together with information on the evaluation methodology used."

Like Christof Heyns, Ben Emmerson argues strongly for much greater transparency around the use of armed drones, especially incidents where there have been reports of harm to civilians Ben Emmerson believes that the UN itself should consider establishing an investigatory body. Drones attacks by the USA raise fundamental questions … If they don't investigate themselves, we will do it for them. The USA is not a signatory to the International Criminal Court (ICC) or many other international legal forums where legal action might be started. It is, however, part of the International Court of Justice (ICJ) where cases can be initiated by one state against another.

Heyns told The Guardian later that his future inquiries are likely to include the question of whether other countries, such as the UK, share intelligence with the USA that could be used for selecting individuals as targets. A legal case has already been lodged in London over the UK's alleged role in the deaths of British citizens and others as a consequence of US drone strikes in

There are three legal paradigms examined in all the reports:

- The law governing the inter-State use of force (the *jus ad bellum*).
- The law governing armed conflict (the *jus in bello*). The capture versus kill debate is of significance because many lawyers claim the state authorities should try to capture the enemy and to kill in last resort only.
- The law governing international human rights (this is referred to by Alston as "the law enforcement model").

The reports have consistently found that the existing legal frameworks—*jus ad bellum*, *jus in bello*, and "the law enforcement model"—are not in issue, but rather the practical application of these paradigms by States, particularly in recent times during the "War on Terror," are stretching widely accepted legal principles and norms.

The next step in the UN debate was a report of the Secretary General on the protection of civilians in armed conflict (United Nations S/2013/689.) submitted on November 22, 2013 (Report of the Secretary General 2013a, b).

It says that in Afghanistan, the first six months of 2013 saw a significant increase in civilian casualties compared with the same period in 2012. The number of deaths increased by 14 % (1319 deaths) and the number of injuries by 28 % (2533 injuries) owing to the more extensive use of improvised explosive devices by anti-government elements and increased casualties caused by ground engagements. This is a growing threat to life and security.

Secretary General has been concerned about the use of drones in Afghanistan and Pakistan. He was uncertain whether there was the compliance with international human rights law and with the international humanitarian law rules of distinction, proportionality, and precaution. He stressed that "the proliferation of drone technology and the increasing resort to such weapons systems will also further sharpen the asymmetry that exists in many conflicts between State and non-State parties. As technology allows one party to become increasingly removed from the battlefield, and the opportunities to fight against it are reduced, we may see technologically inferior parties increasingly resort to strategies intended to harm civilians as the most accessible targets. Moreover, drone technology increases opportunities to conduct attacks that might otherwise be considered unrealistic or undesirable through other forms of air power or the deployment of ground troops."

His final appeal to UN member states was rather bleak and very cautious. Nothing more than the present law prescribes and much less in intellectual terms than the three reports by Alston, Heynes, and Emmerson offered.

The member states were called by Secretary General to do the following:

(1) To ensure the protection of civilians in specific drone attacks;
(2) To track and assess civilian casualties resulting from attacks in order to identify all measures feasible to avoid civilian casualties;
(3) To investigate serious violations of international humanitarian law and international human rights law that are alleged to have occurred during such attacks.

The final report by Emmerson was submitted in 2014 and instead of some answers it proposed a list of legal questions to member states of United Nations (Emmerson 2014).

(a) Does the international law principle of self-defense entitle a State to engage in non-consensual lethal counter-terrorism operations on the territory of another State against a non-State armed group that poses a direct and immediate threat of attack even when the armed group concerned has no operational connection to its host State? If so, under what conditions does such a right of self-defense arise? Does such a right arise where the territorial State is judged to be unable or unwilling to prevent the threat from materializing? If so, what are the criteria for determining "unwillingness" or "inability" to act?
(b) Is the international law principle of self-defense confined to situations in which an armed attack has already taken place, or does it entitle a State to carry out preemptive military operations against a non-State armed group on the territory of another State, without the territorial State's consent, where it judges that there is an imminent risk of attack to its own interests? If so, how is imminence to be defined?
(c) Does the international humanitarian law test of intensity of hostilities (which is one of the criteria determining whether a non-international armed conflict exists) require an assessment of the severity and frequency of armed attacks occurring within defined geographical boundaries? In applying the intensity test to a non-State armed group operating transnationally, is it legitimate to aggregate armed attacks occurring in geographically diverse locations in order to determine whether, taken as a whole, they cross the intensity threshold so as to amount to a non-international armed conflict? If it is possible for a State to be engaged in a non-international armed conflict with a non-State armed group operating transnationally, does this imply that a non-international armed conflict can exist which has no finite territorial boundaries?
(d) Does international humanitarian law permit the targeting of persons directly participating in hostilities who are located in a non-belligerent State, and if so, in what circumstances?
(e) Does the pattern and frequency of the armed attacks currently being perpetrated by al-Qaida, and the various affiliate organizations in different parts of the world that claim allegiance to al-Qaida, satisfy (or continue to satisfy) the criteria of organization and intensity required under international humanitarian law to qualify as a state of armed conflict?

(f) Assuming that a non-international armed conflict exists, does the test of "continuous combat function," as elaborated by the International Committee of the Red Cross ("ICRC") for determining whether a person is a "member" of an armed group (such that they may be targeted with lethal force at any time) reflect customary international law?

All questions asked by Ben Emerson are quite clever, but all proceedings of the UN in this matter are so clumsy and unfocused that one can expect a new convention in the year 2050 perhaps. His hope on consensus on these matters among member states is hopelessly naïve or cynical or both. We may conclude that the debate will go on and the targeted killing as well. What a nice new, brave world, isn't it?

Imagine the Next Stage of the Story: Fully Autonomous Weapons

Current drone computer programs merely advise human operators on the decision to launch an attack. In future, drone computers may be programmed to launch attacks on the basis of preset parameters without the need for a human being to make the real-time decision. Shift of responsibility triggered by fully automated warfare seems to be hardly predictable. According to a 2009 US Air Force report, by 2047 drones will be fully automated. Doubts are expressed by many experts, as to who should be held accountable for possible serious violations of the laws of war. Decisions are usually taken by responsible moral agents capable of rational judgment. The notions of moral agency and responsibility, however, are difficult to reconcile with algorithms. The human agent monitoring the execution of an operation does not take the decision to engage a target at present conditions. Should he be held accountable for having failed to intervene on time to prevent undesirable killings? Should commanders be responsible for having wrongly defined the parameters and rules of engagement? Or should the programmers be held to account for having instilled deficient assessment mechanisms into the machine?

When that 'person' is a machine, a crucial link is missing in the mental representation of responsibility. In the absence of a primary offender, the narrative of death is conceptually altered. Technological advances in unmanned warfare displace the burden of decision making and contribute to outsourcing and distorting responsibility. This is a gloomy future. Instead of more security, it may disperse responsibility and accountability completely. Many experts predict autonomous weapons systems will become the norm in the next 20 years.

Fully autonomous weapons, also known as "killer robots," would be able to select and engage targets without human intervention. It is not certain that fully autonomous weapons do not exist yet, but it is obvious that they are being developed by several countries and precursors to fully autonomous weapons have already been deployed by high-tech militaries. The UK's Taranis combat aircraft, whose prototype was unveiled in 2010, is designed to strike distant targets, even in another continent. While the Ministry of Defiance has stated that humans will remain in the loop, the Taranis exemplifies the move toward increased autonomy (Autonomous weapons 2012). Some experts predict that fully autonomous weapons could be operational in 20 to 30 years.

These weapons would be incapable of meeting international humanitarian law standards, including the rules of distinction, proportionality, and military necessity. The weapons would not be constrained by the capacity for compassion, which can provide a key check on the killing of civilians. Fully autonomous weapons also raise serious questions of accountability because it is unclear who should be held responsible for any unlawful actions they commit. Human Rights Watch calls for a preemptive prohibition on fully autonomous weapons. Human Rights Watch is a founding member of the Campaign to Stop Killer Robots, and currently serves as the campaign's global coordinator.

Growing public concern and mounting opinions of critical experts inspired the Secretary General of the United Nations to include in his final report on protection of civilians in armed conflicts some vague recommendations (Report of the Secretary General 2013a, b).

First his report repeated some basic moral questions—"is it morally acceptable to delegate decisions about the use of lethal force to such systems? If their use results in a war crime or serious human rights violation, who would be legally responsible? If responsibility cannot be determined as required by international law, is it legal or ethical to deploy such systems?" Secretary General proposed only … a debate. He has put it this way: "Although autonomous weapons systems as described herein have not yet been deployed and the extent of their development as a military technology remains unclear, discussion of such questions must begin immediately and not once the technology has been developed and proliferated. It must also be inclusive and allow for full engagement by United Nations actors, ICRC and civil society." It seems to be too little and too late. The Mountain gave birth to a mouse, again. Hopeless weakness of once quite powerful and reasonable organization one may say.

Conclusions

I would like to conclude this essay with some forecasts and recommendations of my own:

1. I endorse the general judgment of John Horgan "The invention that escapes our control, has been a persistent fear of the industrial age—with good reason. Nuclear weapons are too easy an example; consider what cars have done to our landscape over the past century, and it's fair to wonder who's in the driver's seat, them or us. Most people would say cars have, on the whole, benefited humanity. A century from now there may be the same agreement about drones, if we take steps early on to control the risks" (Horgan 2013).

2. Regardless of power game between superpowers on the legal rules of using drones in surveillance and combat missions, there is an obvious need to work on the new treaty arranged for by the United Nations after the final report of Ben Emmerson in 2014. If we do not develop the international law of war, the military technology will make the law obsolete and hopelessly old-fashioned and unable to regulate anything of significance. Do we expect future terrorists or partisans to use military uniforms and to carry a flag in order to be able to order our soldiers to fight against them? Are we to think that terrorist is a civilian protected by the law because he/she is operating in civilian clothes? If humans will not control the technology, the technology will control humans.
3. The beginnings of the British parliamentary debate on this matter show how hard it will be to reach a broader consensus on a new convention. At a parliamentary debate on June 17, 2013, the UK Minister for Counter Proliferation, Alistair Burt MP provided further information on British policy regarding fully autonomous weapons. The Minister stressed that the UK does not possess fully autonomous weapon systems and has no intention of developing them. (UK says killer robots will not meet requirements of international law 2013). By recognizing that fully autonomous weapons "will not" be able to meet the requirements of international humanitarian law, this position provides a significantly stronger barrier to the development of fully autonomous weapons than the government's previously stated position, presented at the UN Human Rights Council on May 30, 2013, that existing International Humanitarian Law is sufficient to regulate the development and use of such weapons. However, a certain ambiguity reappeared when the Minister noted later in the debate that: "We think the Geneva conventions and additional protocols provide a sufficiently robust framework to regulate the development and use of these weapon systems."
4. Many varieties of highly sophisticated and smart weapons should be covered by the future international treaty that must be negotiated as soon as possible.
5. The USA will probably adopt a new law on civilian use of drones in 2015, as we already know. The other states should follow as soon as possible with their own legislation regulating drone flights by private companies, by government agencies, and by individual hobbyists. The new law should require the aviation authorities to allow the safe integration of UAVs into national and international airspace. The civilian market for drones—and especially small, low-cost, tactical drones—could soon dwarf military sales. Drone fever might explode in the USA first and very soon after in many advanced countries.

   Before it happens the clever law makers should do their job. But regulation should not mean prohibition, of course. Privacy, clean sky, personal safety is at stake if we regulate civilian drones in a wrong way.[6]

Schonfield seems to be right that "lawyers and judges now play starring roles both in making national security policy and in overseeing military operations. The

[6]The New York University School of Law professors have established in Autumn 2013 a successful Web site on http://security-related legal thinking/justsecurity.org/.

result is that, when it comes to the American government's efforts to provide for the common defense, a far-reaching legalism has taken hold" (Schonfield 2011). In the USA, "significant intrusion by lawyers and courts into the conduct of national security and warfare is not something the framers of the Constitution even remotely envisioned. Under the arrangement they established, the political branches were to hold the reins in wartime." Lawyers have penetrated every crevice of the US national security machinery; there are more than 10,000 attorneys in the Defense Department alone, and they determine the conduct of war to a degree without any precedent.[7]

No wonder that legal regulation of military duties increased. In 1914, the War Department—fulfilling America's obligations under the Hague treaty—published its *Rules of Land Warfare*, which contained 139 pages of text. Today, many editions later, the Pentagon issued yet another update of its Law of War Manual, which exceeds 1100 single-spaced typewritten pages, with more than 3000 footnotes.

While planning new laws, it is wise to remember that—as Plato said—"The excessive increase of anything causes a reaction in the opposite direction."

References

Alston P (2010) Report of the Special Rapporteur on extrajudicial, summary or arbitrary executions, Addendum, Study on targeted killings. United Nations, General Assembly, Human Rights Council 28 May 2010

Alston P (2011) The CIA and targeted killings beyond borders. Harv Natl Secur J 2:2011

ASIL Keynote Highlight (2010) US legal adviser Harold Koh Asserts Drone Warfare is lawful self-defense under international law. http://insidejustice.com/law/index.php/intl/2010/03/26/asil_koh_drone_war_law

Autonomous weapons—the risks of a management by 'partition'. 10 Oct 2012. http://www.article36.org/other-issues/uncategorized/autonomous-weapons-the-risks-of-a-management-by-partition/. The emergence of armed drones and how we assess the acceptability of new weapons, March 21, 2013. http://www.article36.org/weapons-review/the-emergence-of-armed-drones-and-how-we-assess-the-acceptability-of-new-weapons/

Brennan JO (2011) Remarks "Strengthening our security by adhering to our values and laws. 16 Sep 2011. http://www.whitehouse.gov/the-press-office/2011/09/16/remarks-john-o-brennan-strengthening-our-security-adhering-our-values-an

Brennan JO (2012) Assistant to the President for Homeland Security and Counterterrorism "The ethics and efficacy of the president's counterterrorism strategy". Woodrow Wilson International Center for Scholars, Washington, DC, Monday, 30 Apr 2012. http://www.wilsoncenter.org/event/the-efficacy-and-ethics-us-counterterrorism-strategy

Covert War on Terror (2012). http://www.thebureauinvestigates.com/category/projects/drones/. Visited 12 Sept 2012

[7]On the other hand, there is in the USA the "Guantánamo bar"—the 500 or so lawyers, many of them from top-drawer firms, leading law schools, and civil liberties organizations—who are representing terrorist defendants. Many of them have often gone far beyond providing competent counsel in court.

Emmerson B (2013) Promotion and protection of human rights and fundamental freedoms while countering terrorism. United Nations A/68/389, General Assembly, 18 Sept 2013

Emmerson B (2014) Report of the Special Rapporteur on the promotion and protection of human rights and fundamental freedoms while countering terrorism, 28 Feb 2014, United Nations, A/HRC/25/5

Horgan J (2013) Unmanned flight. http://ngm.nationalgeographic.com/2013/03/unmanned-flight/horgantext?rptregcta=reg_free_np&rptregcampaign=20131016_rw_membership_n1p_intl_ot_c1#. John Horgan's most recent book is *The End of War,* San Francisco: Sweeney's, 2012. He argues that war is neither necessary nor inevitable

Heyns C (2013) Report of the Special Rapporteur on extrajudicial, summary or arbitrary executions. United Nations, General Assembly, Human Rights Council, 9 Apr 2013

Lamentowicz W (2009) Amerykański i europejski wariant atlantyckiej kultury strategicznej. In: Bezpieczeństwo międzynarodowe: wyzwania i zagrożenia XXI wieku, Wydawnictwo Wyższa Szkoła Handlowa, Radom 2009, pp 105–118. I claim that US and Western Europe are two different varieties of Atlantic strategic culture

Losing Humanity. Report (2012) by Human rights watch. http://www.hrw.org/sites/default/files/reports/arms1112_ForUpload.pdf

Luban D (2013) A strange and far-flung war, the dissent magazine October 23, 2013

Obama B (2013) Remarks by the president at the National Defense University National Defense University. Fort McNair, Washington, DC, The White House, 23 May 2013

O'Connell ME (2010) The international law of drones, 14 American Society of InternationalLawInsight. 12 Nov 2010. http://asil.org/files/2010/insights/insights_101112.pdf

Mayer J (2009) The predator war, New Yorker, October 26, 2009. http://www.newyorker.com/reporting/2009/10/26/091026fa_fact_mayer#ixzz2NWZuvk2y

Press Conference by Special Rapporteurs on Executions, Human Rights While Countering Terrorism (2013) 25 Oct 2013, Department of Public Information, News and Media Division, United Nations, New York

Report of the Secretary-General on the protection of civilians in armed conflict (2013) United Nations S/2013/689 Security Council, 22 Nov 2013

Schonfield G (2011) Legalism in wartime, national affairs. Number 7, Spring 2011

The Year of the Drone: an analysis of U.S. Drone Strikes in Pakistan, 2004–2012. New America Foundation. http://counterterrorism.newamerica.net/drones. Last updated 4 Apr 2012

Report of the Secretary-General on the protection of civilians in armed conflict (2013) United Nations S/2013/689 Security Council, 22 Nov 2013

UK says killer robots will not meet requirements of international law (2013). http://www.article36.org/ weapons-review/uk-says-killer-robots-will-not-meet-requirements-of-international-law/. Accessed 18 June 2013

US Policy Standards and Procedures for the Use of Force in Counterterrorism Operations outside the United States and Areas of Active Hostilities is a short brief made available on the White House website. The full document is classified

Zenko M (2013) US public opinion on drone strikes, March 18, 2013, Council on foreign relations, New York. Sources of data used are as follows: Washington Post/ABC News Poll, (1 Sep 2011), CBS News Poll (6–10 Nov 2011) Washington Post/ABC News Poll (1–4 Feb 2012) Pew Research Center (April 2012) Farleigh-Dickinson Public Mind Poll (7 Feb 2013) CBS News Poll (6–10 Feb 2013) YouTube/Huffington Post Poll (6 Mar 2013)

Part IV
Technology Evaluations and Policies

Chapter 13
Technology Assessment in Systems Analysis

Piotr Sienkiewicz and Halina Świeboda

Abstract In every age cumulative innovation accrue from the previous period, which discarded are those that do not meet the new challenges of civilization. Industrial Era brought a "mass society" and bureaucratic organizations. Information Era and emerged "information society" and network and virtual organizations. Always one of the sources of social change was the development of technology: the good and the bad. Therefore, multidimensional (multicriteria multiattribute) systems analysis is required, which is an important element of the valuation of real and potential technologies.

Keywords Systems analysis · Evaluation · Technology assessment

Introduction

Observation of the effects of technological development brings the need for evaluation, such as analysis and evaluation, in order to obtain useful knowledge in systems analysis as the basis of rational decisions. Rationalism in the research system implies the need to formulate both ex post and ex ante in order to anticipate future states and the design and development decisions. If you cannot predict systemic change (development) system, it should be the future shape so as to minimize the likelihood of possible side systems. Adoption of global and holistic perspective of the system seems to be obvious. Difficulties, however, concern the construction of modern instruments of research and description including (Sienkiewicz 1985):

P. Sienkiewicz (✉) · H. Świeboda
Engineering Institute of Security Systems, National Defence University,
Al. A. Chruściela 103, 00-910 Warsaw, Poland
e-mail: p.sienkiewicz@aon.edu.pl

H. Świeboda
e-mail: h.swieboda@aon.edu.pl

L.W. Zacher (ed.), *Technology, Society and Sustainability*,
DOI 10.1007/978-3-319-47164-8_13

- description language with correct terminology and categories,
- methods of observation of phenomena and processes,
- meters properties and attributes of objects (systems, processes),
- methods for the identification of the cause–effect relationship,
- model systems (structure and dynamics of process).

When structured empirical material is rich enough and efficient enough, instruments are efficient for:

- diagnosis (e.g., the degree of impact of technology on social development),
- comparison of the predicted effects of technology development in distinguished areas of society,
- technology to identify attributes that contribute to obtain a higher quality of life,
- predict social impact of socioeconomic development (Sienkiewicz 1983).

Failures related to "strict" forecast of social systems development meant that more important than the absolute accuracy of the specific predictions is to achieve social consensus of strategic directions of system development (the state). Thus, scenarios and techniques such as Delphi method, supported the need for simulation models, are an essential tool for predictive analysis. However, more often "forecasting" is replaced by the "foresight" which could mean an analysis by creating a social vision of the future that is likely to become "self-fulfilling forecast." (Sienkiewicz and Świeboda 2008)

On the other hand, evaluation studies of science and technology are conducted in the context of the impact of specific policies and scientific and technical use of the method called as Technology assessment (TA). Using both elements of "forecasting" and TA, evaluation methodology of technologies in the systems analysis of socioeconomic development has been developed (Table 13.1, Fig. 13.1).

Table 13.1 Technology assessment – nine primary sources (impacts) and benefits (outputs) of social development

	Development resources	Aspects of development	Consequences
1	Society	Sociology	Changes in quality of life
2	Economy	Economic	Changes in welfare
3	Policy	Politological	Changes in state power
4	Education	Educational	Changes in intellectual capital
5	Science	Cognitive	Changes in knowledge
6	Culture	Cultural	Changes in cultural identity
7	Energy	Energetic	Changes in the level of energy security
8	Natural environment	Ecological	Changes in the ecosphere
9	Information infrastructure	Informative	Changes in the value of information assets

Source own

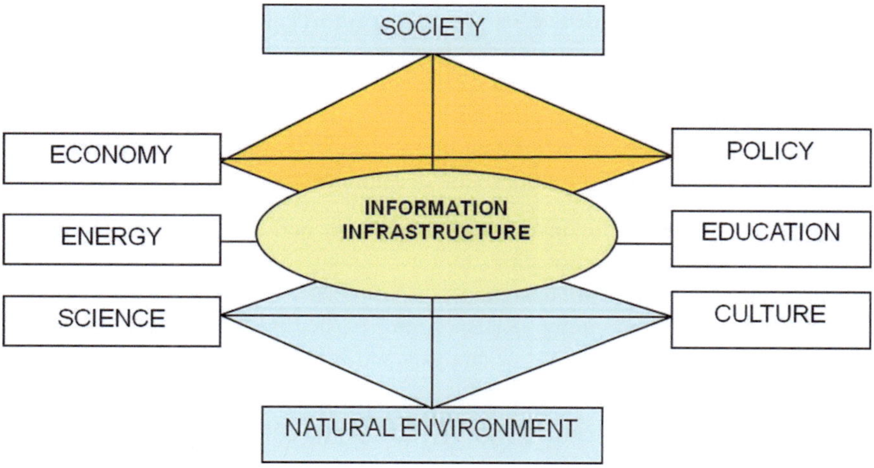

Fig. 13.1 Impact of information infrastructure into separate segments of socioeconomic life. *Source* own

It should be noted that in systems analysis there is a "nonlinear feedback system," which brought a kind of rationality conflict that regards:

1. Time—primary effects are different from late, secondary effects;
2. Spatial and functional—what is good locally, it can be bad globally, conversely.

Globalization in the late twentieth and twenty-first centuries has brought a peculiar coincidence of these various processes, such as (Sienkiewicz and Świeboda 2009):

- economic development of transnational corporations and international investment capital inflows;
- international transfers of technology;
- development of information and communication technologies and the massification of different uses;
- dissemination of effective and standardized production of goods and services;
- development of universal and global availability of electronic media;
- rollout of universal consumer attitudes (also in the sphere of mass culture);
- creation of new sources of danger (globalization of terrorism, cyber-terrorism);
- create new dilemmas of civilization (such as globalism–alterglobalism, freedom–security).

The Evaluation of the Possible Consequences

"Foresight" includes a set of activities "that enable multidimensional determination of future social development directions, based on analysis of current state of science, technology and public awareness as well as their relationships." It is a strategic process designed to (Sienkiewicz and Świeboda 2009):

- organization of public debate on the future (prospective states) and ability to achieve the desired state through the development of science and technology;
- influence on decisions concerning development;
- creation of information resources necessary to develop medium- and long-term vision of development (development trends, investment priorities, etc.);
- gain public acceptance of public social development programs.

Identifying priorities for investment in research and development, support of absorption of technology innovations by the economy, and the reorientation of the country's scientific policy are designed, e.g., to reorient the traditional economy to the knowledge-based economy (KBE). Along particularly distinctive "foresight" of other methods and techniques of analysis and forecasting is the introduction of social dialog in order to turn on specific social groups for discussion and cooperation both with stakeholders such as representatives of science investment (R & D sector) (Sienkiewicz 2009).

"Foresight" can take place at four levels: (1) over-national, (2) national, (3) regional and local levels, and (4) industry, and includes two main stages (Fig. 13.2):

(1) identification of research areas;
(2) determination of the structure of the areas of research.

Methods include following activities:

- expert panel discussions;
- determination of the key technologies for individual panels;
- use of technology scenarios in reports creation.

It is assumed that the introduction of new technology, its modernization, or widening of the scale of appliances of existing technique causes different effects: social, economic, environmental, health, organizational, legal, etc. Therefore, the basic tasks of TA include the following activities:

- predict and determine the systematic identification and analysis of the consequences (impacts) for the latest technology;
- establish policy options to minimize the social impact of technology and comparative analysis of policy alternatives allowed (strategy);
- ensure the efficient implementation of the selected policy (strategy) (Świeboda 2009).

The general methodology of TA includes detailed analysis such as feasibility studies, market analyses, laboratory tests, analysis of the efficiency (cost-benefit, cost-effectiveness), the analysis of the impacts (environmental impact, economic impact), risk analysis, and assessment social acceptability of risk.

TA general methodology consists of the following basic steps (Fig. 13.3):

(1) problem identification,
(2) technology description,
(3) technology forecast,

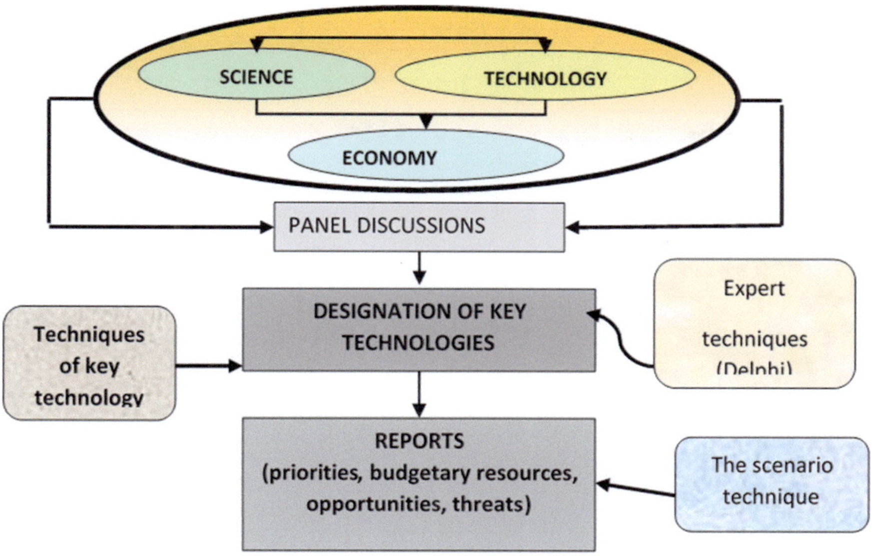

Source: own.

Fig. 13.2 Methods of the "foresight" process. *Source* own

(4) social context description,
(5) social forecast,
(6) impact identification,
(7) impact analysis,
(8) impact evaluation,
(9) analysis of policy and strategic options;
(10) communication of the results.

Methodological Aspects

Suppose there is a set of acceptable technologies, each of which is characterized by a set of parameters (attributes) on the basis of which it defines a set of criteria for assessing the effectiveness (utility) formulated for the purposes of systems analysis. The problem consists in determining a strategy that is optimal for the matter accepted criteria (e.g., Pareto), i.e., that there is no strategy "better" among analyzed (Zacher 1981).

A strategy to be implemented in a specific social environment can have both desirable effects (positive and negative) and side effects. Creation of event scenarios resulting from the strategy and forecasts the development of technology leads to identification analysis of the impact (e.g., using expert techniques of "brainstorming" or Delphi, and especially computer simulation). Following step

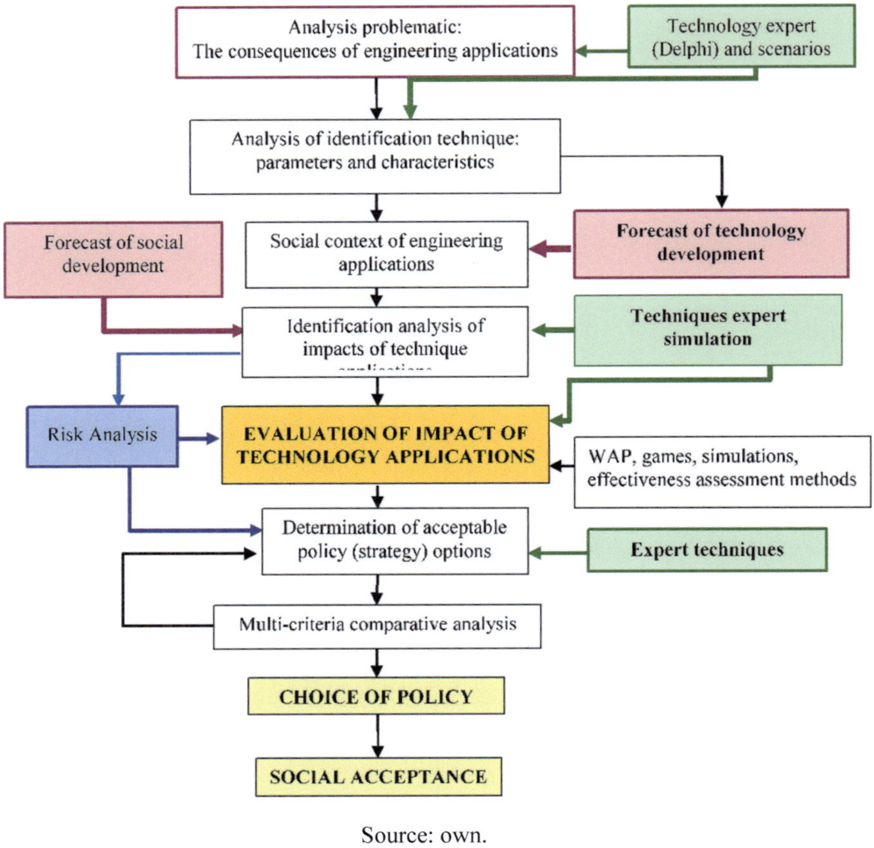

Fig. 13.3 Methods of evaluation techniques. *Source* own

is to conduct an assessment of particular impacts with attribution of numerical values (expressing natural measures or "points"—written by the experts). The final stage of systems analysis applications (development) of a particular technology is to analyze the positive and negative effects and evaluation of its effectiveness (approval or disapproval). The assessment provides grounds of decisions such as "implement or withdraw," "upgrade or replace" ("adopt alternative technology,").

(1) The identification strategy due to the level of modern science and technology (KNOWLEDGE) and scale applications (SCALE).

Different levels of technological development ("retarded" or underdeveloped technology, developed technology, highly developed technology, advanced technology).

(2) Identification of the social environment due to the rate of economic development (quality of life) and the index of social, educational,

cultural, etc. determining the capacity of absorption and assimilation of technology, the "range of influence" (range) for the basic types of environment (underdeveloped, before to modernize, modernizing is developed, highly developed).

It is assumed that the type of technology can produce different effects depending on the type of environment in which it is used (implemented).

(3) A preliminary evaluation of the development of technology in the social environment: identify opportunities and threats to development, in particular risk assessment (as a function of opportunity and risk assessments) uses technologies such as above and a social environment (made by a team of experts using the technique of "brainstorming" or Delphi, or in the process of foresight).

(4) Identification of the social impact of technology applications:

All the possible and probable social impacts caused by the use (implementation) of the technology in a particular social environment are divided into following types (Zacher 1990):

- Key (K)—that arise from forecast and basic functions of technology (i.e., the effects of unintended use and unforeseen).
- Closer in time (B) and further (F), which may show up in the long term.

Each effect can be assessed by means of two values (e.g., utilities) benefit (B) and losses (L) or the difference between these values.

The final assessment of technology in the social environment is done with the utility of technology used in the form of, for example, composite indicator.

References (all in Polish)

Sienkiewicz P (1985) Analysis system. Bellona, Warszawa

Sienkiewicz P (1983) Systems engineering. Warszawa, MON

Sienkiewicz P, Świeboda H (2008) Models for evaluation of quality of life in a research system. BOS'08, IBS PAN Warszawa

Sienkiewicz P, Świeboda H (2009) Systems analysis of the development of the information society: visions and scenarios, opportunities and threats. ZN USz, Szczecin

Sienkiewicz P (2009) Evaluation of the technological development of the global information society, Transformacje, nr 58, Warszawa

Świeboda H (2009) The perception and evaluation of the effects of the development of the information society (in empirical research). Transformacje, nr 58, Warszawa

Zacher L (1981) Technical rationality of human actions. Studia filozoficzne, nr 7–8

Zacher L (1990) Methodology of evaluation techniques (in terms of public acceptance of nuclear energy), Warszawa

Chapter 14
Technology Assessment and Policy Advice in the Field of Sustainable Development

Armin Grunwald

Abstract Sustainable development as a societal vision meets with correspondingly broad approval across all societal groups and political positions, nationally and internationally. The number of nations which have signed and ratified the documents of Rio 1992 and the corresponding follow-up papers and the numerous local or regional activities are impressive. However, the pathway to a more sustainable society needs high effort. Scientific analysis and policy advice is required for monitoring and assessing developments and trends relevant to sustainability as well as for designing political instruments, measures and strategies for sustainability governance. In particular, strategies have to be developed to deal constructively with the enormous uncertainties involved. Therefore, sustainability has become a major issue in technology assessment giving scientific advice to political bodies. In this Chapter I will briefly describe the main motivations and origins of technology assessment and its relations with sustainable development as well. As an institutional case of policy advice, the Office of Technology Assessment at the German Bundestag will be introduced, followed by the presentation of some of its reports on different issues such as energy, tourism, and access to information.

Keywords Policy advice · Technology assessment · Parliament · Uncertainty

---

A. Grunwald (✉)
Institute for Technology Assessment and Systems Analysis,
Karlsruhe Institute of Technology, Karlsruhe, Germany
e-mail: armin.grunwald@kit.edu

A. Grunwald
Office of Technology Assessment at the German Bundestag, Berlin, Germany

L.W. Zacher (ed.), *Technology, Society and Sustainability*,
DOI 10.1007/978-3-319-47164-8_14

Introduction

Sustainability as a societal vision is—at least on the political-programmatic level—not only potentially acceptable, but does, in fact, meet with correspondingly broad approval across all societal groups and political positions, nationally and internationally (Grunwald and Kopfmüller 2012). The number of nations which have signed and ratified the documents of Rio 1992 and the corresponding follow-up papers and the numerous local or regional activities are impressive.

In monitoring and assessing developments and trends relevant to sustainability as well as in designing political instruments, measures and strategies for sustainability governance (Voss et al. 2006), scientific analysis and advice is required to a high degree, in particular for developing adequate strategies to deal constructively with the enormous uncertainties involved (Grunwald 2008). Therefore, sustainability has become a major issue also in technology assessment (TA) giving scientific advice to political bodies (Grunwald 2009).

In this paper I will briefly describe the main motivations and origins of technology assessment (Sect. "Motivations of Technology Assessment") and its relations with sustainable development as well (Sect. "Technology Assessment for Sustainable Development"). As an institutional case of policy advice by TA, the Office of Technology Assessment at the German Bundestag (TAB, Sect. "The Office of Technology Assessment at the German Bundestag") will be introduced, followed by the presentation of some TAB reports on different issues such as energy, tourism, and access to information related with sustainability (Sect. "Sustainability Issues at TAB: Some Examples") and a cross-cutting project on the relation between parliaments and sustainable development (Sect. "Sustainable Development at Parliaments: A New TAB Project").[1]

Motivations of Technology Assessment

In the twentieth century, the importance of science and technology in almost all areas of society (touching on economic growth, health, the army, etc.) has grown dramatically. Concomitant with this increased significance, the consequences of science and technology for society and the environment have become increasingly serious. Decisions concerning the pursual or abandonment of various technological paths, regulations and innovation programs, new development plans, or the phasing-out of lines of technology often have far-reaching consequences for further development. They can influence competition in relation to economies or careers, trigger or change the direction of flows of raw materials and waste, influence power supplies and long-term security, create acceptance problems, fuel technological conflict, challenge value systems and even affect human nature.

[1]This paper provides a summary of several earlier papers of the author and brings together work from the fields of technology assessment, policy advice, and sustainable development.

Since the 1960s also adverse effects of scientific and technical innovations became obvious some of them were of dramatic proportions: accidents in technical facilities (Chernobyl, Bhopal, Fukushima), threats to the natural environment (air and water pollution, ozone holes, climate change), negative health effects as in the asbestos case, social and cultural side effects (e.g., labor market problems caused by productivity gains) and the intentional abuse of technology (the attacks on the World Trade Centre). The experience with such unexpected and serious impacts of technology is central to TA's motivation (Bechmann et al. 2007; Grunwald 2009). Indeed, in many cases, it would have been desirable to have been warned about the disasters in advance, either to prevent them, or to be in a position to undertake compensatory measures. This explains why the methodologically quite problematic term "early warning" with regard to technological impacts (Bechmann 1994) has always had a prominent place in TA discussions from the very beginning (Paschen and Petermann 1992:26).

Early warning is a necessary precondition to make societal and political *precautionary action* possible: how can a society which places its hopes and trust in innovation and progress, and must continue to do so in the future, protect itself from undesirable, possibly disastrous side effects, and how can it preventatively act to cope with possible future adverse effects? Classic problems of this type are, for example, the use and release of new chemicals—the catastrophic history of asbestos use being a good example (Gee and Greenberg 2002)—and dealing with artificial or technically modified organisms (for further examples, cf. Harremoes et al. 2002). In order to be able to cope rationally with these situations of little or no certain knowledge of the effects of the use of technology, prospective precautionary research and corresponding procedures for societal risk management are required, for instance by implementing the precautionary principle (von Schomberg 2005).

Parallel to these developments, broad segments of Western society were deeply unsettled by the "Limits of Growth" (Club of Rome) in the 1970s which, for the first time, addressed the grave environmental problems perceived as a side effect of technology and technicisation. The optimistic pro-progress assumption that whatever was scientifically and technically new would definitely benefit the individual and society was questioned. As of the 1960s deepened insight into technological ambivalence led to a crisis of orientation in the way society dealt with science and technology. Without this (persistent!) crisis TA would presumably never have developed.

Technology assessment (TA) has been developed since the 1960s as an approach first to explore possible unintended and negative side effects of technology, to elaborate strategies for dealing with them and to provide policy advice (early warning, see Sect. "Technology Assessment for Sustainable Development"). From the 1980s on the idea of *shaping technology* by early reflection on possible later impacts and consequences of technology was postulated (Bijker and Law 1994). The adaptation of this social constructivist programme to TA was done within the approach of Constructive Technology Assessment (CTA, cp. Rip et al. 1995). Parallel to this development in the field of TA, the Leitbild of sustainable development became a major issue in public debate and scientific research. Against this background, it is not surprising that TA took up the challenge to start thinking about shaping technology in accordance with sustainability principles (Weaver et al. 2000). Terms such as "transition management" (Kemp et al. 1998) and "reflexive governance" (Voss et al. 2006) were coined in order to demarcate the need for and approaches to embed sustainability assessments into the consideration of the governance of transformation processes toward sustainable development.

New and additional motivations entered the field of TA over the past decades, leading more and more to a shift toward "shaping technology" according to social values (and therefore, building a bridge to the idea of design for value):

- issues of democracy and technocracy, or of democratizing technology (von Schomberg 1999): from the 1960s on there are concerns that the scientific and technological advance could threaten the functioning of democracy because only few experts were capable of really understanding the complex technologies. One of the many origins of TA is to counteract and to enable and empower society to take active roles in democratic deliberation.
- the experience of technology conflicts and of legitimacy deficits and little acceptance of some decisions on technology motivated TA to think about a more socially compatible technology. The very idea was to design technology according to social values.
- in the past decade the innovation problems of Western societies influenced also motivations and driving forces of TA. TA was considered part of regional and national innovation systems (Smits and den Hertog 2007) and expected to contribute to "responsible innovation" (Siune et al. 2009) by taking into account ethical aspects.
- techno-visionary sciences such as nanotechnology, converging technologies and synthetic biology entered the arena. Visions and metaphors mark the expected revolutionary advance of science in general and became an important factor in societal debates (Grunwald 2007).

Compared to the initial phase of TA a considerable increase of its diversity and complexity can be observed. In modern TA, it is often not only a question of the consequences of individual technologies, products, or plants, but frequently of complex conflict situations between enabling technologies, innovation potentials, fears and concerns, patterns of production and consumption, lifestyle and culture, and political and strategic decisions.

Technology Assessment for Sustainable Development

Several experiences have already been made in applying ideas of TA to shaping technology for sustainable development. There many different opportunities, contexts, situations, stage of development of the respective technologies but also different challenges, obstacles and difficulties. Main areas have been the fields of sustainable energy supply technologies, waste disposal, environmental technologies, mobility and transport, and also the exploration of sustainability potentials of new technologies such as nanotechnology and synthetic biology. Recently, an integrative framework for sustainability assessments of technology was proposed (Grunwald 2012).

Technology and Sustainability: An Ambivalent Relation

Generally, a deep-ranging *ambivalence* of the roles of technology in regard to sustainable development can be observed. The relation between technology and sustainable development is usually discussed under contrary aspects: On the one hand, technology is regarded as a *problem* for sustainability and as cause of numerous problems of sustainability, but, on the other hand, it is also and directly considered as a *solution* or at least one aspect of the solution of sustainability problems. This ambiguity is the reason for classifying the relation between technology and sustainability as *ambivalent* (Fleischer and Grunwald 2002).

On the one hand, the use of technologies in modern society has numerous impacts and consequences which conflict with sustainability requirements. This applies for ecological impacts, especially problems with emissions which are harmful for the environment or health and the rapid exploitation of renewable and non-renewable resources. Also in view of social aspects, the technological progress causes sustainability problems, such as the consequence of the technical rationalization for the labor market. At the same time, the *distribution* of both the possibilities and risks of modern technology often objects the claim for justice of sustainability—for example: industrialized countries are often the beneficiaries of technological innovations, while developing countries are primarily affected by the disadvantages. The "digital divide", describing not only the unequal opportunities to use the Internet in industrialized compared to developing countries but also within industrialized countries is an often quoted example (Grunwald et al. 2006).

On the other hand, there are also many impacts and consequences of technological progress which are *positive* in the sense of sustainability. Well-known examples are the prosperity which has been achieved in many parts of the world and the consequential security of livelihood and quality of life, the successful control of many diseases which were disastrous in former times, food security in many (not all!) parts of the world, and the possibility of global information and communication through the Internet. *Innovative* technologies play a key role in the so-called efficiency strategies of sustainable development (cf. e.g., von Weizsäcker et al. 1995). To some extent, modern technologies can already replace conventional technologies and thus contribute to more sustainability (e.g., by fewer emissions and reduced consumption of resources).

This ambivalent relation between technology and sustainability is the starting point for approaches for shaping technology and its societal ways of use (Weaver et al. 2000). These approaches shall be used to realize the positive sustainability effects of innovative technology and minimize or avoid the negative ones in order to contribute through technological progress to a sustainable development in an optimal way. The resulting question is not whether technological progress has positive or negatives effects on sustainability, but how scientific-technological progress and the use of its results has to be designed to achieve positive contributions to a sustainable development. The questions which have to be analyzed in this context include (following Grunwald 2012):

- How and to which extent can research, development, and use of new technologies contribute to sustainability? How do technology's contributions to sustainability influence other contributions (e.g., of changing lifestyles and a "sustainable consumption")? Within which period of time can the impacts relevant for sustainability be expected?
- Which societal framework conditions can serve as incentive for the development, production, and market integration of innovative technology as a contribution to more sustainability? Which political instruments can support this?
- Which methods can be used to assess whether and to which extent the use of technology can result in more or less sustainability? Which sustainability criteria can be the basis for these assessments and how are they justified? Where are methodological new or further developments necessary, e.g., in life cycle analysis?
- Which standards of comparison, weighing principles, and criteria for consideration can be used in situations of contrary effects and conflicts of aims concerning sustainability?
- How reliable or arguable are sustainability assessments of technology? How should be dealt with the unavoidable uncertainty and ambivalence concerning the knowledge on impacts and assessment problems?

The structure of these questions is very similar to that of the types of tasks of technology assessment (TA) (Grunwald 2009). In the end it is about *prospectively* understanding and assessing technology impacts relevant for sustainability—preferably already during the *development* of a technology. The principle of considering such knowledge on presumable or probable technology impacts already in the early stages of decision-making and making it thus usable for the design of technology itself or its societal "embedment" is part of the basic concept of TA. Therefore, the experience of TA can be used to answer the above-mentioned questions of a prospective sustainability assessment of technology (Fleischer and Grunwald 2002).

Transformation of Infrastructures

The transformation of large infrastructures (such as energy supply, water supply, information and communication, and transport) toward more sustainable structures has to be a *system transformation* where singular technologies are only parts of the game but where social issues, acceptance, user behavior, governance, power and control are main elements.

For example, the transformation of the energy infrastructure in conjunction with principles of sustainable development is a considerable challenge. Industrialized countries such as Germany have achieved high standards of energy supply. Energy in the form of electricity, gas, or fuel is reliable and has been more or less available to industrial and private consumers without restriction for decades. Changes in these framework conditions can easily lead to societal controversies. Therefore, transformation processes must always take into account the willingness of customers and users to support these changes and implement behavioral adaptations where required. Sustainability assessments of technology therefore must include the 'social side' of the technologies. The energy supply infrastructure is a *socio-technical system*. It can only fulfill its function if supply and demand are balanced, and if the required changes can be integrated into the existing routines of functioning societal processes, or if new routines can be easily established. Therefore, not only is technical competence necessary for the analysis and design of future (sustainable) energy infrastructures, but so are insights into organizational and societal circumstances such as political-legal framework conditions, economic boundary conditions, individual and social behavior patterns, ethical assessment criteria, and acceptance patterns. In addition, other infrastructures must be co-considered with the energy system: in particular the transport infrastructure (through the development toward e-mobility) and the information infrastructures. The interplay between technical potential, complex social usage patterns, and connected regulation and control processes requires a holistic investigation and an interdisciplinary assessment of the transformation-and-governance strategies aiming at sustainable development. Sustainability assessment of new energy infrastructure elements, therefore, must not be restricted to exploring the supply side and to the provision of technical artifacts (machines, power stations, pipelines etc.). Instead, sustainability assessments must consider also *the societal demand and user side*. Research must bridge disciplinary boundaries between the natural, technical, and social sciences and link technical developments to context conditions of markets, organizational strategies and individual behavior. Multiple interfaces between technical, environmental and social issues have to be taken into account to arrive at a comprehensive and transformational knowledge (www.energy-trans.de).

New and Emerging Science and Technology (NEST)

New and emerging science and technologies (NEST) such as nanotechnology, micro-systems technologies, converging technologies and synthetic biology are *enabling technologies*: they can lead to a lot of applications, even revolutionary developments in many different areas. Therefore, they often show high potentials for supporting strategies of sustainable development—however, most of them are related also with high uncertainties and possible risks. Therefore, there is much more open space for shaping technology compared to the field of infrastructures because of the early stage of development. The main (research) questions for a sustainability assessment of nanotechnologies are, for example (Fleischer and Grunwald 2008): Can nanotechnology development and the application of the resulting products, processes and systems be organized in a sustainable—or, at least, *more* sustainable—manner? How can nanotechnologies and their application paths be shaped in a way that they positively contribute to sustainable development? Are there possibilities of shaping nanotechnologies already in early stages of R&D?

The application of nanotechnology in products and systems is expected to produce a significant relaxation of the burden on the environment: a saving of material resources, a reduction in the mass of by-products that are a burden on the environment, improved efficiency in transforming energy, a reduction in energy consumption, and the removal of pollutants from the environment (Fleischer and Grunwald 2008). A number of studies on precisely the issue of the sustainability of nanotechnology have been published in the meantime (e.g., JCP 2008). However, these developments might have a price. The consequences of the use and release of nanomaterials into the environment are unknown. Although it is not very probable that synthetic nanoparticles in the environment will have long-term effects because of anticipated agglomeration processes, there is no proof available. We do not know about possible long-term effects comparable to the HCFC problem that created the hole in the ozone layer. This situation of high uncertainty and ignorance places a burden of possible risk on future generations while we are exploiting the benefits of nanotechnology today.

Anticipatory assessments of nanotechnology have to cover the entire *life cycle* of the respective technological products or systems. They should include a *temporal integration* and *balancing* of all sustainability effects which might occur during the complete life cycle. For such analyses to contribute to shaping nanotechnologies for more sustainability, they must provide reliable *prospective* life cycle information, such as on health and environmental implications, consumption and production patterns, future developments of lifestyles and markets, and the political and economic framework conditions for the later usage of new technologies. These are only some examples of aspects of the future that need to be known *in advance* in order for reliable life cycle analyses to provide sustainability assessments. A start has already been made toward addressing this challenge for the creation of prospective life cycle assessment. The increasing focus on life cycle assessment as a tool for example in strategy and planning processes, including for long-term issues, and in scenario processes has triggered methodological developments that try combine traditional technology foresight methods with life cycle assessment methods (Schepelmann et al. 2009). Decisive for a comprehensive assessment of nanotechnology or of the corresponding products from a sustainability point of view is that the entire course of the products lifetime is taken into consideration. This extends from the primary storage sites to transportation and the manufacturing processes to the product's use, ending finally with its disposal (Fleischer and Grunwald 2002). In many areas, however, nanotechnology is still in an early phase of development, so that the data about its life cycle that would be needed for life cycle assessment are far from being available. Empirical research on the persistence, long-term behavior, and whereabouts of nanoparticles in the environment as well as on their respective consequences would be necessary to enable us to act responsibly in accordance with criteria of sustainable development.

The Office of Technology Assessment at the German Bundestag

The Office of Technology Assessment at the German *Bundestag* (Büro für Technikfolgen-Abschätzung beim Deutschen Bundestag, TAB) was founded in 1990 and has become a permanent institution of the German legislature.[2] The purpose of the TAB is to provide contributions to the improvement of the legislature's information basis, in particular, of research- and technology-related processes of parliamentary discussion. Among its responsibilities are, above all, drawing up and carrying out TA projects, and—in order to prepare and to supplement them—observing and analyzing important scientific and technical trends, as well as societal developments associated with them (Monitoring). Since 1990, the TAB has been staffed by the Institute for Technology Assessment and Systems Analysis (ITAS) of the Karlsruhe Institute of Technology (KIT; formerly Research Centre Karlsruhe). The 20th anniversary of TAB was celebrated in 2010 in a ceremony conducted by the President of the *Bundestag*.

The TAB is oriented strictly on the German *Bundestag's* and its committees' information requirements. The TAB's principal is the Committee for Education, Research, and Technology Assessment. The choice of subjects for TA projects as well as their delimitation and specification is the *Bundestag's* responsibility. Decisions on the urgency of problems and the scientific advice desired belong on the political agenda. The choice of topics and of problems to be treated is primarily a political responsibility.

The decisions made in the preliminary stages of a TA project have decisive influence on the results obtained and on determining which questions can be answered. With the design of a TA project, it is also determined which aspects of the investigation, which interactions, or which segments of a subject area are relevant for the desired analysis or problem solution, and which are not. Preliminary decisions about the scientific disciplines which are to participate in providing information and on the experts to be consulted also belong in this category. These preliminary decisions are frequently made by TAB, in a close dialogue with the

[2]This section is a modified version of the corresponding section in Grunwald (2003) where TAB's work was analysed with respect to expectations concerning "democratizing expertise".

legislature, in particular with the group of rapporteurs which is in charge of taking care about the many interfaces of the Committee and TAB. In order to support this process, TA investigations are often carried out as so-called pre-studies, the purpose of which—besides an initial appraisal of the state of research—consists in a transparent, comprehensible, and purposeful inquiry into the possible research designs, the formulation of questions, goal directions, etc. The legislature then has the opportunity to decide on a reflective clarification of the subject matter and the design of the respective TA project. Treatment of the topics set in this manner by the legislature is carried out by the TAB in scientific independence and neutrality (Grunwald 2006).

The TAB is a small unit (at present, ten scientists). The various requests and topics are treated by obtaining a number of expert opinions on the respective subject from scientific institutions. The results of this groundwork are evaluated by the TAB-team, are concentrated on the legislature's advisory requirements, and are summarized in the form of a report to the legislature. By means of this networked method of operation, the pertinent competence and knowledge of the science system can be mobilized, case- and subject-specific, for the legislature's purposes in decision-making. The results of TAB studies sometimes lead to *Bundestag* resolutions, and sometimes they have indirect influence on processes of opinion formation and decision-making in the legislature.

The subjects of the TAB's studies stem from all fields of technology. The "classical" TA subjects, such as technology and the environment, energy, and bio- and genetic engineering, predominate. There are also studies on selected fields of science and technology (e.g., on new materials) and on new and emerging technologies such as nanotechnology or synthetic biology. An increasing share of studies on medical technology (Health Care Technology Assessment) and on Information Society issues can be observed in the recent years.

Sustainability Issues at TAB: Some Examples

The leitmotif of Sustainable Development leads to challenges for nearly all fields of policy-making, far beyond being merely a new phrase for environmental protection (Grunwald and Kopfmüller 2012). The following selection of finished or ongoing TAB projects might be a bit arbitrary because sustainability is an issue in most of the TAB studies.[3] The main criterion of selection was to show the thematic diversity of sustainability research at TAB.[4]

---

[3]This section and the following one are based on the corresponding sections in Grunwald (2011).

[4]For more information see the homepage of TAB (www.tab.beim-bundestag.de) which includes English versions of the Work Programme and of the summaries of finished projects. The descriptions given at the following pages are short versions of the information which can be found there. Also contact persons can be identified and approached via TAB's homepage. The full text of finished reports is available for download (German language).

Electric Mobility Concepts and Their Significance for the Economy, Society and the Environment (Ongoing)

Over the last few years, electric mobility has once again been the subject of lively discussions due to new, more powerful rechargeable batteries and the high volatility on the oil markets. Political objectives such as reducing the dependency on oil, more efficient energy transformation, significant CO_2 reductions and lowering local emissions from transport are important drivers of electric mobility. Electric mobility is therefore a prominent topic in sustainability discussions on the future of mobility and of the energy infrastructure.

A series of challenges still have to be solved. The lithium-ion battery which is deemed to be the key technology today still has to be substantially improved with regard to costs, energy density, weight, cycle and calendar life, and speed of charging. A charging infrastructure has to be implemented and business models have to be developed. For electric mobility services to be successfully diffused and used, it is imperative that these are aligned to their users and take their needs and preferences into account. Actually, there is high uncertainty concerning most of these challenges.

Policy support can make a decisive contribution to solving these challenges. Electric mobility is already being promoted by the German government in numerous research programmes and pilot projects. However, there has been no comprehensive compilation of the currently available results up to now. The main task of this TAB project will be providing such an overall picture which is essential for a wide-ranging discussion and evaluation of electric mobility and for any decision about political objectives and measures in this context. Conclusions and recommendations will be derived for the further development and promotion of electric mobility in Germany. The following steps are planned:

1. Establish the foundations for a comprehensive evaluation of electric mobility. This starts with an extensive description and analysis of the technologies and applications regarded. The focus here will be on vehicles with traction batteries (plug-in hybrids and battery electric vehicles). Hydrogen-powered fuel cell vehicles are included for comparison. Promising market penetration scenarios and transport concepts for integrating electric mobility into today's transport systems are described and analyzed. The impacts of electric mobility on local authorities and cities are considered as is their role in introducing electric mobility. Finally, energy scenarios for the improved integration of fluctuating renewable energy sources due to the use of electric mobility are described and analyzed.
2. Evaluate comprehensively technologies, applications and development scenarios with regard to sustainability aspects such as greenhouse gases, air pollutants and noise over the life cycle of batteries and vehicles, critical raw materials, opportunities and risks of shifts in the automotive value added chains and their impacts on Germany as an automobile producing country, effects of electric mobility on employment, road users' acceptance of electric mobility as well as expected impacts on their mobility behavior.

3. Provide an overall assessment of electric mobility and its potentials and draw conclusions. Different support strategies and measures are described and critically assessed. Finally, conclusions and recommendations are derived for the further development and support of electric mobility in Germany.

Ecological Farming and Biomass Production (Ongoing)

The national sustainability strategy of the German government includes the aim of increasing the proportion of ecological farming in the next few years to 20 % of productive agricultural land (from 5.4 % in 2008). At the same time, the share of renewable energy in primary energy consumption shall be increased to 10 % and in gross electricity consumption to at least 30 % by the year 2020. Biomass represents the most important renewable energy source, accounting for around two-thirds of the whole.

Provision of bio-energy depends increasingly on the agricultural cultivation of energy crops. This cultivation also covers areas which were previously used for food production or were part of set-aside schemes. At the same time, ecological food production requires a larger land area per unit produced compared to conventional production. These developments indicate increasing competition concerning land use. Analyses carried out by TAB in the context of the most recently concluded project »Opportunities and Challenges Facing New Energy Crops« have shown that the future development of competition for acreage is dependent on a multitude of factors.

The aim of the TAB project is to examine whether ecological farming and biomass production for energy and material purposes can be interconnected more closely in order to satisfy a growing demand simultaneously in both areas or not. If the latter is true, a priority would have to be set either on ecological farming or on biomass production. There will be the following focus points of investigation:

- Competition between the two sustainability aims of ecological farming and energy crop use as part of renewable energy production (in particularly acreage requirements under different conditions).
- Determining reasons and impediments to changing over to ecological farming in the past few years (for instance, influence of the spread of energy crop cultivation)
- Opportunities for integrating bio-energy production and energy crop use in ecological farming and their effect on acreage requirements and ecological system benefits.

Reduction of Land Use Rate: Objectives, Measures, Impacts (Finished in 2005)

The high consumption of land for settlement and transportation is a significant problem on the route to sustainable development. Soil as an environmental medium is a non-reproducible resource which is one of humanity's finite natural assets fulfilling many essential functions. Land cannot, however, actually be »consumed« in the proper sense of the word, but it can be used in such a way that the spectrum of future opportunities for use is significantly restricted. As the supply of land is limited and non-reproducible, the various types of utilization are in competition with each other. As the ecological value of the soil is not generally reflected in the real estate price, land is frequently consumed in the »ecologically wrong place«. A regionally differentiated approach is required, which takes the different quality, carrying capacity and susceptibility of soils into account. The overarching goals of sustainable development have to be the maintenance of the multifunctionality of soils, the protection of open spaces and the pursuit of a land reserves policy, which also opens up the broadest possible range of possibilities for use to future generations.

The final report on the TAB project »Reduction of Land Use Rate—Objectives, Measures, Impacts« is not concerned with an analysis of the usefulness of the »Target-30-ha« of the German sustainability strategy or with alternative types of land use. The report is more concerned with a critical discussion of instruments and measures with regard to their contributions to the achievement of objectives.

Internet and Democracy: Analysis of Network-Based Communication from Cultural Aspects (Finished in 2005)

This TAB study addresses the impacts of Internet communication on democracy and its cultural foundations. The approach to this diverse topic is shaped by the following questions:

- How is the Internet changing the technological possibilities for political information, communication and participation?
- What cultural changes induced by the Internet are evident with consequences for political communication?
- What visions and potential of the Internet for democracy and what related concerns have been and are being discussed in the academic literature, and how should these be evaluated today?
- How are institutions of the executive and parliament viewing and using Internet communication? How far is this contributing to achieving the democratic potential of the Internet?
- How are civil society actors using the Internet for political communication and what change is this leading to in the political public? How true is the idea of the Internet as a new form of political public (public on the net)?
- Where is there political need and opportunity for shaping this, specifically at the German Bundestag?

The leitmotif of sustainability is touched by these questions mainly by equity issues concerning the access to information and communication technologies (the digital divide issue) as well as via the exploration of the potential of the Internet for approaching some of the social dimensions of sustainable development such as enabling far-reaching participation. In this project, there is a direct link to culture, in particular to cultures of political communication and its changes in the Internet age.

The Internet is accordingly investigated as a possible medium for supporting and strengthening democracy. In dealing with skeptical assessments and concerns on the one hand and high-flying visions and expectations on the other hand, the report (Grunwald et al. 2006) seeks realistic answers to the question of the Internet's impact on political communication and democratic culture. The line of argument in the report is:

- start from the normative paradigm of a deliberative democracy which is solidly justified by democratic theory and rooted in political discourse and culture;
- look at the Internet's possible contributions for achieving these normative concepts;
- review the reality of political Internet communication and study if and how far this potential is already realized;
- identify promotional factors and also obstacles which have so far prevented implementation;
- use this as a basis for identifying political opportunities for action. In particular proposals were made toward the implementation of petitions to the *Bundestag* via the Internet which had remarkable influence in the last years. E-petitions were established at the *Bundestag* and opened up a new window for political communication between members of parliament and citizens.

Internet Communication in and with Developing Countries: The Example of Africa and Opportunities for Development Co-Operation (Finished in 2008)

Since the 1990s Africa has moved back up the political agenda—above all as a result of the United Nations Millennium Declaration and African reform and unification aspirations. On the other hand, the use of information and communications technologies to promote social development continues to remain marginalized in the discussion on international development policy. In this field primarily the sustainability dimension of intra-generational justice is addressed by looking at development challenges, opportunities and strategies.

The topicality of the issue is demonstrated by looking at the current programmes and strategies pursued by African states and regional organizations, at the practice and statements made by the institutions of African civil society, as well as at the local, in part extremely rapid development of the Internet and mobile telephony. German development policy and cooperation will have to continue to occupy itself with this issue and to clarify its own strategy simply because of the interest in Africa itself in the implementation of information and communications technologies for the promotion of development.

The TAB report addresses the reality and potential of Internet use in sub-Saharan Africa giving emphasis to the issues of democratic development, government action and civil society; economic development and trade as well as to education, research and technological development. The focus of the investigation is on Internet use which is not viewed as detached from other conventional (e.g., radio and television) and modern (e.g., mobile telephony) information and communications technologies. The analysis is thus also a contribution to the overall discussion on the use of information and communications technologies for development (ICT4D).

Future Trends in Tourism (Finished in 2006)

Demographic, socio-structural and socio-cultural developments have always led to changes in tourist demand and faced service providers in tourism with substantial need to adjust. These constant challenges have expanded and intensified considerably in the first few years of the new millennium. War and tourism, extreme weather, the ongoing internationalization of tourism and the aging of society (increasingly prominent in public awareness) have emphatically demonstrated the latent vulnerability of tourism as a boom industry. The survival of the tourist industry depends, on the one hand, decisively on recognising relevant trends and allowing for them in good time. On the other hand, tourism is related with many sustainability issues and goal conflicts which make this field to a prominent area of study also with regard to sustainable development.

The focus of the TAB project "Future trends in tourism" was addressing the themes "demographic change", "EU expansion" and "security, crises and dangers". The report

- identifies the relevant trends and their implications for tourism in Germany and by Germans, on the basis of a review and an analysis of current socio-demographic data;
- looks at the impacts of the eastward expansion of the EU and considers what trends in vacation traffic can be expected in and from the new EU nations and to and from Germany;
- describes current and future potential dangers to tourism and discusses possibilities for improving information, prevention and crisis management.

These three themes raise numerous important research issues. For example, interdisciplinary approaches could monitor and analyze the demographic shift periodically, taking into account tourist aspects such as the development of travel motivation and behavior in different age groups. In expectation of growth in tourism from the EU expansion, the consequences of this could be explored in dialogue between politics and science, for example if and how framework conditions could be created which increase travel to Germany. Research into risk and consequences should analyze and evaluate future tourism in the shadow of possible structural global peacelessness, the consequences of climatic change and growth in epidemics. Finally, all these trends and challenges should and could be assessed with respect to sustainable development in order to allow designing pathways to establishing a more sustainable culture of tourism.

These examples show the rich diversity of sustainability issues dealt with at the German Bundestag and its Office of Technology Assessment. More or less, the cultural dimension of sustainability is present in most of these studies. However, mostly this is only the case in form of a cultural "background" of sustainability, while other sustainability aspects such as environmental issues, challenges of future responsibility or equity are in the foreground.

Sustainable Development at Parliaments: A New TAB Project

In this cross-cutting project addressing the question in which way and by which means parliaments can (better) tackle challenges of sustainable development, more emphasis will be given to the relations between sustainability and culture, in particular with respect to political and democratic culture.

Following the UN Earth Summit in Rio in 1992, a great variety of activities were initiated in politics and society all over the world. Quickly, it became clear that the political perception and realization of the goal of sustainability pose far-ranging challenges to governments and parliaments, specifically to their established procedures and to their organization based on a division of labor. The long-term trends of social development and the effects of political measures must be identified and taken into account. Political actions must be coordinated among numerous affected departments. A high degree of coordination with special interest groups in society is necessary if long-term goals affecting more than one area of policy are to be achieved. To this end, procedures as well as forms of institutionalization (e.g., interdepartmental programs, creation of specialized scientific institutes, boards of inquiry, expert committees, communication fora between politics and society) have been established by parliaments and governments in many European countries.[5]

The politics of sustainability has provided new momentum for the long-term orientation of politics and the formulation of corresponding goals in all countries. Awareness of long-term and interdepartmental policy formulation has been sharpened and has led to new forms of » governance". A TAB report (TAB working

[5]The European Parliamentary Technology Assessment institutions constitute the EPTA network. See www.eptanetwork.org for information about the European landscape of parliamentary TA.

report no. 86 published in 2003) confirmed that both the initiative and the further development of sustainability policy are essentially carried by the executive. At governmental level, sustainability policy has led to new administrative structures (specific supervisory responsibilities in the Department of the Environment) and to the installation of new institutions (e.g., sustainability councils and task forces in the ministries such as "Green Cabinets"). The role of parliament, in contrast, seems to lie merely in accompanying and supporting the policy of sustainability. Its role as a forum for public consultation and as a body supervising the executive probably, however, does not exhaust the potential of parliament, according to the results of the TAB study mentioned above.

On the other side, the German *Bundestag* has, compared with many parliaments in other countries, made a considerable contribution to national sustainability policy, in particular by creating the Commission of Enquiry on »Protecting People and the Environment« as well as appointing the parliamentary advisory body on sustainable development (see Ulla Burchardt in this Volume). A range of measures can, however, further strengthen its role in German sustainability policy, ranging, for example, from exercising its budgetary powers more strongly to control governmental programmes and draft legislation with regard to their contribution to promoting sustainable development, to regular plenary debates on the status of German sustainability policy and finally to intensive use of the options for monitoring the sustainability of laws and programmes (Grunwald and Kopfmüller 2007).

In this new TAB report, an overall review and some country studies are planned, the latter as examples that analyze in depth how the parliaments in other countries have initiated institutional innovation and learning processes. These should then be examined with regard to whether they can be transferred to the structures and procedures of the German Bundestag. Against this background, some institutional and procedural options to strengthen the role of the German Bundestag will be identified. Three basic focal points of parliamentary work are of particular interest in this context:

Accompanying the work of the government on the topic of sustainability and supervising the work of the government from the perspective of its contribution to sustainable social development;

Participating in a substantive sense in the further development and implementation of sustainability strategies (definition of sustainability goals and measures, proposals for legislation with particular relevance for sustainability policy;

Stimulating and supporting social discussion about sustainable development by handling sustainability issues in a prominent and public manner and by establishing the principles of sustainability, such as by means of participation and communication in the work of parliamentary agencies and committees (councils, hearings, and commissions of enquiry).

In doing this, a considerable contribution to a "culture of sustainability" shall be made at the level of parliaments.

References

Bechmann G (1994) Frühwarnung – die Achillesferse der TA? In: Grunwald A, Sax H (eds) Technikbeurteilung in der Raumfahrt. Anforderungen, Methoden, Wirkungen. Campus, Berlin, pp 88–100

Bechmann G, Decker M, Fiedeler U, Krings B-J (2007) Technology assessment in a complex world. Int J Foresight Innov Policy 3:6–27

Bijker WE, Law J (eds) (1994) Shaping technology/building society. MIT Press, Cambridge

Fleischer T, Grunwald A (2002) Technikgestaltung für mehr Nachhaltigkeit – Anforderungen an die Technikfolgenabschätzung. In: Grunwald A (ed) Technikgestaltung für eine nachhaltige Entwicklung. Von der Konzeption zur Umsetzung. Edition Sigma, Berlin, pp 95–146

Fleischer T, Grunwald A (2008) Making nanotechnology developments sustainable. A role for technology assessment? J Clean Prod 16(2008):889–898

Gee D, Greenberg M (2002) Asbestos: from 'magic' to malevolent mineral. In: Harremoes P, Gee D, MacGarvin M. (eds) The precautionary principle in the 20th century. Late lessons from early warnings

Grunwald A (2003) Technology assessment at the German *Bundestag*: 'expertising' democracy for 'democratising' expertise. Sci Public Policy 30(3):193–198

Grunwald A (2006) Scientific independence as a constitutive part of parliamentary technology assessment. Sci Public Policy 33(2):103–113

Grunwald A (2007) Converging Technologies: visions, increased contingencies of the conditio humana, and search for orientation. Futures 39:380–392

Grunwald A (2008) Working towards sustainable development in the face of uncertainty and incomplete knowledge. J Environ Plan Policy Manag 9(3):245–262

Grunwald A (2009) Technology assessment: concepts and methods. In: Meijers A (ed) Philosophy of technology and engineering sciences, vol 9. North Holland, Amsterdam, pp 1103–1146

Grunwald A (2011) Sustainable development—an issue of the Office of Technology Assessment at the German Bundestag. In: Banse G, Nelson GL, Parodi O (eds) Sustainable development—the cultural perspective. Concepts—aspects—examples. Edition Sigma, Berlin, pp 279–289

Grunwald A (2012) Sustainability assessment of technologies—an integrative approach. In: Ghenai C (ed) Sustainable development—energy, engineering and technologies—manufacturing and environment. Open access: InTech 2012, pp 35–62

Grunwald A, Banse G, Coenen C, Hennen L (2006) Netzöffentlichkeit und digitale Demokratie. Tendenzen politischer Kommunikation im Internet. Edition Sigma, Berlin

Grunwald A, Kopfmüller J (2007) Die Nachhaltigkeitsprüfung: Kernelement einer angemessenen Umsetzung des Nachhaltigkeitsleitbilds in Politik und Recht. Forschungszentrum Karlsruhe FZKA 7349

Grunwald A, Kopfmüller J (2012) Nachhaltigkeit, 2nd edn. Frankfurt, New York

Harremoes P, Gee D, MacGarvin M, Stirling A, Keys J, Wynne B, Guedes Vaz S (eds) (2002) The precautionary principle in the 20th century. Late lessons from early warnings. Sage, Beverly Hills

Journal of Cleaner Production (JCP) (2008) Sustainable nanotechnology development. J Clean Prod 16(special issue):133–184

Kemp R, Shot J, Hoogma R (1998) Regime shifts to sustainability through processes of niche formation: the approach of strategic niche management. Technol Anal Strateg Manag 10:175–195

Paschen H, Petermann T (1992) Technikfolgenabschätzung – ein strategisches Rahmenkonzept für die Analyse und Bewertung von Technikfolgen. In: Petermann T (ed) Technikfolgen-Abschätzung als Technikforschung und Politikberatung. Frankfurt, New York, pp 19–42

Rip A, Misa T, Schot J (eds) (1995) Managing technology in society. Pinter, London

Schepelmann P, Ritthoff M, Jeswani H, Azapagic A, Suomalainen K (2009) Options for deepening and broadening LCA. CALCAS—co-ordination action for innovation in life-cycle analysis for sustainability. European Commission

Siune K, Markus E, Calloni M, Felt U, Gorski A, Grunwald A, Rip A, de Semir V, Wyatt S (2009) Challenging futures of science in society. Report of the MASIS expert group. Brussels, European Commission

Smits R, den Hertog P (2007) TA and the management of innovation in economy and society. Int J Foresight Innov Policy 3:28–52

Stirling, A, Keys, J, Wynne, B, Guedes Vaz S (eds) Sage, Routledge, pp 49–63

von Schomberg R (ed) (1999) Democratizing technology. Theory and practice of a deliberative technology policy. ICHPA, Hengelo

von Schomberg R (2005) The precautionary principle and its normative challenges. In: Fisher E, Jones J, von Schomberg R (eds) The precautionary principle and public policy decision making. Edward Elgar, Cheltenham, Northampton, pp 141–165

von Weizsäcker EU, Lovins AB, Lovins LH (1995) Faktor vier. Doppelter Wohlstand – halbierter Naturverbrauch. Droemer Knaur, München

Voss J-P, Bauknecht D, Kemp R (eds) (2006) Reflexive governance for sustainable development. Springer, Cheltenham

Weaver P, Jansen L, van Grootveld G, van Spiegel E, Vergragt P (2000) Sustainable technology development. Greenleaf, Sheffield

Chapter 15
Cooperation with Middle and Eastern European Countries in the Field of Technology Assessment: Results and Experiences: A Short Overview

Gerhard Banse

Abstract The basis of the following is the cooperation in the field of technology assessment with colleagues and institutions in Middle and Eastern European countries in the last 15 years. The aim of these activities is, on the one hand, to "watch" institutional and content-related activities in the field of interdisciplinary technology and environmental research (monitoring) in order to create starting points for cooperation opportunities. On the other hand, bi- and multilateral activities should be carried out to concentrate and combine different conceptual as well as methodological knowledge in the fields of technology assessment, technology risk assessment and environmental research. Many activities faced (and partly still do) the following difficulty: political power constellations are changing as the basic economic conditions do, potentials of science were reorganised or newly organised just as the administrative bodies on state and regional level. Thus, both political aims and priorities and opportunities for social interference and action changed as well (occasionally very quickly). Consequently, there was often a lack of time and continuity required for consolidation and differentiation processes. However, one aspect becomes clear: with the transformation processes of the last nearly 20 years in these countries the opportunities in the field of interdisciplinary technology and environmental studies have improved on the one hand, but, on the other hand, they have also deteriorated at the same time: improved, since there is a stronger scientific recognition, social need and political desire for such investigations as a means of policy advice and decision preparation than before; deteriorated, since in the individual countries both the overall industrial and financial conditions and the situation on the labour market

G. Banse (✉)
Institute for Technology Assessment and Systems Analysis, Karlsruhe Institute of Technology, Karlsruhe, Germany
e-mail: gerhard.banse@partner.kit.edu

L.W. Zacher (ed.), *Technology, Society and Sustainability*,
DOI 10.1007/978-3-319-47164-8_15

are generally more unfavourable and thus the funds for considerations preparing and accompanying mechanisation projects in the interest of policy advice and social decision preparation are (probably) very limited. The article will give some examples for it.

Keywords Interdisciplinary technology and environmental research · Middle and Eastern European countries · Technology assessment · Transformation

The Background

In the middle of the 90s, we began research cooperation with colleagues especially in the Czech Republic, Hungary, the Russian Federation, Poland and Slovak Republic. The research fields are Technology Assessment (TA), ethics of science and technology, sustainable development and knowledge-based information society (cf., e.g. Banse 1998, 2007, 2009, 2011; Banse et al. 2000).[1]

Important aims of these "go east" activities—mostly by ITAS—are:

- to "watch" institutional and content-related activities in the field of interdisciplinary technology and environmental research (monitoring) in order to create starting points for cooperation opportunities;
- bi- and multilateral activities should be carried out to concentrate and combine different conceptual as well as methodological knowledge in the fields of technology risk assessment and environmental research.

This mainly concerns

- research on theoretical and methodical aspects of technology assessment, risk assessment and environmental research under the influence of ethical interrelations;
- knowledge transfer in the field of education (related to both that of natural, technological and economic scientists and of social and human scientists).

The background is the "non-knowledge" of this situation especially in Germany, but in other Western European countries as well: Middle and Eastern Europe was

[1]In this time there were differences in terminological and conceptual respects. In Germany, as in France or Great Britain, quite heterogeneous scientific and political, methodological and ethical, participative as well as elitist, institutionalised as well as "free" concepts (which all certainly somehow have something in common), are concealed behind the abbreviation "TA"—if it is used at all. On the other hand, there are many and varied designations—in the German language area alone—from "Technology Impact Evaluation" via "Estimation" or "Judgement of Technological Impact" to "Technology Impact Assessment". Can one expect conceptual or terminological "assimilation" from people in the nations of East Central and Eastern Europe, who have not had the chance to gain enough experience with the "discussion culture" prevalent there? For this reason, we find in those countries some activities which in fact are TA, but are not called "TA", as well as some which are called "TA", but are not.

a "terra incognito"! It was an "unknown land"—regarding the knowledge there of interdisciplinary technology and environmental research, the situation of the environment, restructuring in industry and agriculture, transformations in the field of science or approaches in technology policy.

But in the past years a lot has changed: improved situation regarding information, better understanding of the respective problem situation and joint activities. With the transformation processes of the last 15–20 years in these countries, the opportunities in the field of interdisciplinary technology and environmental studies have improved.

Some remarks in this direction to the past and to the present are in order. Starting point for me was a project of the European Academy Bad Neuenahr-Ahrweiler, Germany, entitled "Technology Assessment und Ethics of Science in Central and Eastern European Countries" (Czech Republic, Hungary, Poland, Slovak Republic), 1997/1999. The aim was a state-of-the-art report on the situation in these countries (a short overview of this project in English is given in Banse 2000):

- topics;
- institutions;
- persons;
- activities;
- forms, etc.

Before this project started, there were two interesting activities:

1. In 7–9 October, 1991, in Prague took place the conference "Technology assessment and its role in processes of society transformation in Central and Eastern European countries"—and one of the organisers was ITAS (cf. Proceedings 1992);
2. 1994–1995 was the project "Transformation of the Central and East European Science Systems", lead by Renate Mayntz, Uwe Schimank and Peter Weingart (cf. Mayntz et al. 1995; cf. also Mayntz et al. 1998; Provazník et al. 1996).

The background for both activities was the following situation: With the process of transformation in East Central and East European states in the 1990s, the chances for progress in the fields of TA and the (practical) ethics of science and technology have, on the one hand, improved, because TA and the ethics of science as a means of advising politicians are—scientifically—much more widely recognised than was formerly the case, are supported by society and are politically accepted; on the other hand, worsened, because the basic industrial as well as the financial conditions and the situation on the labour market in the individual nations show in general a negative trend. For this reason, the (financial) means for considerations which could—in the sense of political advice as decision support in social questions—pave the way for mechanisation are probably quite limited (see also the articles on the background theme "Science and Technology amid Change in Eastern Europe" in: Technology 1993).

But these two activities by ITAS and by Renate Mayntz and colleagues were without sustainable effects!

My first steps showed:

1. There were different activities in research and teaching (e.q.: in Poland by Lech W. Zacher and Andrzej Kiepas, in Czech Republic by Ladislav Tondl, in Hungary by Imre Hronszky; cf. Hronszky and Tibor 1994; Kiepas 1987, 1995, 2007; Tondl 1992; Zacher 1984, 1996).
2. Many of these activities have not the name "technology assessment", but "philosophy of technology", "science and technology studies", "problems of scientific-technological revolution" or "technological prognostic".
3. The solutions and activities were mostly in a specific national manner (based on the national culture, history, political traditions, etc.).
4. That means that the "transfer" of solutions, which were established under other national conditions (eq. in the institutional direction) mostly had no chance of realisation (see the "destiny" of TA in Hungary or of the Prague Institute of Advanced Science [PIAS] in the Czech Republic—cf. Pechan 1996; but the PIAS do not exists since many years…). This was influenced by the national strategy of development—in science, technology and education.
5. The conclusion was that there is a pool of different experiences, of specific knowledge and a good basis for a transfer of ideas not only from west to east, and from east to west too. (Good examples are the three German-Polish Conferences on Sustainable Development and some SD-Meetings in Poznán and Rzeszów; cf. Banse et al. 2011; Banse and Kiepas 2005, 2007.[2])

Some Activities

At first, let us name some cooperation partners:

- for Poland: Silesian University Katowice (Institutes for Philosophy and Cultural Science), University for Social Sciences and Management Tychy as well as the Technical Universities Gliwice, Rzeszów and Wrocław;
- for Romania: University "1 Decembrie 1918" Alba Iulia;
- for Russia: Institute for Philosophy of the Russian Academy of Sciences, Lomonossov-University, International Independent University for Ecology and Politology as well as State Technical Baumann University (all in Moscow);
- for the Slovak Republic: Chair for Ethics and Applied Ethics of the Matej Bel-University Banská Bystrica as well as the Technical University Košice;

[2]The presentations showed interesting cultural differences between the Polish and the German participants: more in an environmental direction versus more in a social and institutional direction, more in the form of case studies versus more conceptual-systematic, more national or regional view versus more international or global view. These activities were the starting point of discussions around the "cultural dimension" of SD.

- for the Czech Republic: Centre for Science, Technology and Society Studies at the Institute for Philosophy of the Academy of Sciences of the Czech Republic, Prague, and the University Plzeň;
- for Hungary: Technical and Economic University Budapest.

Our experience with these institutions (and persons): It is not easy to establish a continuous cooperation (there were many activities like a "mayfly" resp. short time cooperation—based on different interests, financial supports and persons, etc.).

Forms or kinds of cooperation are:

- Initiation and implementation of joint (short-term as well as long-term) activities (in particular research projects, workshops and conferences, publications);
- Mutual participation in (national) scientific events;
- Holding lectures (e.g. on technology risk assessment, on the philosophy and ethics of science and technology, on sustainability and socio-scientific environmental questions);
- Organisation and preparation of translations (e.g. into German, Polish, Russian, Slovak, Czech);
- Preparation of stays of guest lecturers (in Germany, Poland, the Slovak and Czech Republic);
- Signing cooperation agreements (with institutions in Poland, Russia, Hungary and the Slovak Republic);
- (Joint) supervision of doctoral candidates (in Poland and Czech Republic), cooperation in scientific councils as well as in appointed commissions;
- Establish joint "real" or "virtual" research and teaching institutions (like International Centre of Sustainable Development and Information Society at Silesian University, Katowice, Poland, or International Research Centre for Social Consequences of Scientific and Technological Development and Innovation at the Lomonossov University, Moscow, Russian Federation).

All this based on different financial support (mostly on support by DAAD [Deutscher Akademischer Austausch Dienst/German Academic Exchange Service], ERASMUS/SOCRATES programme, DFG [Deutsche Forschungsgemeinschaft/German Research Funding organisation]).

Some specific examples show the different kinds of cooperation:

- Project "Electronic Signature. Cultural Frame and Conditions of a Technological Development" (with participants from Belgium, Czech Republic, Germany, Italy and Poland)—2000/2001 (cf. Langenbach and Ulrich 2002);
- "International Network on Cultural Diversity and New Media"[3] (CultMedia) (with partners from Austria, Czech Republic, Germany, Spain, Hungary, Poland, Russian Federation and Slovak Republic)—founded in 2002, Prague

[3]The basic assumption of the cultmedia network is that the interactions between the development of European nations and the use of means of communication that have existed since the invention of the alphabet still play an important role in the present. The introduction of printing

(Czech Republik)—more than 12 workshops, since 2009 as annual conferences (Cottbus, Katowice, Prague, Karlsruhe, Wrocław), since 2004 an own publication series "e-Culture/Cultural Diversity and New Media" with currently 20 volumes, mostly in German (cf., f.i., Banse 2005; Banse and Krebs 2011; Banse and Metzner-Szigeth 2005, 2012);

- "Forum on Sustainable Technological Development in a Globalising World"[4] (with participants of the founder states Germany, Hungary and the USA, and from all over the world, f.i. Australia, Belgium, Brazil, China, Costa Rica, Mexico, Spain, Switzerland and Ukraine)—founded in 2002, Eger (Hungary), since this time 11 workshops alternately in Europe (Hungary or German) and in Florida, USA, and nine book publications (in English; published in Germany, Hungary and the USA) (cf., as an overview, Nelson 2009; cf. also Banse et al. 2005, 2007, 2011);
- Project "Technology Assessment—Methods and Impacts (TAMI)" (with participants from Belgium, Czech Republic, Danmark, Germany, Netherlands, Poland, Switzerland and UK)—2002–2003 (cf. Decker and Ladikas 2004);
- "Network for Sustainability Strategies, Monitoring and Management in Southern Eastern Europe" (NESSEE)—founded in 2006, Alba Iulia (Romania) (with partners from some South-East European countries);

Footnote (continued)
with movable letters, print media, radio and television has led to cultural revolutions that have been associated with radical economic, political and social changes. With the development of digital media and computer-mediated communication, and due to the importance of such media for the transition of the present European society/societies to a "knowledge-based society", these interrelations remain effective in the current day. In connection with the transformation of their cultural basis through widespread use of new media, the operational conditions of societal subsystems change, together with the working methods of organisations and the possibilities for interaction between individuals. CULTMEDIA deduces the concomitant changes in the relationship between sociality and culturality in three fields of research, namely "Privacy and the Public Sphere" (for the socio-political dimension of the subject), "Identity and the Community" (for its social-cultural dimension), and "Knowledge and the Economy" (for its socio-economic dimension). In addition, there is a cross-sectional topic "Security/Insecurity and Trust".

[4]The basic idea is that because we live more and more through technology, a forum should be developed where a heterogeneous group of people, natural scientists, engineers, engineering researchers, social scientists, philosophers, policy analysts, even policy makers, interested public may meet each other to develop a systematic discourse over topics chosen for the workshops to be realised in each second year. It was also a founding idea that the workshops would be held alternately, one time in Europe and one time in the USA, to express the importance of trans-Atlantic participation by the places where the discussion takes place. The aim was the establishing of a continuous "transatlantic" debate in the field of sustainability, especially concerning technology with the main interest of the different patterns of discussion, argumentation, … in this field. To set up a multi- and transdisciplinary forum developing a wide range discourse and special knowledge on the social role of technological development in the age of globalisation, to examine possible actions based on widely shared consensus and discussions among the technologically leading countries and European countries in transition were further aims.

- German-Russian Academy of Sustainable Development (based on some institutions in Karlsruhe and Moscow)—founded in 2002, Karlsruhe (Germany) (f.i. a series of conferences "Sustainable Development and Modern Civilisation", Moscow 2006 ff.);
- Ph.D. thesis: Krzysztof Michalski (Technical University of Rzeszów, Poland): Ethics and Technology Assessment in the Most New German Philosophy of Technology, 2003), and Petr Machleidt (Czech Academy of Sciences, Prague, Czech Republic): Technology Assessment in the Czech Republic and in Germany, 2007;
- Discussions on the relationships of innovation, security, technology, technology assessment and sustainable development to culture in a broad sense, mostly in Poland and the Czech Republic (cf. Banse 2014a, b, 2015; Banse and Belyová 2011; Banse and Hauser 2010, 2011; Banse and Parodi 2011; Parodi et al. 2011).

Some Results

Gerhard Banse, Christian J. Langenbach, Petr Machleid (eds.): Towards the Information Society: The Case of Central and Eastern European Countries. Berlin a. o. 2000

The background of this publication was the final activity of the project "Technology Assessment und Ethics of Science in Central and Eastern European Countries", the workshop "From an Information Society to a Knowledge Society: Democracy—Participation—Technology Assessment", Prague, 3–5 February, 1999 (The international conference "From Information to Knowledge Society: Reloaded. e-Participation—e-Identity—e-Society", Prague, 15–17 June, 2011, went back to this international conference directly, cf. Banse et al. 2013). More than 40 scientists and experts of various institutions from nine European countries—from universities, academic centres and from various governmental and private organisations—participated in the conference. Approximately two thirds of participants came from EU countries and one third from the Central and East European countries.

The conference program was thematically and organisationally divided into three parts:

- more general and conceptual considerations;
- application areas;
- evaluation process.

Objectives of the conference are

- to express and illustrate various viewpoints and attitudes to the problems of the presented subject matter;
- to make use of the feedback effects, i.e. to attain the exchange of experience and information.

The Prague conference provided a fair overview of the status and perspectives of technology assessment in individual countries of Central and East Europe. However, it also showed the complexity of creation of space for TA type activities in individual countries—for independent activity of both experts and the public based on individual responsibility. As the key role of knowledge in the scope of emerging trends towards the information or the so-called knowledge society we consider knowing the limitations of technical and technological applications, possible risks and potential failures that cannot be avoided. More and more important role here is played by the subjective factor, by human being in the full range of capabilities and interests—this is what we call human capital. This also includes the problems of participation and democracy which need to be further discussed with respect to the Internet and safety of information technologies. Searching for adequate relation to technology in the future assumes more comprehensive attitude to the public, which is much more an active bearer of information technologies development than its target.

In principle, the book came from the contributions presented at the Prague Workshop. However, it reflects the ideas and incentives sounded in the discussion. Therefore, the book consists both of the final versions of the presented papers and the new contributions initiated by the Workshop.

Gerhard Banse, Armin Grunwald, Michael Rader (eds.): Innovations for an e-Society. Challenges for Technology Assessment. Berlin 2002

The international congress "Innovations for an e-society. Challenges for technology assessment" took place in Berlin, 17–19 October, 2001, with participants from 24 European and non-European countries, among them from Czech Republic, Poland, Russian Federation and Slovak Republic (cf. Banse et al. 2001).

The congress provided a forum for technology assessment and innovation and technology analysis to scientifically handle and discuss these existing challenges due to currently available or already foreseeable information and communication technology. The plenary sessions and the presentations to working sessions during the congress contributed towards the following goals:

- Investigation of the potential effects and implications of Information- and Communications Technologies in political, economic, societal, cultural and environmental respects.
- Analysis of the institutional assumptions and basic conditions which are necessary or desirable to shape a sustainable and democratic "e-society".
- Proposal and discussion of procedures, through which society's demands for this project can be determined and implemented.

- Improvement in the interface between scientifically fuelled TA on the one side, and the expectations of its "systems of application" (industry, politics) on the other. For this reason, there were discussions on how TA can contribute to shaping products and processes within companies.
- Illumination of the potential for shaping within existing scenarios for further technological development and discussion of concrete options for decision-making and action, for instance, in the fields of e-commerce, e-governance, e-health or data protection and privacy in international comparison.
- Enabling mutual learning processes and contributions to improving mutual understanding beyond cultural differences in handling technology and technisation, especially concerning culturally different concepts of an e-society.

The overall aim was to determine, analyse and propose solutions for the challenges of an e-society which arise at the interfaces between science, technological development, the economy, politics and the general public. The theoretical and empirical analyses presented in the congress were oriented towards constructively shaping the e-society. The various methods of implementing technology assessment in politics and in the economy attempted to identify the preconditions for "best practice" and for "good practice" in advising politics and in industrial technology planning. It was important to consider both types of addressees for technology assessment: the political system—which is responsible for setting the framework for an e-society—and the economic system, in which concrete products, systems and services are developed for this e-society.

Gerhard Banse, Armin Grunwald, Imre Hronszky, Gordon Nelson (eds.): Assessing Societal Implications of Converging Technological Development. Berlin 2007

This book is the result of the third workshop of the Forum on Sustainable Technological Development in a Globalising World, which took place in Budapest, 08–10 December, 2005.

In it are contributions from authors of 11 countries (among them from Germany, Hungary and Ukraine).

Perhaps the most comprehensive challenge in the early twenty-first century is the promise of the so-called converging technological development based on the recent development of nanotechnology and realising more and more on this base some sort of fusion with such basic technologies and sciences as biotech, informatics and cognitive science. The usual emphasis in assessing this process is made with these elements, in abbreviation the NBIC developments. But it is obvious that what is at stake is the social and environmental change that becomes possible through this new technological wave. All the social sciences and economics are needed to be heavily involved in the assessment of this overarching process that not only promises a deeper penetration of all social relations than any earlier technological wave, but as a result of accelerating trends of each of its constituents, it may cause this penetration in an accelerated way in a very short time. All social sciences/humanities have an essential role in this assessment.

Concerning the convergence characteristics typical of technological development today and especially the NBIC convergence, the present book concentrates on the following elements:

1. It goes back and overviews what happened as converging technological developments in ICT, information and communication technology occurred.
2. It brings cases how nanotechnologies were developed and utilised for such important issues as environmental problems.
3. It brings a rather systematic assessment of the utopias and fictions around nanotech and converging technologies.
4. It gives an overview of research in some countries from which we know less in this respect, China, Mexico, the Ukraine.
5. It gives an overview of research supported at NSF (National Science Foundation), discusses the European Committee 6th Framework program, and efforts by UNESCO to focus on the possible social impacts of nanotechnology.
6. Finally, it closes with articles dealing with the ethical problems in the field.

The content includes papers in the following directions:

- Natural Sciences and Medicine on Nanotech-Based Converging Technologies;
- Visions and Utopias;
- Utopian Aspects of the Debate on Converging Technologies;
- Development of Nanotech in Different Countries;
- Social and Ethical Aspects of Converging Technologies.

Gerhard Banse (ed.): Technological and Environmental Policy. Studies in Eastern Europe. Berlin 2007

This publication presents results of interdisciplinary studies in the field of technology, innovation and environment in countries of Central and Eastern Europe with contributions from Czech Republic, Finland, Poland, Romania and Russian Federation.

The necessity for interdisciplinary technology, innovation and environmental studies in these countries also depends on the following "boundary conditions":

(a) there are enormous ecological and economic technology-induced problems and burdens from the contaminated sites resulting from technology utilisation in the fields of energy generation, chemical industry, agriculture as well as transport;
(b) there are a number of decisions to be made regarding technical solutions, which can modify, supplement or substitute the currently used technology or which new solutions are to be developed and utilised;

(c) there is a need for overview and orientation knowledge as a basis for technology-political decisions in politics, economy and science (especially against the background of the restructuring of the whole industrial basis);
(d) it is necessary to sensitise the general public regarding the consequences of technical developments and their utilisation (also against the background of previously refused opportunities of participation and discussion).

The contributions summarised in this book refer to this situation—to a varying extent—and partly give detailed descriptions and mainly offer suggestions for solutions. They are primarily "insiders", who, if they are not actors, they are at least observers, but definitely they are "concerned" by the transformation processes which took or are still taking place.

The content is divided into two parts:

- Technology Assessment and Sustainability;
- Innovation and Economic Transformation.

The book begins with a contribution on the importance of scientific knowledge in and for the modern age. In the texts it becomes obvious that in the individual countries own ways have to be found, independent solutions have to be found and experience must be gained. At the same time this indicates the limits of the transformability of knowledge generated elsewhere—i.e. also under different economic, political, social and cultural conditions—or—in other words—the relationship of general approaches and concrete, specific national conditions.

Closing Remarks

1. In a more conceptual direction: In the 1990s were conceptualisations in a specific national manner (eq. topics, relationship research/teaching interests or scientific/public debates), in the present mostly the same topics or debates like in Western Europe (EU membership!)—examples are the relationship between description and evaluation or the role of norms and values in technology assessment, kinds of participation, … It brings in a broad scientific potential and a broad spectrum of questions, answers and solutions—technology assessment in general, in specific technologies (environmental technologies, sustainable development, new media, …).
2. In a more practical direction: Political power constellations are changing as basic economic conditions change, potentials of science are reorganised or newly organised just as the administrative bodies on state and regional level. Thus, both political aims and priorities and opportunities for social interference and action change as well (occasionally very quickly). Consequently, there was (and is) often a lack of time and continuity required for consolidation and differentiation processes in TA.

Cooperation with countries in Middle and Eastern Europe is a "normal" process today! There is a "pool" for cooperation in TA.

References

Banse G (ed) (1998) Technikfolgenbeurteilung und Wissenschaftsethik in Ländern Ostmitteleuropas, vol 2. Bad Neuenahr-Ahrweiler

Banse G (2000 The genesis of the project "TA-East". In: Banse G, Langenbach CJ, Machleidt P (eds) Towards the information society. The Case of Central and Eastern European Countries. Berlin a.o., pp 9–19

Banse G (ed) (2005) Neue Kultur(en) durch Neue Medien(?). Das Beispiel Internet. Berlin

Banse G (ed) (2007) Technological and environmental policy. Studies in Eastern Europe. Berlin

Banse G (2009) Technology assessment & sustainable development: information society & eastern Europe. In: Nelson G, Hronszky I (eds) Sustainability 2009: the next horizon. Conference proceedings. American Institute of Physics, Melville, pp 83–92

Banse G (2011) Cooperation with middle and eastern European Countries in the field of technology assessment: results and experiences. In Banse G, Grunwald A, Hronszky I, Nelson GL (eds) On prospective technology studies. KIT, Karlsruhe, pp 187–194 (KIT scientific reports 7599)

Banse G (2014a) Nachhaltige Entwicklung und Kultur – Anregungen zur Diskussion. In: Humanities and Social Sciences (Rzeszów), vol XIX/21, no 3 (July–September), pp 9–24

Banse G (2014b) Von der sozialen zur kulturellen Technikbewertung. In: Věda, kultura, veřejnost. Mimořádne číslo k 90. narozeninám prof. Ladislava Tondla. In: Teorie vědy. Časopis pro meziobarová zkoumání vědy, vol 36, pp 133–154 (in German)

Banse G (2015) Technisches und Kulturelles. Historisches und Aktuelles. In: Banse G, Rothkegel A (eds) Neue Medien: Interdependenzen von Technik, Kultur und Kommunikation. Berlin, pp 21–41

Banse G, Belyová L (2011) Safe innovations, innovative safety. In: Loudín J, Hochgerner J (eds) Social and cultural dimensions of innovation in knowledge societies. Prague, pp 67–80

Banse G, Hauser R (2010) Technik und Kultur – ein Überblick. In: Banse G, Grunwald A (eds) Technik und Kultur. Bedingungs- und Beeinflussungsverhältnisse. Karlsruhe, pp 17–40

Banse G, Hauser R (2011) Culture and culturality. approaching a multi-faceted concept. In: Banse G, Nelson GL, Parodi O (eds) Sustainable development—the cultural perspective. Concepts—aspects—examples. Berlin, pp 31–50

Banse G, Kiepas A (eds) (2005) Nachhaltige Entwicklung: Von der wissenschaftlichen Forschung zur politischen Umsetzung. Berlin

Banse G, Kiepas A (eds) (2007) Nachhaltige Entwicklung in Polen und Deutschland. Landwirtschaft—Tourismus—Bildung. Berlin

Banse G, Krebs I (eds) (2011) Kulturelle Diversität und Neue Medien. Entwicklungen—Interdependenzen—Resonanzen. Berlin

Banse G, Metzner-Szigeth A (2005) Veränderungen im Quadrat. Computervermittelte Kommunikation und moderne Gesellschaft. – Überlegungen zum Design des europäischen Forschungs-Netzwerks "Kulturelle Diversität und neue Medien". In: Banse G (Hg.) Neue Kultur(en) durch Neue Medien (?). Das Beispiel Internet. Berlin, pp 17–46

Banse G, Metzner-Szigeth A (2012) Cultural diversity and new media—their interaction as an element of european integration: elaborating a european research network. In: Rothkegel A, Ruda S (eds) Communication on and via technology. de Gruyter Mouton. Berlin, Boston, pp 217–257

Banse G, Parodi O (2011) Kulturowe uwarunkowania rozwoju zrównoważonej gospodarki opartej na wiedzy. In: Poskrobko B (ed) Uwarunkowania rozwoju zrównoważonej gospodarki opartej na wiedzy. Białystok, pp 31–46

Banse G, Langenbach CJ, Machleidt P (eds) (2000) Towards the information society. The case of central and Eastern European Countries. Berlin a.o
Banse G, von Berg I, Rader M (2001) Der Beginn einer neuen Tradition? Der Internationale Kongress "Innovations for an e-Society. Challenges for Technology Assessment" (Tagungsbericht). In: TA-Datenbank-Nachrichten, vol 10, no 4 (December), pp 161–172
Banse G, Hronszky I, Nelson G (eds) (2005) Rationality in an uncertain world. Berlin
Banse G, Grunwald A, Hronszky I, Nelson G (eds) (2007) Assessing societal implications of converging technological development. Berlin
Banse G, Janikowski R, Kiepas A (eds) (2011a) Nachhaltige Entwicklung – transnational. Sichten und Erfahrungen aus Mitteleuropa, Berlin
Banse G, Nelson GL, Parodi O (eds) (2011b) Sustainable development—the cultural perspective. Concepts—aspects—examples. Berlin
Banse G, Hauser R, Machleidt P, Parodi O (eds) (2013) Von der Informations- zur Wissensgesellschaft. e-Society – e-Partizipation – e-Identität. Berlin
Decker M, Ladikas M (eds) (2004) Bridges between science, society and policy. Technology assessment—methods and impacts. Berlin a.o
Hronszky I, Tibor L (eds) (1994) Bevezetés a technológia hatáselemzésbe. Budapest
Kiepas A (1987) Wprowadzenie do filozofii techniki. Katowice 1987
Kiepas A (1995) Ekologiczne wartościowanie techniki. Zarys koncepcji. In: Migula P, Nakonieczny M, Dąbrowska E (eds) Problemy środowiska i jego ochrony, vol 2. Katowice, pp 51–61
Kiepas A (2007) Technology assessment in Poland. Tradition, current initiatives and experiences. In: Banse G (ed) Technological and environmental policy. Studies in Eastern Europe. Berlin, pp 75–82
Langenbach CJ, Ulrich O (eds) (2002) Elektronische Signaturen. Kulturelle Rahmenbedingungen einer technischen Entwicklung. Berlin a.o
Mayntz R, Schimank U, Weingart P (eds) (1995) Transformation mittel- und osteuropäischer Wissenschaftssysteme. Länderberichte. Opladen
Mayntz R, Schimank U, Weingart P (eds) (1998) East European Academies in transition. Dordrecht
Nelson G (2009) Introduction. In: Nelson G, Hronszky I (eds) Sustainability 2009: the next horizon. Melville, pp vii–xiv
Parodi O, Ayestaran I, Banse G (eds) (2011) Sustainable development—relationships to culture, knowledge and ethics. Karlsruhe
Pechan P (1996) PIAS—the Prague Institute of advanced studies. In: TA-Datenbank-Nachrichten, vol 5, no 4 (December), pp 50–51
Proceedings (1992) Institute for Theory and History of Science, Czechoslovak Academy of Sciences; Department for Applied Systems Analysis, Nuclear Research Center (eds) Technology Assessment and Its Role in Processes of Society Transformation in Central and East European Countries. Proceedings of the International Seminar, Prague, 1991. Prag/ Karlsruhe (Institute for Theory and History of Science, Czechoslovak Academy of Sciences/ Department for Applied Systems Analysis, Nuclear Research Center)
Provazník S, Filácek A, Krízová-Frýdová E, Loudín J, Machleidt P (1996) The transformation of the research system in the Czech Republic. In: Science, technology and society, no 2, pp 311–331
Tondl L (1992) Sociální hodnocení techniky. Plzeň
Zacher LW (ed) (1984) Społeczne wartościowanie techniki. Przegląd zagadnień. Wrocław/ Warszawa
Zacher L (1996) Ewaluacja techniki jako element makrozarządzania. In: Transformacje, December, pp 23–28

Part V
Approaches to Sustainability—Some Examples

Chapter 16
Sustainability as Growth

Tsvi Bisk and Piotr Bołtuć

Abstract According to a good old definition, sustainable is *what can be sustained or maintained.* The only way to attain sustainable balance between civilization and nature is through technological development. The term *sustainable development* implies development and therefore growth. We discuss the two kinds of growth—*vampire growth* that sucks human and environmental resources, and *mutually beneficial growth.* The latter involves public and private choice that allows a free market/socially responsible society. The authors design a development paradigm that enhances the natural environment. We specify the *bottlenecks of civilizational development*: food, water, energy (including transportation) and waste disposal. We focus on water conservation and production (e.g., desalination), food production (e.g., vertical farms), energy saving (negawatts) in material science (bucky paper, piezo-crystals) and transportation (hyper-loop, online electric vehicles) to show how those bottlenecks can be tackled and resolved using recent developments in science and technology. Their solution leads to progress on second-order civilizational problems such as global warming. In this scenario the negative environmental footprint decreases when development occurs.

Keywords Bottlenecks in development · Mutually beneficial growth · Hyper-loop · Negawatts · Bucky paper · Environmental blueprint

Let us think of the word *sustainable*—it implies balance, longevity and proportion. The 1828 Webster's dictionary defines sustainable as 'That may be sustained or

T. Bisk (✉)
Center for Strategic Futurist Thinking, Kfar Saba, Israel
e-mail: bisk@futurist-thinking.co.il

P. Bołtuć
Department of Philosophy, University of Illinois at Springfield, Springfield, IL, USA
e-mail: epetebolt@gmail.com

L.W. Zacher (ed.), *Technology, Society and Sustainability*,
DOI 10.1007/978-3-319-47164-8_16

maintained'.[1] Some people view sustainability as the opposition of growth; such definition predominates in the more recent dictionaries. This view implies that the only way to attain balance is by scaling back. But, empirical studies should determine which strategies achieve balance in a given historical, environmental and technological situation. The authors of this paper contend that sustainability today can best be attained by technological growth.

We define two kinds of growth. The first we call 'vampire growth'; growth that sucks human, social and ecological resources out, leaving in its wake depleted natural resources, destroyed ecosystems, dysfunctional societies and impoverished human lives.[2] Some old-style advocates of capitalism would claim that such 'vampire capitalism' does not exist, that in the long run, in a free market economy it all evolves into rational growth.

We disagree! Even if such a vision of the invisible hand would be correct at some very long-term *philosophical* level, in fact there are no societies guided by perfectly competitive markets. There is always some political regulation and redistribution as well as some economy of scale or natural monopoly that limits the number of competitors. Indeed, the very dynamic of corporate capitalism is to crush competition. Every great capitalist from Rockefeller and J.P. Morgan onward strove to destroy competition. The irony is that Laissez Faire is hardly ever practicable; in reality it leads to oligopolies and therefore may even undermine the market economy and create mini-command economies. It can be demonstrated through game theoretical simulations (conducted by Dr. Jason Alexander from the London School of Economics), that unavoidable divergence from perfectly competitive markets prolongs the works of the 'invisible hand' into several generations. This means that to get many environmental or social problems resolved just by free market forces would take several generations, which would be an unacceptable human cost since many people would live during those unnecessarily prolonged 'transitional periods'.

The second kind of growth we define as 'mutually beneficial growth'; growth that produces a dynamic balance, or a positive sum game among various stakeholders, including society and individual human beings as well as the environment and natural resources. This is why *public*, as well as *private*, *choice* is important within a free market capitalist society. The existence of this second alternative, therefore, demonstrates how a free market society, geared toward growth, can be a socially responsible option and in fact, as we argue below, is the most sustainable social alternative.

A similar view is shared by many persons, from Larry Summers, former Secretary of the Treasury, to Presidents Clinton and Obama, all the way to the Pope. Here is an excerpt of a recent Tom Ashbrook radio interview with Summers

---

[1]http://1828.mshaffer.com/d/search/word,sustainable. We want to thank graduate students at the University of Illinois Springfield: Roel Ybarra, Daniel Rush and especially James McGennis for their helpful comments and suggestions on different parts of the paper.

[2]Vampire growth is different than social Darwinism since the point is not its very competitive nature but rather sucking out the resources, even those needed for an enlightened selfish individual or group.

(Larry Summers's Stagnation Warning 2014, http://onpoint.wbur.org/2014/02/06/larry-summers-economy-federal-reserve-stagnation):

> Ashbrook: "Do you see a contradiction between renewing growth and addressing climate change".
>
> Summers: "I actually think that's a very, very dangerous idea. I think it is dangerous because if this was ever accepted people are going to give up on environmental improvement because they are not going to give up on the idea of improving economic standards of living, they are not going to give up on economic growth".
>
> Very consistently as countries get richer their environment improves (…) China's environment will improve more rapidly if they have the economic growth. (…) There are far more trees here in the New England than there was in the 1900s. If you look at smog in American cities it is 90% down. That is a reflection of the fact that we have become a more prosperous society and we can afford environmental investments and we demand environmental investments. Environmental investment and economic growth go together.

Surprisingly, a similar message can be found in the encyclical *Caritas in Veritate*. It rejects, on one hand, attempts 'to entrust the entire process of development to technology' yet 'on the other hand (…) an upsurge of ideologies that deny *in toto* the very value of development, viewing it as radically anti-human and merely a source of degradation.' The encyclical continues 'This leads to a rejection, not only of the distorted and unjust way in which progress is sometimes directed, but also of scientific discoveries themselves, which, if well used, could serve as an opportunity of growth for all. The idea of a world without development indicates a lack of faith in humankind and God. It is therefore a serious mistake to undervalue human capacity to exercise control over the deviations of development or to overlook the fact that human beings are constitutionally oriented toward 'being more'. Idealizing technical progress, or contemplating the utopia of a return to humanity's original natural state, are two contrasting ways of detaching progress from its moral evaluation and hence from our responsibility' (Benedict XVI 2009).

What unites the ideas of Summers, the Vatican and the authors of this article, is clear understanding of two things: 1. that civilizational and economic growth is a necessary aspect of human life; 2. that in order for growth to benefit human beings it needs to be under some level of scrutiny. This second point runs counter to the view of radical free market technocrats. More importantly for the gist of this paper, the first point is incompatible with current main stream understanding of sustainable growth.

Paradigms of Growth and Decay

Modern society, and its entire socioeconomic structure, is based on the idea of progress understood as growth. Our pension systems are a prime example. They rely on the idea of paying it forward (where future generations, assumed to be richer and more numerous, cover the pensions of current payees). National debts are viewed as sustainable because we assume that the economy will grow and such growth will decrease the proportion of national debt to GDP. But the impact of our belief in progress-as-growth is most profound in human psychology and philosophy.

When we feel joy when a human being is born, when we feel that a doctor or a fireman does something good by saving human lives, this is due to the basic belief that human life is a good, in fact that it is *prima facie* good in itself. On the other hand, when we feel sad when a person dies, especially a young person, when we oppose the killing of humans, and especially genocide, it is primarily because we view human lives as *prima facie* something good. Alternative explanations have been developed, for instance that human life should be respected because it involves one's projects in progress, or because we need to respect a social contract for various practical reasons. Yet, as Dancy (1993; also Hooker and Little 2000) rightly argued people disagree about moral justifications, whereas we tend to agree on basic moral intuitions about actions and judgments. (For instance, people would agree that we should not beat somebody up randomly but they would provide very different reasons why). This is due to benevolence toward others and basic moral goodness[3] (Haidt 2001).

If we reject the idea of human life as valuable, and if we follow radical ecology and view human beings as *scum on the face of the earth*, we run counter to the very idea that saving human lives is, in some generic sense, a good thing to do. In a civilization that abandons this idea, there would be no obvious societal praise for saving lives as opposed to killing people. The life of an individual would not be viewed as a good that does not require further explanation. Such faux-Nietzschean re-evaluations of all values would put our civilization and its underpinning of humanism into question.

There is a healthier approach: the approach of the Greeks and the Jews, of the Renaissance and the Enlightenment. In fact, this is an approach shared by Adam Smith and Karl Marx, though rejected by Pol Pot; the idea that human progress is closely linked to civilizational and technological growth. Some claim, as Aquinas did, that God created human persons, and the world, in such a way that there is a natural fit between technological and moral progress. Others, such as major figures of the French Enlightenment, argued that the human mind can solve all problems. Still others believed that we evolved to tackle all those things; in fact, those approaches are surprisingly compatible in practical terms. Even cautious skeptics of progress, such as Edmund Burke, trusted enough in the natural course of events (while distrusting any ideas that would interfere with it) not to try to reverse the natural human desire to thrive; they just wanted to slow it down a bit.

[3]Recent moral psychology discovered that there are five different 'moral domains' or moral senses embedded in various parts of the brain coming from different evolutionary periods: benevolence, justice, the difference between what's clean and unclean (in some more or less metaphoric sense), loyalty, obedience to the authority and liberty. We share most of those domains at least with some animals, and their application tends not to require much intellectual reflection, but it is deeply evolutionarily engrained in what we are.

We have good reasons to believe—epistemic reasons detailed in the sections below and pragmatic ones detailed above—that the paradigm of human progress, broadly understood, is the right one to choose. The pragmatic reasons, including business reasons as well as the most natural human longings known to psychology, are in surprising agreement with the most profound philosophical ones, with the very core of humanism of all stripes.

The epistemic reasons to believe that human beings are not, and should not be viewed as, the scum of the Earth, are twofold: historical and future oriented. In looking back, we perceive many doomsday predictions that seemed reasonable at the time, yet proved to be shortsighted in the course of time. People feared that agglomerations would not be able to develop because of the tremendous accumulation of horse manure in big cities. This was a very reasonable prediction when all transportation was completely dependent on the horse: individuals on horseback, small groups in horse-drawn carriages and masses of people on horse-drawn trams. The natural tendency would have been to limit both transportation and city growth to *sustainable* levels. In fact, it would have been a good short-term strategy, until human ingenuity had found a better way.

What really happened was that the development of cars and trains took care of this problem—just like effective new heating methods took care of the London fog. Limitations and self-restraints tend to be a necessary short-term strategy able to provide short-term fixes but the true solutions, and therefore the main effort, should always be focused on solving the problems that lie behind those difficulties. We call them **bottlenecks of civilizational development**. The rest of this article is devoted to the discussion of such bottlenecks. We also draft rather obvious, yet often disregarded, solutions that go beyond the temporary calls for constraints and restraint.

Faces of Environmentalism

Environmentalists might be divided into three categories:

1. *Human Centered (anthropocentric)*—A point of view which centers on the value of human beings and advocates for a clean environment because it is good for human beings. This is close to our viewpoint, which we call *humanistic environmentalism.*
2. *Nature Centered (biocentric)*—A point of view which centers on the value of non-human species, processes and ecosystems, the proponents of which sometimes denigrate policies based on a primary concern for human beings. We call this *pagan environmentalism*—in extreme versions it becomes a latter-day version of nature worship.
3. There is a third category which we call *Marxist environmentalism*. This is a position held by sullen and morose Marxists who are resentful that human history has turned its back on Marxist historicism but still wish to influence, if

not take control of, economic policy by maintaining that American consumerist capitalism is the root cause of environmental degradation. This of course ignores the environmental desolation left behind by the former Soviet Union—devastation several times greater than the capitalist West without the saving grace of delivering a high standard of living. It also ignores the historical fact that present-day environmentalism is a major social movement derived from western middle class consumerism—the bourgeois realization that we all are 'consumers' of our environment and thus we want that environment to be of the highest quality. It has been bourgeois civil society that has taken the lead in this development—not the so-called working class.

Anthropocentric environmentalism trumps *biocentric* environmentalism at least 90% of the time from a strategic point of view in terms of practical environmental policy making. This is because in democracies, arguments for quality environmental policies such as those pertaining to the health of one's children are most effective. There are exceptional cases such as saving the pandas or European bisons, or some picturesque environments, but the more pragmatic arguments usually win the day.

These categories are further divided into two attitudes to life whether one trusts in the human capacity for technological growth solving problems or one is skeptical.

The *cornucopian attitude* believes in what Ramez Naam calls *The Infinite Resource* (Naam 2013)—of which Ray Kurzweil, wrote 'Naam shows that innovation is the only force equal to the global challenges that face us, and that we can prosper if we harness it.' Julian Simon called human imagination *The Ultimate Resource* (Simon 1998), i.e., the infinite creative ability of the human imagination coupled to the human spirit to solve problems related to scarcity of energy and natural resources, pollution and population.

Or as Henry George wrote in *Progress and Poverty*, 'That the earth could maintain a thousand billions of people as easily as a thousand millions is a necessary deduction from the manifest truths that, at least so far as our agency is concerned, matter is eternal and force must forever continue to act.' The following is a good example of the cornucopian attitude (www.omg-facts.com/Interesting/The-Entire-World-Population-Could-FitIn/55348#11bXouCAQCYQqVpt.99):

> There are 6.8 billion people on Earth…if…everyone on Earth lived on a space with the same population density as New York City; we could fit everyone in about 666,265 square kilometers…every man, woman, and child (on Earth) could fit pretty comfortably within the perimeter of the state of Texas…we (would) have…the entire world except Texas, to farm and ranch for our food supply…we need 350 billion liters of water per day to properly hydrate 6.8 billion people…the Columbia River alone could produce that amount in less than a day…the Columbia River is the U.S.'s fourth largest river…that leaves the rest of the world's water supply open and ready to serve…we're not really overpopulated. We just need to be better at managing our resources.

The *catastrophic attitude* expresses itself in prophecies of doom that see humanity headed for global catastrophe due to climate change, population growth, shortages of water, deforestation, food shortages and the end of oil; all of which are converging to destroy the planet and modern human civilization as we have known it. "We must stop economic growth in order to 'save the planet' and survive as a species! James Lovelock (of Gaia fame) has said, 'The whole idea of sustainable development is wrongheaded. We should be thinking about sustainable retreat.' Environmental activist George Monbiot says that the campaign against climate change '... is a campaign not for abundance but for austerity... not for more freedom but for less... it is a campaign not just against other people, but against ourselves.' We must forgo our growth fetish and addiction to unfettered consumerism and embrace the halcyon simplicity of years gone by. Monbiot says this policy must be achieved by 'political restraint'—i.e., enforced by the police power of the state; in effect criminalizing innovation and creativity (Bisk 2012).

The authors of this essay believe that we are capable of designing a development paradigm that enhances the natural environment. We also think that calls to 'de-develop' will result in a subsistence economy that will actually degrade the environment. Development, unlike stagnation and crisis, can enhance the environment, and it already does so. There are reasons to believe that by the 2099 the planet earth will be capable of sustaining 12 billion people with an American standard of living and a fraction of the negative environmental impact of today.

History matters! It teaches us that prophets of doom—from Malthus to Ehrlich to 'The Club of Rome'—have been proven wrong time after time by that 'infinite resource'—the human mind. This essay challenges the assumptions of the 'catastrophe industry'. We believe that growth in *value* (of GDP) is not synonymous with growth in the *volume* of raw materials—that we can grow value while using less of our natural resources, and that the 'wealth of nations' can increase as the human burden on the carrying capacity of the planet decreases.

Bottlenecks to Environmentally Sound Development

Human beings require three things to survive; two of which we have in common with all other living things: food, water and energy (uniquely human). Modern civilization requires three other things in order to survive and prosper: waste disposal (sewage, garbage, trash and industrial refuse), education and health maintenance.

The four bottlenecks to environmentally sound development are food, water, energy and waste disposal. All are inter-related, and thus solutions must be multi-dimensional and multipurpose. Present methods of production and distribution of the first three are *the* major threat to public health maintenance around the globe. Agriculture generates 20% of anthropogenic greenhouse effects (more than all the vehicles in the world) as well as numerous harmful particulates. Two billion people are without clean drinking water which is the greatest threat to health maintenance in the world. These are the same people without adequate sewage and garbage disposal. So we see that water, food production, waste disposal and health maintenance are inter-related.

Underdevelopment is the main cause for human suffering. Women and children in the underdeveloped portions of the world spend up to 80% of their time gathering wood for energy as well as going to the well for water. This colossal waste of human energy not only denudes the world's forests and jungles; it is one of the major causes of chronic poverty. To the extent that children take part in this, it interferes with their education. To the extent that children are debilitated by intestinal worms because of the poor quality water, we know that this also interferes with their education as well as health and proper development. The cheapest resources are the resources you do not use, water is a prime example (Wenz[4] 2015). If world was as water efficient as Israel, as we show below there would be no water problem. Following are some solutions to the water problem.

Efficiency of Water Use

1. Eliminating Non-Revenue Water (NRW)—NRW is water produced and lost during the water cycle. A large amount of water is lost from leaking water mains, service line pipes and connections, as well as pipe breaks. A considerable amount of water waste can be prevented with leak detection technologies that *already* exist. NRW is a global crisis, with upwards of 60% water loss in some water systems. Efficiency represents a huge new global water resource.
2. Recycling—Israel treats 80% of its sewage producing 400 billion liters a year of added water which is reused as irrigation water for agriculture and public use. The second largest waste water reclamation program in the world is in Spain, where only 12% of the nation's waste is treated. Recycling represents a huge new under-exploited global water resource. The technologies already exist.
3. Efficient agricultural use of water (sophisticated irrigation technology)—the USA uses 80% of its water for agriculture, while Israel only 40%. The waste of irrigation water in the undeveloped part of the world is enormous. Recycling represents a huge new global water resource. Once again, the technologies already exist.
4. Collecting runoff from winter rains in drainage basins and directing them into recharge ponds that can seep or be pumped back into the aquifer. These are ancient methods. Water collected in this way may be marginal compared to the previous three but in many developing countries can have a significant mitigating effect on water shortages (unfortunately, in some places, including some states in the US, collecting such water may still be illegal, see Johnson 2009).

[4]The author views sweet water as the main bottleneck.

Production of New Water

1. Seawater desalination—is still expensive (and energy intensive) at about 70 cents a cubic meter. But recent developments have reduced energy requirements (which account for 40% of the total cost of desalination and which constitute the most polluting input). A new technique being researched at MIT and Johns Hopkins (and elsewhere) revolves around Microbial fuel cell (MFC) technologies which can not only desalinate water while producing *excess* electrical power but can treat waste water while doing it—thus also eliminating the energy inputs for wastewater. A win-win-win trifecta for both the environment and human well-being (Mehanna and Logan 2010; Kim and Logan 2012).
2. Brackish water desalination from large underground deposits is much cheaper than seawater desalination but while marginal compared to other sources can have a significant immediate mitigating effect on water shortages in the desert areas of undeveloped countries.
3. Water from Air. We quote from the *Science Daily*:
   Research scientists at the *Fraunhofer Institute for Interfacial Engineering and Biotechnology IGB* in Stuttgart working in conjunction with their colleagues from the company *Logos Innovationen* have found a way of converting air humidity autonomously into drinkable water. The process is based exclusively on renewable energy sources such as thermal solar collectors and photovoltaic cells, which makes this method completely energy-autonomous (www.sciencedaily.com/releases/2009/06/090605091856.htm).
4. Genetic engineering to enable irrigation from brackish or even seawater. This would solve two problems: food and some of the need for water. One example would be based on putting to further use existing evolutionary adaptation that exists in mangrove trees. There are two strategies of bioengineering salt-tolerant plants: 1. grafting genes from mangrove trees onto salt-sensitive plants (Meera et al. 2013). Engrafting mangrove trees with productive (e.g., fruit producing) plants (Boltuc P, Matt Wheeler M—the University of Illinois team, work in progress.).

Food Production and the Environment

A new paradigm for producing food (the urban vertical farm) is being developed. This is a concept popularized by Prof. Dickson Despommier of Columbia University (Despommier 2011). A 30-story urban vertical farm located on 5 square acres could yield food for 50,000 people. We are talking about high-tech installations that would multiply productivity by a factor of 480: 4 growing seasons, times twice the density of crops, times 2 growing levels on each floor, times 30 floors = 480. This means that five acres of land could produce the equivalent of 2600 acres of conventionally planted and tended crops. Just 160 such buildings occupying only 800 acres could feed the entire city of New York. Given this calculus, an area the size of Denmark could feed the entire human race.

Vertical farms would be self-sustaining. Located contiguous to or inside urban centers, they could also contribute to urban renewal. They would be urban lungs, improving the air quality of cities. They would produce a varied food supply year-round. They would use 90% less water. Since agriculture consumes two-thirds of the water worldwide, mass adoption of this technology would solve humanity's water problem. Food would no longer need to be transported to market; it would be produced at the market and would not require use of petroleum intensive agricultural equipment. This, along with lessened use of pesticides, herbicides and fertilizers, would not only be better for the environment but would eliminate agriculture's dependence on petroleum and significantly reduce petroleum demand. Despite increased efficiencies, direct (energy) and indirect (fertilizers, etc.) energy use represented over 13% of farm expenses in 2005–2008 and have been increasing as the price of oil rises (Sands and Westcott 2011).

Many of the world's damaged ecosystems would be repaired by the consequent abandonment of farmland. A 're-wilding' of our planet would take place. Forests, jungles and savannas would re-conquer nature, increasing habitat and becoming giant CO_2 'sinks', sucking up the excess CO_2 that the industrial revolution has pumped into the atmosphere.

Countries already investing in the adoption of such technology include Abu Dhabi, Saudi Arabia, South Korea and China; countries that are water starved or highly populated.

Energy

1. Negawatts is a term coined by Amory Lovins to compare with megawatts. Megawatts is what we produce, Negawatts is what we save (from negating watts). The energy we do not use is the cheapest and most plentiful; we can—with present economically viable technology—halve the energy consumption of buildings. Since buildings consume 40% of energy in USA and 70% of electricity, this would be an important new source of energy.
2. Energy-autonomous buildings are possible if we implement a combination of household, local and regional geothermal heat pumps, *depolymerization* of garbage and sewage, household solar and wind as well as existing lighting systems that can cut electricity use for same amount of lighting up to 70%.
3. By the end of the century, space/solar energy could become ubiquitous (in space solar is 24/7)—transferred to the planet by laser or microwave it could eliminate all other sources of energy.

Transportation

1. Revolution in material science (Bucky paper) will make everything that moves super strong and super light requiring much less energy to move it. Light and smart materials will obviate the need for energy and smart materials (such as piezo-crystals) will produce energy.
2. Space Solar linked to OLEV (online electric vehicles) magnets under roadbeds will be a cheap and easy to implement as it simply retrofits an existing infrastructure rather than requiring an entirely new infrastructure.
3. The old idea of hyper-loop, renewed by Elon Musk, is transportation by individualized cars electronically guided in groups within a system of pipes, it was first developed in the 1800s and re-conceptualized in the 1960s. The reason why it is slow to develop is that it is viable only in heavily populated areas, and present-day human centers are not sufficiently urbanized. Such transportation would only make sense within and in the areas surrounding Holland; near Tokyo, NYC and maybe a few new metropolises. Not enough market for mass production. In the current, relatively sparsely populated environment, Space Solar linked to OLEV (online electric vehicles) magnets under roadbeds may be a cheaper and easier to implement solution as it just retrofits an existing infrastructure rather than requiring an entirely new.

As we wrap up, it may strike an attentive reader that we do not refer directly to the topic of global warming. While we have no need to pitch in on the debate to what degree human activity contributes to the greenhouse effect, since this topic should be left to the scientists, the points we make show that hothouse gases in the atmosphere would be reduced as civilization progresses. More broadly, development is sustainable as long as environmental blueprint decreases when development occurs.

Conclusion

The current crisis of human civilization is a consequence of many economic and political factors, including the incompetence and dearth of imagination of politicians and intellectuals, and especially of the power of inertia. The 'limits to growth' advocates are partly right: if we continue to run our societies and economies as we have been doing, extrapolation shows that we are headed for environmental and civilizational catastrophe. But there is no reason to continue to run our societies and our economies as we have. Nor is there any reason to deprive ourselves and our progeny of the material comforts that many stages of the industrial revolution have provided us, in the name of 'saving the planet'. Instead we can envision and in fact facilitate, a civilization totally committed to enabling the self-actualization of every single human being on the planet; a civilization that will have banished want and hunger for all time; a civilization dedicated to realizing the potential of the human spirit.

References

Benedict XVI (2009) Caritas in Veritate, Sec. 14, Vatican
Bisk T (2012) No limits to growth. World Future Review, Spring
Dancy J (1993) Moral reasons. Blackwell, Oxford
Despommier D (2011) The vertical farm: feeding the world in the 21st century. Picador
Haidt J (2001) The emotional dog and its rational tail: a social intuitionist approach to moral judgment. Psychol Rev 108(4):814–834
Hooker B, Little MO (2000) Moral particularism. Clarendon Press, Oxford
Johnson K (2009) It's now legal to catch a raindrop in Colorado? New York Times, June 28. http://www.nytimes.com/2009/06/29/us/29rain.html
Kim Y, Logan BE (2012) Microbial desalination cells for energy production and desalination. Desalination. doi:10.1016/j.desal.2012.07.022
Larry Summers's Stagnation Warning (2014) On point with Tom Ashbrook: February 6, 10:00 AM minute 15:30–17:0515:30. http://onpoint.wbur.org/2014/02/06/larry-summers-economy-federal-reserve-stagnation
Meera SP, Sreeshan A, Augustine A (2013) Functional screening and genetic engineering of mangrove salt responsive genes: a review. Bioscholar 2(12) http://link.springer.com/article/10.1007/s00468-014-1034-x
Mehanna M, Logan BE (2010) JHU Global Water Magazine. 8 Nov
Naam R (2013) The infinite resource UPNE (9 April)
Paul VI (1971) Octogesima adveniens. Apostolic letter, Vatican
Sands R, Westcott P (2011) Impacts of higher energy prices on agriculture and rural economies. USDA Economic Research report August
Simon J (1998) The ultimate resource. Princeton University Press, Princeton
Wenz P (2015) Functional inefficiency: the unexpected benefits of wasting time and money. Prometheus Books, Amherst

Chapter 17
What Can We Do Better for Sustainability in an Uncertain Future?

Li Xu and Talia Raphaely

Abstract Sustainability is a significant challenge confronting a changing world. With an increasingly uncertain future ahead for human wellbeing, achieving, social–ecological sustainability is more than just a simple goal. The new imperative for natural resources management sheds is how to avoid the collapse of social–ecological systems as a result of external shocks triggered by climate change and anthropogenic perturbations. Building up resilient social–ecological systems is therefore an urgent issue for sustainability science. Using water resources management as an example, this paper discusses the need to introduce resilience thinking into sustainability science, how such a thinking should be incorporated into sustainability management for adapting to the growing uncertainties, and how social–ecological resilience can be enhanced.

Keywords Sustainability · Social–ecological systems · Resilience thinking · Perturbations, water, management, uncertainty, climate change

Introduction

Sustainability (or sustainable development) was introduced as a concept in the 1980s and directed people's consideration towards environmental health and human development. Although definitions abound, one of the most enduring is "development that meets the needs of the present without compromising the ability of future generations to meet their own needs" (WCED 1987). However, global environmental issues—such as water scarcity, food security, peak oil, climate change and natural catastrophes including increasing instances of unexpected external and internal shocks such as earthquakes, extreme climate events

L. Xu · T. Raphaely (✉)
Curtin University, GPO Box U1987, Perth, WA 6845, Australia
e-mail: T.Raphaely@curtin.edu.au

L.W. Zacher (ed.), *Technology, Society and Sustainability*,
DOI 10.1007/978-3-319-47164-8_17

and tsunamis—have become an inevitable truth and a barrier to achieving the goal of sustainability (Barnosky et al. 2012; UNEP 2012). The realisation of social–ecological sustainability is not a simple aspiration but a huge challenge, the achievement of which is necessary for human well-being in the face of the changing world.

Since sustainability is not a "steady state" or "fixed target", achieving the goal of sustainable development requires continuous adjustments that respond to changing conditions, knowledge and priorities (Dale et al. 2013). Integrated natural resources management needs to find optimal ways for effective actions in order to avoid social–ecological systems collapse directly resulting from external shocks triggered by nature and human-induced perturbations. Building social–ecological resilience, by improving the ability of the system to withstand such shocks without changing its original state or domain of attraction, could enhance the likelihood of successful sustainability in an uncertain future (Walker et al. 2004; Adger et al. 2005; Folke 2006; Xu et al. 2015). This chapter discusses why resilience thinking is needed to address sustainable development and how we should use this thinking to build social–ecological resilience for water resource management in an uncertain future.

Sustainability and Uncertainty in Water Resource Management

Water resource sustainability is "the ability to use water in sufficient quantities and quality from the local to the global scale to meet the needs of humans and ecosystems for the present and the future to sustain life, and to protect humans from the damages brought about by natural and human-caused disasters that affect sustaining life" (Mays 2007, p. 4). Freshwater is essential for survival of the living world (Wetzel 2000; Long et al. 2003) and is a prerequisite for the continuity and advancement of human societies (Postel and Carpenter 1997). However, the growing severity of freshwater scarcity has become an increasing threat with freshwater systems directly impacted, damaged and depleted by human activities and anthropogenic climate change. For example, at the turn of this century, about 80 % (at the time almost 5 billion people) of the world's population lived in areas where either incidental human water security or biodiversity threats exceeded the 75th percentile (Vörösmarty et al. 2010). This situation has worsened although no specific figures are available. In China, two-thirds of this country's 669 cities are facing water shortages and 80 % of lakes are affected by eutrophication (Chinese Academy Science 2007; Liu and Yang 2012). Due to increasing demand, high pollution levels and the resulting decline in freshwater ecosystems (Johnson et al. 2001), limits of water availability and related considerations of water security have become major threats in the 21st century (Bismas 1991; Vörösmarty et al. 2010).

The highly uncertain future of water is caused mainly by human and climate-related impacts and changes. The complexity of social–ecological systems gives rise to a variety of projection variables on climate change. These in turn increase uncertainties regarding impacts and consequences of the interaction between the internal mechanisms of social–ecological processes and the impact of external influences (changes) on these systems. The changing impacts of climate extremes on water systems—including floods, droughts and storms—depend not only on changes in the characteristics of climate-related variables but also on water-relevant non-climatic stressors, management characteristics and adaptive capacity (IPCC 2012). For example, climate change has the potential to impact on river flood characteristics by changing the volume and timing of precipitation or by changing evaporation and hence accumulated soil moisture deficits. However, there is considerable uncertainty in the magnitude, frequency and direction of these changes. For freshwater adaptive management, Folke (2003) advocates a shift in thinking arguing that resilience needs to be strengthened to secure and provide the possibilities for prosperous societal development. Folke reasons that active management should be undertaken to help maintain the essential role of freshwater in dynamic landscapes faced with uncertainty and shock (moving from command-and-control to complex systems thinking).

Resilience Thinking for Sustainability

The concept of resilience for ecosystems and social–ecological systems is one of the declared focussed research areas within the sustainability discourse (Levin et al. 1998). A bibliometric analysis on resilience thinking shows it is a dominant approach within the sustainability paradigm, especially when it comes to climate change adaptation and dealing with human impacts and disturbance issues (Xu and Marinova 2013), and its value has been proven in the past decade. To better integrate the concept of resilience into sustainable management of resources facing an uncertain future, the rest of this chapter seeks to provide a definition of resilience processes, the ways in which resilience can be incorporated into the sustainability discourse using water resource management as an example, and, finally, to outline some of the possible future directions such an approach might take.

Defining Resilience for Water Resource Sustainability

Resilience was introduced by Holling into ecological systems theory in 1973 and is generally defined by the Resilience Alliance as “the capacity of an ecosystem to tolerate disturbance without collapsing into a qualitatively different state that is controlled by a different set of processes” (Resilience Alliance 2012).

Social–ecological resilience is the capacity of the system to absorb regular perturbations or uncertain disturbances from natural hazards—such as floods, typhoons or hurricanes—by retaining their essential functions, structures, processes and feedbacks (Walker and Salt 2006; Adger et al. 2005). Applying resilience thinking to sustainability requires a definition of resilience tailored for the specific system being studied. The first question to answer is "resilience *of what* and *to what*" namely *what* system state is being considered and to *what* disturbances does resilience apply. Also important is defining resilience over *what time period*, *to whom* and *at what scale* (Carpenter et al. 2001). As sustainability encompasses three main pillars (environment, economy and society), there is a need to consider the concept of resilience in these three contexts before defining *of what* and *to what* for water resource sustainability. Ecological (or environmental) resilience describes the ability of an ecosystem to absorb environmental disturbances as well as its capacity for renewal, reorganisation, learning, adaptation and development. It includes the degree to which the system is capable of self-organisation and the degree to which the system can build the ability for learning and adaptation to the external perturbations (Carpenter et al. 2001; Folke et al. 2002). Economic resilience refers to "the ability of the system to withstand either market or environmental shocks without losing the capacity to allocate resources efficiently (the functionality of the market and supporting institutions), or to deliver essential services (the functionality of the production system)" (Perrings 2006, p. 418). Social resilience emphasises the time it takes to recover from stress and, more importantly, the access of a community to critical resources such as water (Langridge et al. 2006), land, finances and human skills.

Three key words can be captured from the definition of resilience above. These are *capacity, disturbance* and *state*. *Capacity* is the ability of a system to absorb external shocks and mainly encompasses renewal, reorganisation, learning and adaptation when coping with disturbance. *Disturbance* is the different sorts of undesirable or unpredictable changes or perturbations to a system caused by nature and human activities and includes natural shocks such as floods, storms, earthquakes and hurricanes and human-induced perturbations such as engineering constructions, timbering, land reclamation and rangeland. *State* is the responses of a system to the disturbance. Resilience requires the system to be able to maintain a desirable state and not change to a qualitatively different state when facing with the disturbances.

Accordingly, resilience of sustainable water resource management can be defined as

> The ability of water resources systems to withstand uncertainty and disturbance without shifting into an undesirable state by maintaining abilities of renewal, reorganization, learning and adaptation, to provide sufficient quantities of good quality water to meet the needs of humans and ecosystems for both current and future generations.

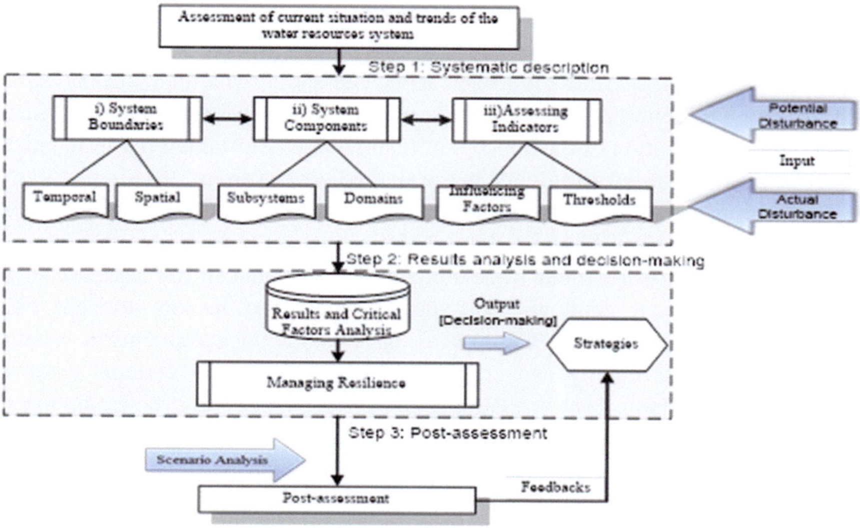

Fig. 17.1 Procedure of incorporating resilience thinking into the sustainable management of water resource systems. *Source*: Own Graphic

Incorporating Resilience into Sustainable Water Management

The most important step for incorporating resilience thinking into sustainable water management is to identify and understand the current circumstances and trends of social–ecological systems. This requires identification and assessment of potential and actual disturbance and external shocks based on their impacts on the sustainability of the specific system being investigated. Once this is accomplished, optimal management strategies can be explored and recommended.

The process of incorporating resilience thinking into the sustainable management of water resource systems is shown in Fig. 17.1.

As Fig. 17.1 shows, the procedure of incorporating resilience into the sustainable management of water resource systems consists of three main steps:

Step 1: *Systematic description*; in this step, three significant interacting characteristics of water resources systems are described from a systematic perspective. One of the primary tasks for resilience analysis is to define resilience over a specific time period, a specific scale and resilience for whom. The critical questions which need to be answered before achieving sustainability are *over what space and time* is sustainability to be achieved? (Bell and Moore 2008, p. 14).

1. The description starts with defining the boundaries of the studied water resource system on a spatial and temporal scale. For example, when assessing the social–ecological sustainability of a lake, the spatial scale can be defined as

the scale of the area which should be assessed in conjunction with the lake, or the areas in which users of the lakes resources live. In addition, the time scale over which the assessment is to be carried out should also be clarified during this step. The definition must be made on the basis of certain specified criteria including a definition of the concept of resilience and sustainability or the average service life of infrastructures in the studied water system.

2. The second part of step 1 is to identify the various components of the system in question. This can be done through defining what subsystems are involved in the specified system and what domains are included in the assessment. In general, ecological, economic and social systems are the key domains used in terms of sustainability. Ecological systems include components relating to environmental quality and ecological health. Economic systems relate to those sectors which have a relationship with production and consumption of the specified resource. Social systems are usually communities and people that have direct interaction with the specified resource.
3. The third and final part of the description is to develop a set of indicators for sustainability within resilience thinking. Two sorts of disturbances need to be considered in this step: actual disturbances (disturbances that have already occurred) such as engineering constructions, and potential disturbances (that may or are likely to occur) such as extreme floods. To do this, the factors influencing sustainability within the system should be identified. That is, what kinds of factors could affect the state of the system and what are the main forces that control these factors? In addition, there is also a need to identify whether there are any tipping points (thresholds), especially critical ones, which determine or could create shifts in the state of system. It is known that ecological systems have tipping points within their components. Whilst the socioeconomic system tipping points, components and causes are more difficult to determine and sometimes less recognised, they are also critically important considerations to consider and include. The likelihood of system transition may gradually increase as the system approaches a tipping point whereupon a minor trigger can invoke "a self-propagating shift to a contrasting state" (Scheffer et al. 2012). Unpredictable external shocks and disturbances increase the possibility of these changes. Certain generic indicators may be useful for identifying the tipping points of a system and detecting if the system is close to the critical tipping point (Scheffer et al. 2012).

Step 2: *Results analysis and decision-making*; this step focuses mainly on managing social–ecological resilience around water resources systems through strategy planning and policy design. Based on the *systematic description* comprising step 1, step 2 focuses on analysing key factors affecting the state of sustainability of the system and critical thresholds that should be considered when confronting external disturbances and strategies for enhancing systems' resilience. Certain planning approaches could be useful in achieving beneficial sustainability outcomes here. For instance, multi-objective planning (MOP) could be one of the options encouraging systematic consideration of multiple objectives including environmental, social, regional, and economic and others (Major 1977). Specific objectives must be defined prior to applying MOP in order to optimise strategies designed for enhancing social–ecological resilience. It is important to recognise any constraints that may create obstacles in achieving the defined objectives. The constraints are identified and determined by disturbance variables, critical influencing factors and tipping points of the system as well as by the conditions of the different components within the overall system.

Step 3: *Post-assessment*; this final step evaluates if the strategies are useful for the enhancement of social–ecological resilience and sustainability of the system. This can be done by observation or simulation. Whilst observation is an effective evaluation option, it is time-consuming and costly due to the lengthy timeframes typically needed in strategy implementation and outcomes. This is particularly true of restoration plans. Consequently a simulation approach, such as a scenarios analysis or computer-based method, is highly recommended as an alternative. This post-assessment is necessary because it can assess the anticipated performance of strategies. It is also a good way to provide feedback to decision-makers for proposed strategy adjustments.

Building Up Social–Ecological Resilience for Water Resources System

Structured scenarios and active adaptive management are two useful tools for building resilience in social–ecological systems. This includes stimulating building resilience in social–ecological systems, creating open institutions for learning and flexible collaboration and directing actions towards building adaptive capacity. Further, motivating the development of indicators and warning signals of gradual change and loss of resilience and thresholds, and encouraging learning and incorporation of ecological knowledge into institutional structures in multi-level governance (Folke et al. 2002; Adger et al. 2005) are important. Social–ecological resilience can also be built up by policy design that strengthens understanding of humanity and nature as interdependent.

Many studies have been conducted regarding building social–ecological resilience, mostly centred on initiatives enhancing collective actions through engaging stakeholders (Folke et al. 2002), co-management (Tompkins and Adger 2004) and legislation (Olsson et al. 2004). Additionally, building indigenous ecological knowledge-based systems, multi-level governance and polycentric institutions are proving to be a helpful means of facilitating institutional and social learning and multi-level governance through education and training (Adger et al. 2005; Silici et al. 2011). However, more research is required to better understand how to manage resilience for sustainability. This includes further exploration of multi-scale effects, further evaluation of environmental, social and economic trade-offs, enhanced monitoring and evaluation strategies and continuing engagement with stakeholders.

Building social–ecological resilience for water resource systems should follow seven general principles described by Folk et al. (2000):

1. Designing management strategies based on traditional local ecological knowledge. Local ecological knowledge may expand sources of information for ecosystem management (Becker and Ostrom 1995). Local water use knowledge, including biological knowledge and knowledge of ecological processes, may complement and enhance scientific knowledge;
2. Designing management systems that "flow with nature";
3. Developing local ecological knowledge for understanding cycles of natural and unpredictable events;
4. Enhancing social mechanisms;
5. Promoting conditions for self-organisation and institutional learning;
6. Rediscovering adaptive management; and
7. Developing values consistent with resilient and sustainable social–ecological systems. Moreover, attempts to build resilience for social–ecological systems should capture and address slow variables that affect resilience rather than trying to control disturbance. This is because change can be inevitable or unobservable but still potent as is the case with climate change, nutrient stocks and soil properties (Folke et al. 2002).

This requires improved understanding of social–ecological systems dynamics and the incorporation of knowledge obtained from and by local users to gain greater insight into how systems respond to potential tipping point shocks (Berkes and Folke 1998; Carpenter et al. 2001; Folke et al. 2002). It also requires efficient management interventions. It is, however, important to realise that management interventions can either build or destroy resilience depending on how the social–ecological system is able to organise itself and respond to management actions. Therefore, ongoing assessment is advised to establish if strategy or intervention adjustments are appropriate.

The Way Forward

Social–ecological sustainability is essential for human well-being in an increasingly uncertain future. The urgent issue for natural resources management is to prevent social–ecological systems from collapsing in the face of external shocks triggered by climate change and anthropogenic perturbations. Establishing and enhancing social–ecological resilience for the sustainable management of natural resources could address this urgent challenge. However, the research on resilience and sustainability is still in the exploratory stage with more attention needed on integrating the abundant ecological evidence with socio-economic aspects and the role of human activities in shaping ecosystems. Future research around resilience and sustainability could focus on questions such as how to identify and manage the key drivers and elements of resilience within social–ecological systems, how to monitor and evaluate whether adopted resilience building strategies are working and how to identify and engage with stakeholders when building social–ecological resilience.

More specifically, it is important to identify and quantify the tipping points (thresholds) for key elements of social–ecological systems and to find the drivers which affect these elements and thereby the state of the system. Appropriate indicators may be useful for achieving this, and they can detect if the system is close to the critical tipping point (Scheffer et al. 2012).

Another important issue for sustainability management into the future concerns how to build flexible institutions with the ability to adjust to changing environmental conditions. This is increasingly significant in a world of growing uncertainty and shock and requires consideration of the dynamics of affected social–ecological systems when considering the sustainable use of water and any other resource. Long-term observation of vulnerable systems needs to be established including frequent monitoring of environmental conditions. Information feedback to institutions should be monitored and assessed. Appropriate indicators (especially early warning indicators) should be developed to ensure the long-term resilience of systems under observation. Perhaps most critically for sustainability, local stakeholders need to be involved in any policy development and management programme to ensure the best result for both the people and the environment.

Acknowledgments The authors would like to thank the reviewers of this article for their helpful comments which helped improve the quality of the manuscript. The authors are also thankful to the book's editors for their invitation and feedback.

References

Adger WN, Hughes TP, Folke C, Carpenter SR, Rockström J (2005) Social ecological resilience to coastal disasters. Science 309:1036–1039

Barnosky AD, Hadley EA, Bascompte J, Berlow EL, Brown JH, Fortelius M, Getz WM, Harte J, Hastngs A, Marquet PA, Martinez ND, Mooers A, Roopnarine P, Vermeij G, Williams JW, Gillespie R, Kitzes J, Marshall C, Matzke N, Mindell DP, Revilla E, Smith AB (2012) Approaching a state shift in Earth's biosphere. Nature 486:52–58

Bell S, Moore S (2008) Sustainability indicators: measuring the immeasurable, 2nd edn. Earthscan, London. ftp://tp.geo.puc.cl/Pub/Apaulsen/TEXTOS/INDICADORES%20DE%20SUSTENTABILIDAD%20BELL.pdf. Accessed 31 Oct 2015

Becker CD, Ostrom E (1995) Human ecology and resource sustainability: the importance of institutional diversity. Annu Rev Ecol Syst 26:113–133

Berkes F, Folke C (eds) (1998) Linking social and ecological systems: management practices and social mechanisms for building resilience. Cambridge University Press, Cambridge

Bismas AK (1991) Water for sustainable development in the 21st century. Int J Water Resour Dev 7(4):219–224

Carpenter SR, Walker BH, Anderies JM, Abel N (2001) From metaphor to measurement: resilience of what to what? Ecosystems 4:765–781

Chinese Academy Science (2007) China sustainable development strategy report 2007—water: governance and innovation. Science Press, Beijing (in Chinese)

Dale VH, Efroymson RA, Kline KL, Langholtz MH, Leiby PN, Oladosu GA, Davis MR, Downing ME, Hilliard MR (2013) Indicators for assessing socioeconomic sustainability of bioenergy systems: a short list of practical measures. Ecol Ind 26:87–102

Folke C (2003) Freshwater for resilience a shift in thinking. Philos Trans R Soc Lond B Biol Sci 358:2027–2036

Folke C (2006) Resilience: the emergence of a perspective for social–ecological systems analyses. Glob Environ Change 16:253–267

Folk C, Berkes F, Colding J (2000) Ecological practices and social mechanisms for building resilience and sustainability. In: Berkes F, Folke C (eds) Linking social and ecological systems: management practices and social mechanisms for building resilience. Cambridge University Press, Cambridge, pp 414–436

Folke C, Carpenter SR, Elmqvist T, Gunderson L, Holling CS, Walker B (2002) Resilience and sustainable development: adapting capacity in a world of transformations. Ambio 31(5):437–440

Intergovernmental Panel on Climate Change (IPCC) (2012) Managing the risks of extreme events and disasters to advance climate change adaptation. IPCC special report on climate change. Cambridge University Press, Cambridge, UK

Johnson N, Revenga C, Echeverria J (2001) Managing water for people and nature. Science 292(5519):1071–1072

Langridge R, Christian-Smith J, Lohse KA (2006) Access and resilience: analysing the construction of social resilience to the threat of water scarcity. Ecol Soc 11(2):18

Levin SA, Barrett S, Aniyar S, Baumol W, Bliss C, Bolin B, Dasgupta P, Ehrlich P, Folke C, Gren I-M, Holling CS, Jansson A, Jansson B-O, Māler K-G, Martin D, Perrings C, Sheshinski E (1998) Resilience in natural and socioeconomic systems. Environ Dev Econ 3(2):221–262

Liu J, Yang W (2012) Water sustainability for China and beyond. Science 337:649–650

Long J, Tecle A, Burnette B (2003) Cultural foundations for ecological restoration on the White Mountain Apache Reservation. Ecol Soc 8(1):4

Major DC (1977) Multiobjective water resource planning. Water resources monograph series, vol 4. American Geophysical Union, Washington, DC

Mays L (2007) Water resources sustainability. McGraw-Hill Education, New York, NY. http://accessengineeringlibrary.com/browse/water-resources-sustainability. Accessed 31 Oct 2015

Olsson P, Folke C, Berkes F (2004) Adaptive comanagement for building resilience in social–ecological systems. Environ Manage 34(1):75–90

Perrings C (2006) Resilience and sustainable development. Environ Dev Econ 11:417–427

Postel S, Carpenter S (1997) Freshwater ecosystem services. In: Daily G (ed) Nature's services: societal dependence on natural ecosystems. Island Press, Washington, DC

Resilience Alliance (2012) Resilience. http://www.resalliance.org/index.php/resilience. Accessed 31 Oct 2015

Scheffer M, Carpenter SR, Lenton TM, Bascompte J, Brock W, Dakos V, van de Koppel J, van de Leemput IA, Levin SA, van Nes EH, Pascual M, Vandermeer J (2012) Anticipating critical transitions. Science 338(6105):344

Silici L, Ndabe P, Friedrich T, Kassam A (2011) Harnessing sustainability, resilience and productivity through conservation agriculture: the case of Iikoti in Lesotho. Int J Agric Sustain 9(1):137–144

Tompkins EL, Adger WN (2004) Does adaptive management of natural resources enhance resilience to climate change? Ecol Soc 9(2):10

United Nations Environment Program (UNEP) (2012) GEO5: global environment outlook. Environment for the future we want. UNEP, Valletta

Vörösmarty CJ, McIntyre PB, Gessner MO, Dudgeon D, Prusevich A, Green P, Glidden S, Bunn SE, Sullivan CA, Liermann CR, Davies PM (2010) Global threats to human water security and river biodiversity. Nature 467:555–561

Walker BH, Salt D (2006) Resilience thinking: sustaining ecosystems and people in a changing world. Island Press, Washington, DC

Walker BH, Holling CS, Carpenter SR, Kinzig A (2004) Resilience, adaptability and transformability in social–ecological systems. Ecol Soc 9(2):5

World Commission on Environment and Development (WCED) (1987) Our common future. Oxford University Press, Oxford

Wetzel RG (2000) Freshwater ecology: changes, requirements and future demands. Limnology 1(1):3–9

Xu L, Marinova D (2013) Resilience thinking: a biliometric analysis of socio-ecological research. Scientometrics 96:911–926

Xu L, Marinova D, Guo X (2015) Resilience thinking: a renewed system approach for sustainability science. Sustain Sci 10(1):123–138

Chapter 18
Economic Prosperity and Sustainability in China: Seeking Wisdom from Confucianism and Taoism

Xiumei Guo, Sandra Krempl and Dora Marinova

Abstract China has enjoyed miraculous economic growth in the past three decades. The country now has the world's second largest economy. However, China's economic expansion has been accompanied with increasing environmental deterioration as the country has become the top global greenhouse gas emitter. There is desperate need for China to adopt a sustainability course, which would allow the right balance between economic prosperity, social justice, and environmental protection. However, is the concept of sustainability new to China? What does the country's rich history and tradition say about sustainability? This chapter explores the wisdom of Confucius and Lao Tzu, respectively, the founders of Confucianism and Taoism, for references to sustainability. It draws on the implications of some of their many teachings to help discuss possible solutions to problems related to climate change. Absorbing nourishment from the Chinese cultural genes of Confucianism and Taoism is important in efforts aimed at educating Chinese citizens on how to improve their environmental awareness. Their relevance is becoming more apparent as both continue to gain influence in China. Confucian and Tao philosophy can provide useful guidelines for achieving sustainability, particularly for changing the course in the country's economic development. The chapter provides a strategic model for the proper interaction and integration of the principles of Confucianism, Taoism, and the modern concept of sustainability. China's rich heritage of Confucian and Tao culture and its recent policy shift towards sustainability will help the nation to alleviate global environment pressure and cooperate with the rest of the world in tackling climate change.

Keywords Sustainability · Confucianism · Taoism · Confucius · Lao Tzu · China · Climate change · Wisdom · Environmental/ecological awareness · Environment

---

X. Guo (✉) · S. Krempl · D. Marinova
Curtin University Sustainability Policy (CUSP) Institute, Curtin University,
GPO Box U1987, Perth, WA 6845, Australia
e-mail: X.Guo@curtin.edu.au

L.W. Zacher (ed.), *Technology, Society and Sustainability*,
DOI 10.1007/978-3-319-47164-8_18

Introduction

As the most populous country in the world, China is enjoying unprecedented economic growth, but this development has caused serious degradation of the country's physical environment. The resulting environmental problems are particularly aggravated by China's greatly expanding urbanization and ageing population. This chapter explores the country's ecological deterioration issues in order to seek a balanced economic developmental strategy model. It examines the influences of population change and economic growth on the environment with the purpose of providing insights and lessons for achieving a better quality of economic growth and sustainability. Changes need to be made to ease the double burden of restoring the deteriorating ecosystems and improve people's quality of life. We refer to Confucian and Taoist teachings to help support the arguments put forward.

Confucius (551–479 BC) was an educator and philosopher who founded and operated a private school. More than 2500 years ago, he advocated the unity of humankind with nature and other ideas to encourage the harmonious relationship between human activities and the natural environment. Nowadays, Confucian ideas are influencing the basics of life in China and other East Asian countries (Yao 2003a). It can be said that Confucius' thought aligns very well with what has become the basic concept of modern sustainability in the West. However, Mao's slogan during the Cultural Revolution "People must be able to conquer nature" has led China to pursue economic prosperity without paying attention to the natural environment. As the most populous country in the world, China contributes greatly to the global environmental problems, particularly as they relate to climate change. Under the influence of global warming and the pressure from international communities, China is desperately seeking solutions.

In 1988, 74 Nobel Prize winners appealed in Paris: "If human beings want to live in peace and prosperity in the 21st century, they must look back 2500 years and seek the wisdom of Confucius" (Pan 2013; Zhang and Li 2000, p. 1). Building current civilization's ecological awareness must be achieved through seeking the wisdom of Confucius and Lao Tzu, the two most famous Chinese philosophers (Kong 2009; Berthrong and Berthrong 2000; Cordier 1912). There were many differences between these two influential Chinese philosophers in terms of both their ways of thought and approach to life. Lao Tzu saw life on earth within its context in the wider cosmos and within the view that there is interdependence between all things animate and inanimate (Cordier 1912; Chiu 2011). While Confucius also spoke of the importance of this interdependence in terms of harmony, he focused on the ethical standards for people to live together (Yao 2003a, b). He introduced a way for people to learn to become more human and developed a code for effective leadership (Littlejohn 2005). Both Lao Tzu and Confucius referred to the *Dao*, which means the truth or the way. Jaspers (1957, p. 59) states "it is often held that Lao Tzu conceived the *Dao* as beyond good and evil, while Confucius moralised the *Dao*". It could be said that Lao Tzu focused on our spiritual well-being and our connectivity with Earth and the universe, while Confucius focused on the ethical governance and management of a nation.

Is the Concept of Sustainability New to China?

Aitken (1908) argued that Confucius grew up a place where the natural environment—the mountains, rivers, land, heavens, sun, moon and stars—was held as sacred. This connectively and high respect for nature would have been the same for Lao Tzu as their lives were almost in the same period. Lao Tzu taught many ways of maintaining the connectivity with nature including bringing animal movements into ritualistic dances. Confucius accepted all that to be natural; even in the appreciation of nature he prescribed an order.

The wisdom that both gave to China and to the world is relevant today, as we struggle to address ways to mitigate global climate change and as we address the mismatch between systems that we need to comply with for economic well-being, what our spirit wants and needs, and how we should live in harmony and interdependence with the living and nonliving things of Earth. We have gone through half a century in the world's development with what seems to be a shift away from the spiritual wisdom of philosophers, such as Confucius and Lao Tzu, towards an economics-based rationale for living. In relation to fostering economic growth, Lebow (1955, p. 3) states that "we convert the buying and use of goods into rituals, that we seek our spiritual satisfactions, our ego satisfactions, in consumption." Spirituality was perceived as the major competing interest to growth, which needed to be overcome. Jackson (2009, p. 39) refers to gross domestic product (GDP) as the means by which nations sum up market exchanges including household, community, business and government expenditure on goods and services. He states that this expenditure in the GDP is taken as a proxy for measuring satisfaction and well-being. Tied in with this, Jackson summarizes Sen's point of view as follows: "The baseline for social functioning is always the current level of commodities. The avoidance of shame—a key feature of social flourishing—will drive material demand forward relentlessly" (2009, p. 147).

The trend of placing economics as the underpinning of national growth started to discard ancient wisdom and successfully replace spirituality with ego-driven consumerism. It seems that the state of the world today has reached a point where a growth-based economy is considered conventional—the norm, and that this is unsustainable (Jackson 2009). How do we bring sustainability back into our thinking? Is there wisdom in ancient philosophy that can guide us?

Confucius saw that customs were key to establishing order in a nation and he himself sought to learn from past wisdom. He said that "a man born in our days who returns to the ways of antiquity is a fool and brings misfortune upon himself" (Jaspers 1957, p. 44). The ways of antiquity were a priority for Confucius; however, his process in finding solutions, according to Jaspers, was to ask:

"What is the old? How can we make it our own? How can we make it a reality?" (1957, p. 43).

Self-modelling what he thought was a starting point for Confucius. He developed what he saw as the essential qualities for a community and national leader and also the qualities for everyday people to transform themselves to be the best they could be. His teachings did not take anything for granted and included details of how to walk, greet, behave in different situations, how to bury the dead, how to administrate and work, how to be a family. If we apply this thinking today, perhaps we could be asking: what would be the equivalent customs for a sustainable person? How would they travel? What would they eat? How would they engage with diversity? How would they create work cultures in harmony with nature? "A man of humanity does not strive for life at the cost of injuring humanity" (Jaspers 1957, pp. 56–57).

Confucius promoted civic sense saying that a good citizen would not neglect their neighbours. They would honour the worthy and be able to work with diversity. High on his agenda was respect for parents. "If respect is absent, wherein should we differ from the beasts?" (Jaspers 1957, p. 46). To follow this family code of conduct might require restructuring of our modern day life imperatives where economic pressures prevail and are the highest on the list of priorities.

What Is the Wisdom of Confucius and Lao Tzu for Sustainability?

The greatest strength of Confucianism is the wisdom regarding government. He advised that there were "aspects of government that had to be developed and there were aspects that had to be nurtured" (Jaspers 1957, p. 47). This nurturing can be interpreted as very early notions of a developmental, participatory approach to assist with change management and capacity building. Linked to the ability to achieve change, Confucius stated that a good government was led by a person who engendered the confidence of the people and provided opportunity for education. In his lifetime, he did not find a leader (a prince) with the qualities he saw as necessary but he did not give up on his teachings. Confucius did not lower his standards—as some thought he should. He founded a school for future statements, trusting in the long-term. Lao Tzu is said to have disapproved of Confucius' planning process, his study regime and demand that people remain impartial. However, we today can draw from the strengths of both of these great Chinese philosophers and see that both elements are essential.

Both sages, Lao Tzu and Confucius, revered nature and the cosmos as sacred. It was with this foundation that Confucius developed his teachings of good citizenship, good leadership and how to run a successful nation. The same focus was used by Lao Tzu to develop his teachings and processes for engagement with nature and the celestial spheres. This underlining valuing of nature and the cosmos as scared is a critical point of difference between Confucius' wisdom on governance and management and our current corporate systems, which seems to disconnect people from the spirituality and sacredness of the physical world. This detached objective approach is part of a package often included in what is seen as the western paradigm where consumerism has replaced spirituality and ethics, as was the intention in the Lebow (1955) plan.

As part of his teachings, Confucius called for impartiality in the leader, allowing for an open mind to absorb new ideas. This, again, can be argued as the birth of community development or participatory planning processes, with the impartial leader guiding the discussion, but not imposing the way forward. The impartiality is related to the objectivity required in corporate planning. However, it is not separated from strong ethical foundations in its teaching or its application. Lao Tzu lived in the world of nature and the cosmos. His response to Confucius according to Jaspers is:

> To make up one's mind to be impartial is in itself a kind of partiality … You had best study how it is that Heaven and Earth maintain their eternal course, that the sun and moon maintain their light, the stars their serried ranks, the birds and beasts their flocks, the trees and shrubs their station. Thus you too shall learn to guide your steps by Inward Power, to follow the course that the Way of Nature sets; and soon you will reach a goal where you will no longer need to go round laboriously advertising goodness and duty …. All this talk of goodness and duty, these perpetual pin-prinks unnerve and irritate the hearer …. The swan does not need a daily bath in order to remain white (1957, p. 59).

Confucius, however, had this inner power and referred to it often as an essential ingredient of his teaching. His focus was the ethical–political state, the qualities and customs of a good citizen of the Earth and a good leader of a nation. Lao Tzu's focus was to create time, space and methods for one to commune with nature and the cosmos.

Tu (2001, p. 253) refers to New Confucian humanism as a source of inspiration for human flourishing in the twenty-first century. According to Tu (1998a), the age in which we live is one where the modern West is the most dynamic and transformative ideology in human history. He sees the achievements we have today—science and technology, industrial capitalism, market economy, democratic polity, mass communication, research universities, civil and military bureaucracies, and professional organizations, as part of the whole Enlightenment mentality. The Confucian Golden Rule: "Do not do unto others what you would not want others to do unto you" seems to rest forgotten, so is his understanding that: "In order to establish myself, I have to help others to enlarge themselves" (Tu 1998a, p. 7). His teachings reinforced the need for society to have ethical foundations and values of a consciousness that goes beyond anthropocentricism to that which embraces the whole of the Earth and beyond.

Confucianism is originally a philosophy, a way of life—not a religion. However, it provides a good ethics underpinning ordering of communities that is compatible with the foundations of many world religions. According to Tu (1998a, b), Western, Eastern and Indigenous spiritualities have informed the Enlightenment mentality while the Confucian tradition avoids anthropocentrism (human-centeredness) in favour of anthropocosmism (or seeing humans as part and parcel of the cosmos). A re-examining of the link between modern religious traditions in the West is needed to position the matter/spirit, body/mind, sacred/profane, human/nature, creator/creature realities in the new mentality and to go beyond anthropocentrism (Tu 1998a).

At the core of Confucianism is the appreciation that human beings are sentient, social, political, historical and metaphysical beings (Tu 1998a). Confucianism refutes any reductionist models that collapse people into anything less; however, the human being has to learn how to be human through transformation in communion with self, community, nature and the transcendent (Tu 1998a).

According to Moller (2011, p. 56), Confucianism and Taoism both view nature, human beings and the cycles of nature as a holistic system in which humans must not only be in harmony with other beings, but also with nature itself. They see people as part of a network of ecological and social relationships undergoing constant change. Being part of the ecology, humans should not use it according to their own interests. This concept leads to a notion of nature-based sustainability and helps to ensure the protection of nature and its resources. When developing its economy, has China's past practice followed these principles?

Economic Prosperity and Environmental Deteriorations in China

Figure 18.1 shows China's share of global gross domestic product (GDP) adjusted for purchasing power parity (PPP dollars) from 2003 to 2013. In 2013, China's share of GDP was estimated at 16 %.

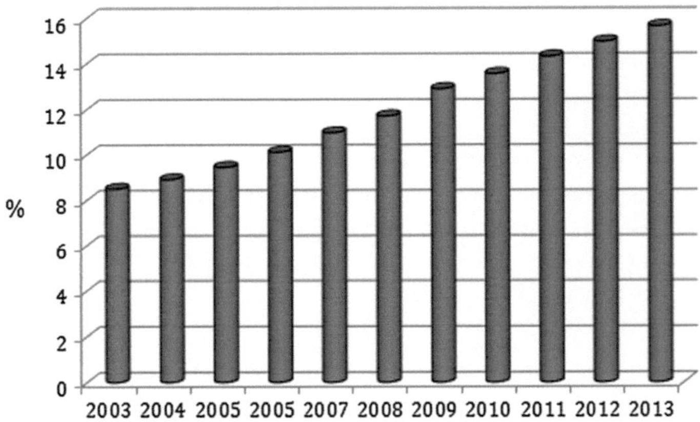

Fig. 18.1 China's share of global GDP (2003–2013). *Source of data* IMF (2013)

As the world's second largest economy, China's average annual GDP growth rate has been over 10 % between 1979 and 2010 with the highest real GDP growth reaching a historical 14.2 % in 2007 (Morrison 2013; Statistia 2013). China has been contributing to the world economy significantly with its GDP (PPP dollars) being 8.5 % in 2013 and projected to jump to 19 % in 2018 (Statistia 2013). Within the next decade, China may overtake the USA to become the largest global economy. However, this economic expansion has been accompanied by an enormous increase in the country's ecological footprint—from less than 1 global hectares (gha) per person in 1961 to 2.1 gha in 2008, significantly higher than the available per capita biocapacity of 1.8 gha (Gaodi et al. 2012). Additionally Gaodi et al. (2012) report that the populations of more than ten flagship and keystone animal species have undergone a marked decline in China. This was particularly severe before the 1990s. Since 2003 the most recent dominant driver of China's ecological footprint, and particularly its carbon component, has been increasing individual consumption (Gaodi et al. 2012). As a result, since 2008 the country has been singled out as the highest greenhouse gas emitter in the world. Furthermore, the 2012 environmental performance index by Yale University ranks China 116th out of 132 countries. These standings are likely to have been unimaginable for China's ancestors.

Miraculous economic expansion in China for the last three decades has been associated with the degradation of its natural resources (e.g. serious water, land and air pollution); both of which are due to the country's increasing industrialization and urbanization. For example, some deforestation and improper land use practices have become major problems in China (UNEP 2002). The current urban air pollution and city smog have been attracting both domestic and international attention (Xinhuanet 2014). In addition, green algae pollution in major rivers of China is a frequent occurrence due to mismanagement of water (People's Daily Online 2011; UNEP 2007). There is also evidence that arable land in China is shrinking due to industrialization and urbanization (China Daily 2011).

Another consequence of a rapidly growing economy is the rapid growth in car ownership. For example, according to car sales in 2009 China has already overtaken the USA and by 2010, it had almost double the number of cars sold in USA (China Daily 2011). Air and water pollution has resulted in problems affecting the whole country. Water shortage is already a big problem in China. For example, the country's per capita water resources availability will drop significantly by 2030 when China's population is expected to reach 1.6 billion with a water-scarcity limit at 1700 m^3 (Wan and Zhou 2009).

As the world's fastest growing developing country, China has attracted global attention for the extensive pressure from its increasing population, consumption of limited resources as well as its role in deteriorating its ecological environment. China's resource, energy and environmental concerns have become a serious constraint for its economic development, social advancement and ecological protection. To sustain development into the future, China must improve its environmental performance and restore the health of the natural environment. This can be achieved through investing in environmental management, restructuring the economy and making the shift to green technologies. The Chinese government has concentrated on economic growth without paying enough attention to environmental deterioration. The issues are particularly aggravated by China's quickly expanding urbanization. Due to these issues, there is concern regarding whether or not China can maintain quality economic growth while satisfactorily preserving its water and other limited natural resources. Solving China's environmental problems, such as carbon emissions, land erosion and degradation, air and water pollution, will add more pressure on the country's population. In order to be able to successfully sustain its growth, China should establish a more effective system and strategy of resources development and the government should strengthen the supervision of resources management and protection.

The teachings of Confucius can be applied in this endeavour. For example, in regard to ecological awareness and environmental protection, Confucius advocated the unity of humankind with nature, for understanding and respecting nature. Confucius believed that a superior person admires mostly the Mandate of Heaven taking the suggestions of great predecessors and listening to what the sagas advice (Lau 1979, 16.8). Furthermore, Confucius valued and treasured mountain and water resources, and he believed that clever people aspiring to be intelligent and honest must love such resources. Through doing so, they can be happy and live long lives (Lau 1979, 6.23). Confucius believed a superior person should not ask too much from life and there is no need to eat full and to live in a comfortable house (Lau 1979, 1.14; Ware 1955).

As part of educating to protect the natural environment, Confucious himself also set a good example for other people. He was known to eat a meal normally consisting of rice, vegetables, meat and a drink, without the meat and drink only water. Also, he was happy to bend his arm as a pillow (Lau 1979, 7.16). In order to maintain ecological balance and environmental protection, Confucius suggested that a hook should be used for fishing instead of a net and that a nest of sleeping birds should not be shot at with an arrow. This demonstrates his adoration, passion and respect for wildlife (Ware 1955). Thus, the knowledge of ancestors can be seen as a useful source when developing guiding principles for the management of the natural resources required for economic growth.

How Can We Draw on the Wisdom of Confucius and Lao Tzu in Our Contemporary Challenges Regarding Climate Change and Environmental Sustainability?

According to *Confucius*, one should not only love his or her parents, but also spread this love to nature. Furthermore, he also advised that one should control personal selfish desires and should not sacrifice the environment for achieving economic needs. Nowadays' people should be more philosophical in their treatment of natural resources and seek to adhere to a more harmonious relationship between human and nature. Confucius, Lao Tzu and other Chinese philosophers and ideologists treated people as equals to nature.

The Confucian thoughts of kindness and responsibility can thus be seen to be based on the notion that humans should see themselves as equal to other natural lives and resources. This means that people should not try to conquer other living entities but should instead protect them. Therefore, people should not try to overcome nature, but rather they should love it as much as they love their parents and relatives including humans and other creatures and resources.

Lao Tzu suggested in the Tao Bible (Daodejing.org 2013) that everything on earth is produced by Dao, the natural rules, and he believed that Dao produces one; one produces two; and two produces three. Therefore, humans have the responsibility to take good care of the other creatures and resources of nature. Protecting and caring for nature are thus viewed as the best of human moral expressions. According to Lao Tzu in Tao Bible (Daodejing.org 2013), Dao, Heaven, Earth and Human are the four biggest things in the universe and he believed that humans depend on the Earth to live, the Earth depends on Heaven, Heaven follows Dao, and Dao eventually follows the principles of nature. This means that humans are the children of nature who have to respect Heaven and Earth. Their actions or behaviours need to follow the natural rules (Dao). Humans are allowed to improve the natural conditions because they are not passive; however, they have to act under humane and righteous motives (Pan 2013).

In general, Confucius advocated that people should treat nature nicely and should exercise self-control in relation to their desires, by applying appropriate and environmentally friendly behaviours. According to Confucius, when dealing with the natural environment and its resources, if righteousness and benefit conflict each other, the former should be treated as the greater imperative. When we are not able to achieve both, Confucius advised that people should forgo benefit and follow the right rules (Dao) of the earth. People should follow this notion of treating the natural environment and exploiting their needed life materials under the condition of not damaging the ecology. The benefits are obvious for all involved. People should follow and respect the natural principles of the environment and balance the economic and personal choices in order to achieve any material needs in the best way. According to Confucius, people should be able to satisfy themselves with enough resources if they treat nature well and follow wisely the natural rules (Pan 2013).

Conclusion

The environmental damage caused by China's pursuit of unsustainable GDP growth is becoming more serious, and the country needs to be prepared to face these difficulties. In particular, there is need for the government to seek a balanced developmental strategy model to achieve this. China's environmental pollution and deterioration may challenge and delay future economic growth. Immediate actions are thus needed to adjust the current growth pattern. This includes achieving the goal of building a harmonious society, outlined by the government, including reducing corruption and enhancing environmental protection laws, regulations and education (Morrison 2013). Much of Confucianism and Taoism can be seen as being important in informing how to achieve these.

The ideas of Confucius, Lao Tzu and the other ancient philosophers can be used to direct and empower people of this day and age to deal with the environmental crisis, while still gaining access to needed resources. Confucius believed that the Earth was big enough to secure life with the condition that humans must treat nature with care, following the natural rules (Pan 2013). China, as the largest developing country, can set a good example by demonstrating that it is actively working to tackle climate change, restore and maintain the health of the natural environment. There is a lot of useful wisdom regarding sustainability that can be derived from ancient Chinese ideologies such as Confucianism and Taoism.

Indeed, it is the case that Confucianism has already been accepted not only in China but across the world (Berthrong and Berthrong 2000; Hsü 2005). People need to remember to look back more often to seek the wisdom from their ancestors when developing economic and environmental policy. They need to pay more attention to the finite nature of many key resources and the planet itself. The sense of sustainability is not a new concept for China. Instead, the root of the sustainability concept was planted more than five thousand years ago. The nation's ancestors set a good example in respecting and taking care of nature, which has sustained human life for many generations. The thoughts of Confucius and other Chinese ancient philosophers informed today's basic sustainability concept, of keeping the balance between economic development and environmental protection to achieve harmony.

References

Aitken CF (1908) Confucianism. The New Advent Online Catholic Encyclopaedia. http://www.newadvent.org/cathen/04223b.htm. Accessed 29 Oct 2015

Berthrong JH, Berthrong EN (2000) Confucianism: a short introduction. One World, Oxford, UK

China Daily (2011) 2010 auto sales driven to a record. http://usa.chinadaily.com.cn/china/2011-01/11/content_11836930.htm. Accessed 29 Oct 2015

Chiu L (2011) Daoism in China. http://chineseculture.about.com/od/religioninchina/a/Daoism.htm. Accessed 29 Oct 2015

Cordier H (1912) Taoism. The new advent online catholic encyclopaedia. http://www.newadvent.org/cathen/14446b.htm. Accessed 29 Oct 2015

Daodejing.org (2013) Tao bible. http://www.daodejing.org/25.html. Accessed 6 May 2013

Gaodi X, Shuyan C, Qisen Y, Lin X, Zhiyong F, Boping C, Shuang Z, Youde C, Liqiang G, Cook S, Humphrey S (2012) China ecological footprint report 2012: consumption, production and sustainable development. World Wildlife Fund Beijing Office, Institute of Geographic Sciences and Natural Resources Research China, Global Footprint Network, Institute of Zoology China and Zoological Society of London, Beijing

Hsü LS (2005) The political philosophy of Confucianism: an interpretation of the social and political ideas of Confucius, his forerunners and his early disciples. Routledge, London

International Monetary Fund (IMF) (2013) China: share of global gross domestic product (GDP) adjusted for purchasing power-parity (PPP) from 2008–2018. http://www.statista.com/statistics/167632/china-share-on-gdp-ppp-of-the-world/. Accessed 8 Jan 2013

Jackson T (2009) Prosperity without growth? The transition to a sustainable economy. Sustainable Development Commission, London. http://www.sd-commission.org.uk/data/files/publications/prospe-rity_without_growth_report.pdf. Accessed 29 Oct 2015
Jaspers K (1957) Socrates, Buddha, Confucius, Jesus. In: Arendt H, Manheim R (eds) The great philosophers, vol 1. Harcourt Brace, New York, NY
Kong SL (2009) Confucian wisdom for the 21st century: a selected rendition. The Chinese University Press, Hong Kong
Lau DC (translator) (1979) Confucius: the analects. Richard Clay, London
Lebow V (1955) Price competition in 1955. Journal of Retailing. Spring. http://www.gcafh.org/edlab/Lebow.pdf. Accessed 29 Oct 2015
Littlejohn R (2005) Daoist philosophy. Internet encyclopedia of philosophy. http://www.iep.utm.edu/daoism/. Accessed 29 Oct 2015
Moller JO (2011) How Asia can shape the world: from the era of plenty to the era of scarcities. Institute of Southeast Asian Studies and ISEAS Publishing, Singapore
Morrison WM (2013) China's economic rise: history, trends, challenges and implications for the United States. http://www.fas.org/sgp/crs/row/RL33534.pdf. Accessed 29 Oct 2015
Pan SD (2013) Learning the wisdom of ancestors and understanding the truth of harmony (in Chinese). http://www.chinavalue.net/General/Blog/2013-3-1/959751.aspx. Accessed 29 Oct 2015
People's Daily Online (2011) Green algae invading China's tourist destination Quingdao. http://english.cri.cn/6909/2011/07/08/2821s647306.htm. Accessed 29 Oct 2015
Statistia (2013) http://www.statistia.com/statistics/263616/gross-domestic-product-gdp-growth-rate-in-china/. Accessed 20 Jan 2014
Tu W (1998a) A Confucian perspective on human rights. http://www.newbooks-services.de/MediaFiles/Texts/7/9789814374477_Excerpt_001.pdf. Accessed 29 Oct 2015
Tu W (1998b) Beyond the enlightenment mentality. In: Tucker ME, Berthrong J (eds) Confucianism and ecology: the interrelation of heaven, earth and humans. Harvard University Centre for the Study of World Religions, Cambridge, MA
Tu W (2001) The ecological turn in new Confucian humanism: implications for China and the world. Daedalus 130(4):243–264
United Nations Environment Programme (UNEP) (2002) State of the environment and policy retrospective, 1972–2002. http://www.unep.org/geo/geo3/english/pdfs/chapter2-1_socioeconomic.pdf. Accessed 28 Oct 2015
United Nations Environment Programme (UNEP) (2007) Wetland biodiversity conservation and sustainable use in China. http://www.undp.org/content/dam/china/docs/Publications/UNDP-CH-EE-Publications-Mainstreaming-Wetland-Biodiversity-Conservation.pdf. Accessed 28 Oct 2015
Wan YX, Zhou CY (2009) The goals, challenges and countermeasures for the harmonious development between human beings and water. Yellow River 31(6):18–19, 118
Ware JR (1955) The sayings of Confucius. The New American Library, New York, NY
Xinhuanet (2014) China focus: air pollution chokes as smog migrates, http://news.xinhuanet.com/english/china/2014-01/22/c_133065998.htm. Accessed 23 Jan 2014
Yao X (2003a) Routledge Curzon Encyclopedia of Confucianism A-N (edited). Routledge Curzon, New York
Yao X (2003b) Routledge Curzon Encyclopedia of Confucianism O-Z (edited). Routledge Curzon, New York
Zhang Y, Li T (2000) Economic lessons from Confucius for the new century. Culture Mandala Bull Centre East-West Cult Econ Stud 4(1):3

Chapter 19
Threats to Sustainable Development

Andrzej P. Wierzbicki

Abstract The paper presents conclusions to the theme of sustainable development drawn from the recent works of the Committee of Future Studies "Poland 2000 Plus" on the *Report Poland 2050* and related to the issue of global threats; both these themes take into account and stress the importance of technological perspective. It is shown that network technologies and other high-tech developments, while bringing many benefits to the mankind, bring also threats related to the speed of change and destruction of existing social structure. Thus, one of the main challenges of mankind in the future will be not *limits to growth,* but *how to limit growth and speed of change.*

Keywords Global threats and challenges · Threats to sustainable development · Benefits and threats of technology

Introduction

The concept of *sustainable development* of H.G. Bruntland, contributed at the end of twentieth century (Bruntland 1987), is especially important, since it transgresses typical short-term economic rationality. The threats to sustainable development are typically understood as (1) *threats of human (or technological) pollution and distortions of natural environment*, thus leaving to next generations much worse environmental conditions. Technology, however, can be used to limit environmental pollution and distortions but, in fact, creates or contributes to the creation of quite different threats to sustainable development discussed below. On the

A.P. Wierzbicki (✉)
National Institute of Telecommunication, The Committee of Future Studies,
Polish Academy of Sciences, Szachowa 1, 04-894 Warsaw, Poland
e-mail: A.Wierzbicki@itl.waw.pl

L.W. Zacher (ed.), *Technology, Society and Sustainability*,
DOI 10.1007/978-3-319-47164-8_19

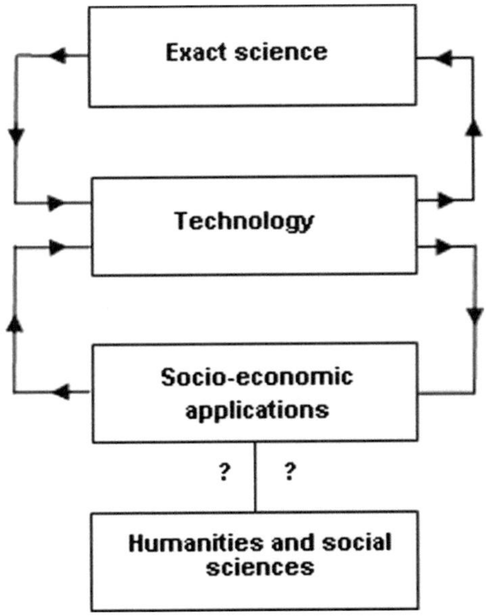

Fig. 19.1 Two positive feedback loops between science, technology and their socio-economic applications

other hand, diverse (not only technological) human anthropogenic pressure on natural environment can be serious in coming 40 years, particularly in countries of rich flora (India, Brazil, etc.) which, if destroyed, could perturb the ecological equilibrium of the world.

However, there are more threats to sustainable development. They are generally related to (2) *an increasing speed of change.* The positive feedback loop between science and technology only stimulates a much stronger positive feedback loop between science and technology on the one hand and market and competitive development of socio-economic applications on the other hand which uses the results of science and technology and co-finances their further development. This is represented in Fig. 19.1, quoted after Wierzbicki (2011).

This leads, however, to an avalanche-like development, additionally stimulated by demographic development. There are small chances of limiting the resulting speed of development, at least until the population of the world stabilizes in numbers (which is expected after 2050) and the economic diversification of the world is limited (which also, we hope, can be achieved around 2050). However, exponential type of growth is not sustainable in the longer perspective, since the Earth—until we start a colonization of other planets—is a closed system. In the *Report Poland 2050* we warned": We are driving with great speed along a slippery and curved road, further acceleration threatens loosing the ability to react if a new, unexpected obstacle arises".

In global development, such obstacles will appear in the form of some new (or already developing) conflicts of economic, socio-economic, socio-technical, etc., nature. From a methodological perspective, such conflicts can be seen as the results of collisions of diverse megatrends, observed already today but usually without a deeper analysis of their possible collisions and resulting threats. We shall give below only one example of such analysis, but start with an enumeration of such threats to sustainable development of the world that could result in serious perturbations (such as mass protests, intensive terrorist activity, revolutions).

(3) *Economic conflict between the virtual and the real economy* where the financial sector of economy practically cut off its former two-sided relation to the real economy and started to dictate what should be done not only in real economy, but also in socio-political matters, threatening democracy;

(4) *Socio-economic conflict of precariat* (described shortly in the "*Poland 2050 Report*"), caused by the collision of two megatrends: an universalization of tertiary education and the destruction of classical permanent places of employment of proletariat resulting from automation and robotization of production, jointly causing the emergence of a new class of highly educated people without permanent employment;

(5) *Three-sided socio-economic conflict about the property of knowledge*, between corporate property knowledge, individual knowledge—both explicit and tacit—and intellectual heritage of humanity constituting public knowledge (the conflict was exemplified by recent clash about ACTA);

(6) *Socio-technical conflict between computers and their human users*, neo-luddism, growing anti-digital attitudes;

(7) *Lurking in the future socio-technical conflict of humans with superhuman cyborgs.*

Another approach is to list the most obvious effects of *informational revolution,* such as universal growth of the access to information and—to some extent—to knowledge; the growth of importance of knowledge as an essential productive factor, gradual transformation towards knowledge-based economy; automation and robotization leading to the so-called *third industrial revolution* (see *The Economist*, issue 21–27 April 2012) and the destruction of the class of proletariat; globalization of life in many dimensions: market, information, travel, etc.; a general acceleration of socio-economic development. Each of these positive processes has a reverse, negative side of threats and conflicts (see, for example, Wierzbicki 2012), but these conflicts confirm essentially the list of threats listed above. Each of these conflicts requires a longer analysis; in this paper, an outline of such analysis will be presented for *the conflict of precariat*; however, in analysing this conflict we shall also point its relations to other threats.

The Third Industrial Revolution

We already mentioned the concept of *third industrial revolution. The Economist* argues that it is possible today to build fully automated and robotized factories, controlled by computers and with negligible labour costs; thus, it could be possible to respond to the challenge of newly developed economies such as China and to take over manufacturing of new products back to new factories in Great Britain. It is true that during last fifty years we accumulated enough knowledge to build such factories (the idea of CAD–CAM, computer-aided design–computer-aided manufacturing, was originated in the Laboratory of Servomechanisms of MIT around 1958) and the economic motivation will result in building such factories, but *The Economist* does not note at all the depth of socio-political consequences of such development.

The third industrial revolution will result in the final destruction of proletariat. It was already noted by Tofflers in *The Third Wave* (Toffler 1980) that the automation and robotization of industrial production will result in such a destruction, but the diminishing of the numbers and power of this class was slow, first restricted to highly developed countries where the employment of labour switched to services. The employment of proletariat in manufacturing was in a sense exported to countries with lower labour costs, but the third industrial revolution will gradually destroy the proletariat even in China. And that information society can be developed only in democratic market society; thus, Toffler (1980) indirectly predicted the fall of the socialist (the so-called communist) block; however, only very few people believed them. Exceptions were Ronald Reagan, who used high technology to weaken that block, and also some leaders of that block, who read the book *The Third Wave* (I know this personally because of trying to popularize the concept of information society at that time), knew the prediction and thus agreed to relatively peaceful transition to democratic market society. Thus, the thesis that the fall of the so-called communist system was unpredicted, a *black swan* event (see Taleb 2007) is simply incorrect.

However, the typical economic forecast that services will provide enough employment to compensate its losses in manufacturing turned out to be imprecise, for two reasons. First, the unemployment in highly developed economies is systematically and inevitably growing; twenty years ago a "natural" unemployment rate was 2–4 %, today it is rather 6–15 %; thus, the demand of labour in services does not compensate the loss of employment in manufacturing. Second, the employment in services turned out to be highly impermanent; the employers exploit the surplus of labour supply and advantageous (for them) labour laws to choose such employees that will not protest when suddenly dismissed. As a result, a new class emerged instead of proletariat: *the class of precariat*, of employees working in highly impermanent positions in services. The emergence of this class leads to numerous conflicts, see next point. Here we must only note that the third industrial revolution will not result in universal happiness, if left to itself or to market forces alone.

The third industrial revolution is inevitable, even if it will take a long time. However, it is the result of an accumulation of human knowledge, developed mostly for public money. The question is: should it be used to enrich a few entrepreneurs, or to provide deserving living conditions for all mankind? We could specify here two extreme cases:

1. *If market will dominate over democracy and human rights,*[1] as it happened during last twenty years (see, for example, Streeck 2011), then this knowledge belonging to public will be privatized, and the profits from third industrial revolution will enrich a narrow stratum of society, leading to a further social stratification and to the growth of precariat class. The consequences of such development are discussed in the next section
2. *If people will decide that human rights should dominate both markets and states*, then this knowledge should be used for the benefit of all people in the world: for abolishing hunger and poverty, for strengthening democratic societies while defending following basic human rights equally for all people:

   - personal security together with health service and social security;
   - freedom of conviction, speech, religion;
   - broad access to education, information and knowledge;
   - relatively permanent work.

The economic organization of such democratic society should be based on market forces, but without believing in their absolute character: market competition is only a tool of economically efficient action, but as every tool it can be distorted or corrupted, and then a correcting or regulating action must be taken. Such a role must be played by the state, but the basic human rights must be sovereign, dominating both the state and the market.

The basic human rights are listed here (and the list is only approximate, certainly not complete) because the forces of informational revolution and third industrial revolution combined with market forces, if left unwatched, can endanger all of them. See, for example, Bard and Söderqvist (2002) for an analysis of the threats of information revolution to democracy; observe how current conflicts about intellectual property (see, for example, Lessig 2004) endanger actually basic freedoms of people; note the arguments of neoliberal economists that knowledge is a commodity and should be privatized, while a full privatization of knowledge would actually mean a new slavery (individual tacit knowledge is the determinant of human individual personality); etc. However, the right to relatively permanent work is a new postulate and should be commented here.

Informational revolution implies a faster change of professions and the need of life-long education, see, for example Wierzbicki (2011). Therefore, no employment can be absolutely permanent in future. Moreover, the third industrial revolution can free people from heavy work, give them more time for leisure and

[1]It is sometimes argued that human rights should be understood as they were formulated in eighteenth century and thus this concept is outdated today; however, such an argument shows ignorance of historical developments. Both Roman society and later feudal society relied on a specific tacit understanding of human rights, appropriate to their level of socio-economic development; without such tacit agreement, these societies would not be stable. New possibilities brought by informational revolution require a new understanding and new formulation of human rights.

entertainment; historically, people always had the aspect of *homo ludens*, and the informational revolution both strengthens and exploits this aspect. However, the first industrial revolution strengthened also another aspect of people, that of *homo faber*, to such an extent that the participation in a relatively permanent work has become a necessary condition of human dignity. It is telling that a vast majority of crimes of affection recently in Polish families are committed by men who lost employment. If we want to use benefits of information revolution and third industrial revolution to provide for deserving living conditions for all mankind, then the right to relatively permanent employment should be treated as a basic human right. Even today, there are societies in the world (e.g. Japan, Sweden) that perceived or anticipated this postulate; Japanese care for relatively permanent work and, if told that this reduces their economic growth, respond that they care more for the quality of life than economic development.

The Conflict of Precariat and the Second October Revolution

If we fail to defend this basic human right, then the class of precariat will grow in numbers, while its social situation will worsen. We already observe the emergence of this class, with the following consequences:

Uncertain of future employment, young people delay forming families and having children, worsening demographic trends of ageing societies. Note that these demographic trends concern mostly developed countries where the conflict of precariat already developed.

Impermanent employment does not guarantee repayment of bank loans—and banks use all possible advertisements and inducements to increase the number and amounts of loans beyond the possibilities of real economy, because of another megatrend of *virtualization of financial economy*, severing its relation with real economy. This was, in fact, the reason of the crisis 2007–2009 and of its continuation today.

The new class of precariat becomes increasingly better educated, because of yet another megatrend of increasing the spread of university education. Families are determined to pay for a better education of their children, because this is the best investment for the uncertain future in knowledge-based economy. However, an equilibrium of demand and supply on the labour markets for educated specialists will be never attained, partly because of this megatrend, but also between others because of 4–5 years delay between starting and finishing studies (thus, instead of equilibrium only educational cycles of 16–20 years duration are observed, see, for example, Wierzbicki 2011). Much more probable is an oversupply of highly educated people.

Already today, young unemployed or impermanent employed people start social protests—in London, New York, Madrid, etc. What will happen, if precariat grows in numbers and, at the same time, becomes better educated, with an access to an organization through Internet, etc.? This is a recipe for a *second October revolution*—this time not in underdeveloped country, but threatening all highly developed countries, and fought with modern weapons including cyber-attacks.

Naturally, such a development will occur only if we neglect the socio-political consequences of the third industrial revolution, leaving it to market forces alone. However, such a development might be one of the most dangerous effects of a too high speed of change, endangering the sustainable development of mankind.

The Speed of Change and Its Dangers

The actual speed of change of economic and technological development is today tremendous, as illustrated by Fig. 19.2 (quoted after *Report Poland 2050* of the Committee of Future Studies of Pac.Sc.) that presents an approximation of the growth of global GDP during recent 250 years. On the one hand, such exponential growth is an obvious systemic result of the double-positive feedback loop presented in Fig. 19.1; one the other hand, it is even more obviously not sustainable even until 2050.

We could benefit to some extent from high speed of change, but only if it would be used for diminishing economic and social disparities on the global scale; left to market forces alone, this growth resulted until now rather in increasing the disparities. Moreover, as indicated by the discussion of the third industrial revolution and the conflict of precariat, this high speed of change results in a correspondingly fast destruction of social structure.

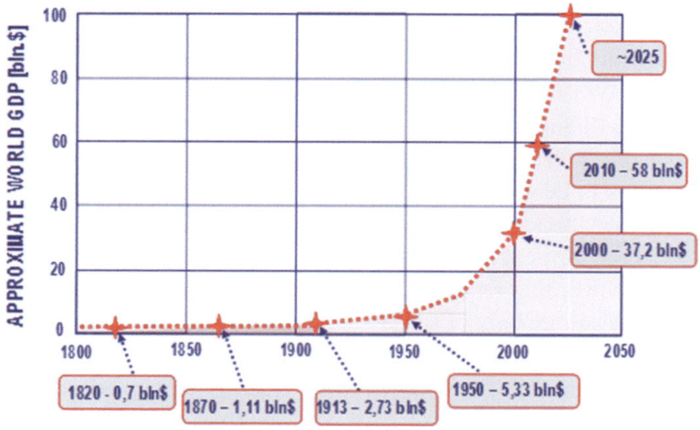

Fig. 19.2 An approximation of the growth of the global GDP in US dollars over recent 2050 years (after *Report Poland 2050*, based on data from Maddison 2004)

The situation is aggravated by an ignorance *what socio-economic or socio-political effects will result from already existing but yet underdeveloped technological breakthroughs.* The point is that all essential technological developments (such as, recently, television, computers, cellular telephony, Internet, WWW) were devised much earlier—usually 30–50 years earlier—than their broad socio-economic penetration, see Wierzbicki (2011). Thus, we can only guess what already existing but not yet fully used technological developments (e.g. robots as companions of people, artificially intelligent implants) will start broad socio-economic penetration in the next 40 years, what benefits and threats they will bring. That requires deeper social studies, but such studies must be based on a realistic, informed understanding of technological realities. Such an understanding can be obtained only if all faculties of social sciences and humanities include in university curricula at least three technological subjects: informatics with its social consequences (called recently *social informatics*), robotics with the concept of feedback, and biomedical engineering. The lack of such understanding is symbolically indicated in Fig. 19.1 by question marks: what is the impact of social sciences and humanities on this double-feedback loop if they do not even understand well the concept of feedback? There are many examples of such lack of understanding, the most notable of them being the treatment of the so-called *vicious circle* as a paradox while it is an example of feedback (all memory elements in computers are based on such feedback; thus, they are vicious circles, see Wierzbicki 2011).

The high speed of development does not permit such ignorance; it is like putting an unschooled driver in a race car on a slippery road. And if the speed of change increases much more, the danger is too fast destruction of social structure; any conflict might lead then to a blow-up, and the conflict of precariat discussed above is only one of possible reasons. Another fact indicates that we might be close to a dangerously high speed of change; this is the observation of *eerie silence* (Davies 2010): after 50 years of using radio signals to indicate the existence of intelligence on Earth, we did not obtain any reply. One of the possible explanations of this fact is that civilizations such as ours are doomed to extinction if they develop too fast, thus are rare ephemerids in the universe.

For the reasons indicated above, *the real threat to sustainable development is too fast speed of change, and the essential question is how to limit it?* An obvious answer seems to be to limit first the dominance of classical, in particular neo-classical economic thinking: countries most criticized by such thinking, such as Japan and Sweden, turn out to be leaders in quality of life. However, this is only a part of the full answer that requires a deeper study. Other parts might be changing the goals of the developments of science and technology from fully commercial to more responsible ones; insisting on the dominance of human rights over markets, corporations and states; etc.

Conclusions: How to Imagine a Sustainable Development that Will Be also Durable?

The real threats to sustainable development are not *threats of human (or technological) pollution and distortions of natural environment*, leaving to next generations much worse environmental conditions. Much *more dangerous threats result from unchecked feedbacks between the developments of science and technology on the one hand and their socio-economic commercial applications on the other hand*, leading to an unbalanced, too fast speed of change. Such too fast speed of change can be explosive because of many possible socio-economic conflicts, mostly resulting from market exploitation of science and technology, thus can leave to next generations a destruction of human civilization as we know it now.

Thus, the challenge to social sciences and humanities is to imagine a sustainable development that will be also durable, that is, slower and more responsible than today. In order to realistically imagine such a development, social sciences and humanities must also change themselves: become more objectively informed about actual possibilities and developments of science and technology. However, slowing down economic development while concentrating on quality of life is possible, as shown by examples of Japan and Sweden. There are also many other aspects of making sustainable development durable, such as insisting on the *dominance of human rights over markets, corporations and states.*

References

Bard A, Söderqvist J (2002) Netocracy: new power elite and life after capitalism. Bookhouse Publishers, Stockholm (Polish translation 2006: Netokracja. Nowa elita władzy i życie po kapitalizmie. Warszawa: Wydawnictwa Akademickie i Profesjonalne)

Bruntland G (ed) (1987) Our common future. The World Commission on Environment and Development, Oxford University Press, Oxford

Davies P (2010) The eerie silence: renewing our search for alien intelligence. Harcourt, Houghton Mifflin

Lessig L (2004) Free culture: the nature and future of creativity. Penguin Books, London (wyd. polskie 2005, Wolna kultura. Wydawnictwa Szkolne i Pedagogiczne, Warszawa)

Maddison A (2004) Growth and interaction in the world economy; the roots of modernity. The AEI Press, Washington, DC

Raport Polska 2050 (English version *Report Poland 2050* also available). Polska Akademia Nauk, Komitet Prognoz, Polska 2000 Plus, 2011 Warszawa

Streeck W (2011) The crisis in context: democratic capitalism and its contradictions. Max Planck Institute fűr Gesellschaftsforschung, Köln. Discussion Paper 11/15

Taleb NN (2007) The black swan: the impact of the highly improbable. Random House, New York

The Economist (2012) The third industrial revolution. Issue April 21–27, 2012

Toffler A (1980) The third wave. William Morrow, New York

Wierzbicki AP (2011) Techne$_n$: Elementy niedawnej historii technik informacyjnych i wnioski naukoznawcze. (in Polish: Techne$_n$: Elements of recent history of information technologies and epistemological conclusions, Springer, 2015). Komitet Prognoz "Polska 2000 Plus" PAN i Instytut Łączności, Państwowy Instytut Badawczy
Wierzbicki AP (2012) Cyfrowa przyszłość: wizje, przewidywania, strategie (in Polish: Digital future: visions, forecasts, strategies). In Zacher L (eds) Nasza cyfrowa przyszłość (in Polish: Our digital future). Komitet Prognoz "Polska 2000 Plus" PAN

Chapter 20
Sustainable Development—Sustainable Man (External and Internal Conditions for Sustainability)

Wiesław Sztumski

Abstract For several years, the idea of sustainable development has become fashionable and is being implemented. The term "sustainable development" is inflected at all possible ways. One creates the necessary conditions for the implementation of such a development, which is believed to be a panacea for economic crises, environmental degradation and premature full exploitation of material resources of our planet. However, internal conditioning—the role of individuals—their mental and spiritual states, value system and environmental awareness—are ignored. The objective of this article is to draw attention to the fact that the success of sustainable development depends mainly on the internal and subjective conditions, especially from the people who implement it. The concept of sustainable development can be fully realized only by sustainable people.

Keywords Sustainable development · Sustainable man · Contemporary life milieu · Golden rule · Inivironment · Environment

A Very Short Foreword

Genesis of the concept of sustainable development is well known. This is an idea that was born at the end of the twentieth century in the minds of the intellectual elites, aware of the risks that result from uncontrolled production and consumption as well as from the fact that the finite Earth is not able to effectively meet the infinite needs of human beings. At the core of sustainable development is social consensus on most important matters of human life on a global scale.

W. Sztumski (✉)
Silesian University Katowice, Okólna 89 F, 42-200 Częstochowa, Poland
e-mail: ws34@op.pl

L.W. Zacher (ed.), *Technology, Society and Sustainability*,
DOI 10.1007/978-3-319-47164-8_20

Reaching such an agreement requires consistent attention of the fundamental principles of the human existence, proclaimed by the philosophical environmentalism: the principle of tolerance and the principle of synergy. Respecting of the first principle reflects the desire to live in peace in a rapidly growing condensation of the environment and the second—the striving to realize the fundamental interest of humanity; it is the preservation of life and the survival of the human species in a progressive degradation of our natural and social environment. A lot of publications about sustainable development appeared in the last time. There are likewise many definitions of this development. Here, I refer to most well-known and frequently repeated definitions, particularly to these, which emphasize the balance in nature and society as well as the harmony between these two components of the world. Authors consider predominantly how one can implement the idea of the sustainable development and with which means and activities one could achieve the goals, which are determined by the strategy of the sustainable development in different areas of social life. Nevertheless, they do not research the problem of sustainable human, without which the full implementation of the idea of the sustainable development seems to be impossible. The sustainability of human is for me the most important matter for life in contemporary world, in the ever more changing, unstable, uncertain and unbalanced life milieu. Furthermore, I think there are three conditions necessary for realizing the idea of the sustainable development: balanced people, reasonable lifestyle and enlightened society.

It is true that some authors draw attention to importance of the subject factors and the role of human in the implementation of the idea of sustainable development. This relates above all anti-ecological activities of individuals and their responsibility for unreasonable politics and accelerated economy, for the use of any technical innovations which can contribute to the degradation of the environment of human life, to the deterioration of the nature and to worsening of the quality of human life. They do not see, what is evident: the idea of the sustainable development can be more effective implemented only by a sustainable individual.

What Is the Sustainable Development?

There are many approaches, dimensions and interpretations of the sustainable development: political, philosophical, anthropocentric, biocentric, economic, ecological etc. Sustainable development can be treated as a political idea. "The idea is working in politics, law, and strategy as well as in political, economic and social programs. It is the center of attention starting from the United Nations by such international organizations as the European Union, the nation-states, local governments and NGOs." (Papuziński 2011). In the anthropocentric interpretation "Sustainable development functions as a password which snatches up the masses to fight with environmental threats, especially with the growing social disproportion in respect of income and wealth, which destroys solidarity, and is contradictory to the current sense of justice and humanity. It seems that the leveling this disproportion should make people happy, and the world—more perfect." (Sztumski 2008). And according to the biocentrists, the sustainable development is "a vision of new human community embracing the whole of humanity. The differentia specifica is in ecological awareness, showing to the man his place in the world that is not distinguished and non anthropocentric. The traditional anthropocentric attitude to the environment is replaced by an ecocentric (biocentric) one, which requires treating the man as only one of elements of cosmic ecosystem having equal rights and playing equal part as other elements." (Gawor 2006). One can quote many other approaches to the concept of sustainable development. Hence, the concept of sustainable development requires a many-sided, multifaceted and multi-dimensional interpretation due to different spheres of life and science. Thus, one cannot reduce it to only one interpretation or to the one system, in which one want to implement it. Therefore, some authors postulate rightly that "The concept and practice of the sustainable development should be seen within the system and spheres in which they operate" (Zacher 2008). Therefore, it would be best to consider sustainable development in a holistic and systemic way. However, as yet—as I know—there is no such systemic approach to the sustainable development.

The conception of sustainable development results from the ideology of consumption and consumerism and it contributes also to the success of this ideology. It aims to reconcile two contradictory interests. On the one hand, it has to ensure the life in balanced and growing prosperity and luxury; on the other hand, it ought to reduce the consumption of material and energetic resources. Thus, it seems to be a solution of the present and future dilemmas:

- How more produce from what is less and less?
- How to be rich, to not become overrich?
- In which way such limited enriching should motivate the people to increasing productivity?
- Could such limitation be the driving force of the economy and the civilization in the future?

In the "Agenda 21", the sustainable development is defined as a process of developing (of countries, cities, business, communities, etc.), which connects unconditionally the needs of the present generation with the ability to meeting the needs of future generations, as well as the needs of one group of people with the needs of others groups. Sustainable development is defined in Polish legislation as a "sequence of changes, in which the use of resources, the structure of investments, orientation of technological progress and institutional structures have to be realized in such a way that they cause no contradiction between the future an present needs". (Prawo Ochrony Środowiska 2001). In these definitions, one speaks about "the intergenerational ecological justice," which was the basis of all international agreements in order to the protection of the environment. All definitions of sustainable development include "uncontradictory of interests". It relates above all to the time dimension (it concerns the successive generations, starting from today), that is in the dimension of history, and also the people living in the time of one generation, that is in the dimension of topicality. So, the sustainable development relates to the in the diachronic and synchronic interests of people. It refers also to the spatial dimension because it concerns the compatibility of the regional and first of all the global interests. And common interests of majority of humanity are:

- To be more rich,
- To live in peace,
- To overcome the global environmental threats.

Commonly one accepts well-known definition of sustainable development: "Sustainable development is development that meets the needs of the present, without compromising the ability of future generations to meet their own needs." Sustainable development is the one that meets today's needs in such a way as to give a chance of life for future generations. This sounds beautiful even if the sustainable development is a very promising idea more for the future generations than for the present one. The goal of sustainable development is to counteract the trend to far-reaching devastation of the natural environment, in order to leave something from it still in the future. Nature does no longer be a system which could itself reaches equilibrium states because its homeostasis is destroyed by the unreasonable activity of humans. Therefore, it is not able to defend itself. On the contrary, it needs human assistance. No one has proven that nature, and the more a society is at the present time a self-regulating system. Instead, one should assume that they are self-deregulating systems in consequence of the increasing growth of chaos and turbulence mainly caused by not reasonable human activities.

The concept of the sustainable development can be interpreted in different ways and it has many interpretations. However, what joins all they, is to care for the equilibrium between all elements of the dynamic social system, in which it ought to be realized. That seems to be main and most essential thought rooted in the ecological awareness of social masses. People, the inhabitants of wealthy countries see at the idea of sustainable development what good is for them, what allows to live further in abundance, and the inhabitants of poor countries see at it a possibility to catching up the wealthier countries. Both of them don't be aware that the equilibrium between the components of social systems which are characteristic for sustainable development needs a lot of renouncements and economical, political and ecological limitation. Functioning under various ecological prohibitions is the fundamental principle of sustainable development. Those prohibitions are not being limited to the natural environment. They are also about the social and cultural environment because they relate to meeting the needs of all people in existing and future communities, promoting personal development, well-being, social cohesion and social inclusion, and creating equal opportunities for all people. The idea of sustainable development is future-oriented; it focuses on the good of the people, which will live in the far future. Hence, no doubt that its implementation requires some sacrifices on the part of people living at this time. Present generation's life quality must be reduced in some degree and this reduction should be higher, the better life condition it wants to create for future generations. In the balanced world and life as in all nature nothing is for free—one must pay for everything and it is always something for something. It is worth to note that the implementation of the concept of the sustainable development does not transform the contemporary "homo oeconimicus" in the future "homo ecologicus", but it balances these two human models at best.

A Sustainable Man: Who This Is?

In the ancient Greek concept, sustainable was such a man, in which a healthy body was in balance with a healthy spirit. The "balanced man" was the personification of perfection being in emotionally, physically, intellectually, and spiritually harmony with himself and with his natural, social and cultural environments. Now, the sustainable man is perceived similarly: he lives in inner harmony without too much deviation from any cultural and social standards or norms, he cares for the unity of his body and spirit and he is emotionally stable. The sustainable man is characterized by a high measure of self-control and equanimity. Therefore, the excessive and unnecessary stress is foreign for him. He tries always to find equilibrium between the pairs of opposites, which influence the human life: good and evil, poor and rich, fair and unfair, cleverly and stupid. But that does not mean that a sustainable man has to be an average, standard or indistinctive man. The balanced man can expose his individual characteristics, is differentia specifica which distinguish him from all other. The question of finding the balance between all oppositions inside and outside of us is the same as the well-known question on finding the Golden Rule in Aristotle's ethic: Treat others as you want to be treated. As it turns out, there is an amazing analogy between the ethics based on the Golden Rule and the concept of the sustainable development. "The Golden Rule is a solid basis for a responsible applied ethics in both theory and practice." (Zecha 2011). It should be added to this statement: in the practice of implementation of the idea of sustainable development in all spheres of the social life. The Golden Rule is in reality a consistency principle that says that our actions should be in harmony with our wishes, desires and expectations. The use of it is a precondition to being a sustainable person. However, to achieve a balance within our interior, we need some rest time, relaxation, periods of silence and the knowledge about what in us is—in our body and soul—as well as we ought to know what the highest value in our life is. And that requires the work on each of us in the calm and contemplation. External conditions, such as for example fixed schedule during the day, contribute to this. One needs time and silence, in order to hear the internal voice—the voice of the heart and the intuition. The normal, i.e., the natural way of functioning of an organism is always optimal, harmonious, balanced and, what is significant, conductive to the health. A really healthy man is mentally and physically balanced men, who found owing to the Golden Rule his internal peace and he friendly (harmoniously) co-exists with his environment. Harmony and order of the external world of a person depend in a rather large measure on the order and equilibrium of his interior world. A regular and balanced man wants to live also into the same environment. Therefore, it forms his external world like himself as a balanced one. Sustainable development can be treated as a reflection (projection) of sustainable human being on the external world. That is one of the other manifestations of the interdependence between these two worlds. Therefore, the balancing of environment should begin to start with restoring the balance of people's invironments—each separately. One should notice: there is no sustainable development of economy and other spheres of social life without the participation sustainable humans in it. In summary, the sustainable person is physically, mentally, emotionally and spiritually balanced. Inner harmony, friendly attitude to other people and to the nature are his characteristic features.

But this is not all. The sustainable person is not only an internally balanced person. He cares not only for his own personal sustainability, but also for the sustainable interior development of the persons of further generations; also they should have possibility to functioning as internally balanced persons, as sustainable people. Therefore, he should think, he ought to leave them sufficient environmental reserves—raw material and climatic reserves, as well as own and social reserves. That is to say, as consumer he must be able to be led by the basic rule of thriftiness regarding the natural and social capital; it should be a frugal man. For this reason during his individual development, he has to preserve as long as possible his vital forces for the other—for the family and society, for now and later. And it should be at least in such degrees future-oriented as present-oriented man. However, contrary to appearances, people do not so much care for the future. This is because they live mostly for today and become less and less frugal. Contemporary man—especially in developed countries—is more and more prodigal and wasteful.

Life in the modern world, dominated by the paradigm of the economy, of increasing efficiency and profit, forces humans to manage most economically with the natural resources, social capital, and with the forces of their own organisms. Under the influence of a free market economy aiming at profit maximization, rationality was connected with the economy, and homo rationalis—a wise man—became ever more also homo frugi (frugal man). Not only the economic, but also the ecological rationality should lead to the formation a homo frugi. And people act exactly on the contrary as if guided by the principle of contrariness. On the one hand, thanks to the ecology they know that all resources are limited and they are running out faster and faster with progress of the civilization. Therefore, they have invented the idea of the sustainable development and they promote it. And on the other hand, they behave carefree, as if these resources can be fully rebuilt, in accordance with the illusory principle: what people will destroy, they are able to repair. Meanwhile, resources reduction process, like all natural processes, is irreversible despite the partial renewal of their and recycling. And economic growth based on consumption growth requires increasing destruction of natural and human resources. Thus one of the fundamental contradictions of the modern world develops. From the biological nature of man follows, that he must live only at the cost of his environment. So, willy-nilly, he must exploit environmental resources. If he wants to live, he must destroy their environment. And if he wants to live better, he must destroy them even more. This causes another contradiction of the modern world: between the pressures on an increasingly prosperous life and the desire to preserve environmental resources. It seems to me that sustainable development is not able to eliminate it in spite of all.

Humans would like to live getting better, and therefore they use everything and itself excessively and in an accelerated tempo and that for many reasons. Because of this, they will consume more and more and in consequence more destroy their environment. Despite the best intentions and different efforts, one cannot be remedied them. As a result of intentional steering of the economy by the ruling elites of the world market, the economy becomes more unstable and dynamic and at the same time less stable. This is proved by the often repeated crises in different countries, even in the most developed ones. The preference of the overconsumption and overproduction guarantees the economic growth, but in the larger degree it contributes to increasing waste. Increasing waste somehow incorporated into the model of the modern economy is a fact and an alarming phenomenon. The sustainable development does not aim to efficient management of energy and material resources, but to the replacing natural raw materials by artificial and traditional sources of energy (coal, wood and oil) by alternative (wind, water, sun). So, it is not about resignation from excessive consumption of energy resources (such as the unnecessary illuminations, advertising) or from excessive consumption of materials for production of more and more superfluous goods. Therefore, it is unlikely that the idea of sustainable development could stop the rush to waste. Prodigality and waste goes hand in hand with the progress of civilization and the increase in living standards. It manifests itself in different forms and concerns a variety of goods. The most visible and outrageous is the waste of equipment footwear, clothing, food and paper.[1] The progress of Western civilization and permanent stupidization caused that modern man is not able to defense himself against the consumptionism and against the economy feeding on overproduction and overconsumption. Thus, foolishly and without a sense of responsibility in everyday life, people squander all that they can and as much as possible to fulfill their cravings and to feel fully valuable. While earlier rationality was connected with saving, today it is associated with wastefulness. A frugal man—the product of a rational economy—already transformed in a wasteful man. This transformation of humans is continued because it does not show anything to it that it will different go, at least in the foreseeable future.

[1]According to research due in 2011 by the Swedish Institute for Food and Biotechnology in the international program Global Food Losses and Waste Foot under the auspices of FAO, is lost each year about 1.3 billion tons of food, which accounts for one third of the total global annual production. In industrialized countries, retailers and consumers throw away for various reasons 40 % of food fit for consumption. That could feed 12 billion people. Research due in March 2012 at the University of Stuttgart have shown that in one year in Germany wasted 11 million tons of food, of which 60 % of households, i.e. about 82 kg food per capita. In the last 40 years consumption of the paper in the world has quadruple increased despite—or possibly because of—the computerization.

Conclusion

On the present stage of the civilization progress, our environment and we ourselves are threatened as never before in consequence of the uncontrolled and accelerated overproduction and overconsumption. This is because people consumes too much and lives in a world of increasing number of consumer goods, in the world ruled by a huge increase of needs, stupidity, avarice and greed. It remains for us still 10–15 years time, in order to prove that we as human species are survivable, survive willing and also survive worthy. A lot of people is aware this situation and takes seriously the pessimistic forecasts for the next not-so-distant future. At the present time, humanity stays before the dramatic choice: either to live sustainably or to dead in a short time (Umwelltbrief 2013). Therefore, sustainability is more a prerequisite for survival than the wishful thinking of people. However, this condition can be fulfilled above all by sustainable people. Only a sustainable person is able to transform the economy—and in consequence all other spheres of social activity—according to the principle of sustainability following the common sense, but also science, reason, intuition and emotions, as well as the supreme value, which is the preservation of his species. This value is generally due to biological instinct. Therefore, a supportable man is relatively strongly resistant to dictates or stereotypes of the culture and fashion. He cannot allow that culture and artificiality prevailed on biology and naturalness. He should counteract the rapid development and spreading of the stupidity (which is due to the social progress and which is subordinated to the principle of the unlimited economic growth), because this reduces our instinct for surviving to such an extent which can cause probably the premature extermination of the human species. Sustainable person can effectively resist various temptations of consumptionism thanks to highly developed environmental awareness and because he feel his own significance and dominance over what is foolish and ephemeral. One thing is certain: without existence and increased activity of sustainable people, one cannot efficiently implement the concept, principles and recommendations of sustainable development on a global scale. In other words, sustainable development cannot be implemented by unsustainable people.

The sustainable human is not only necessary for the implementation of the sustainable development, but also for economic growth at all. The advantage of such human is also that he is more efficient in the mental work and probably also in the physical work than an unsustainable human. Relevant studies could confirm that an internal unsustainable man and in addition stressed by the work tempo and other factors of the contemporary, enormously dynamic and uncertain environment really does not use fully his production capacities and creativity. There is no doubt that the carnal, intellectual and spiritual inner balance of a man contributes to an increase of human capital no less than the education, health, cultural values, standard of living and other factors (Šlaus and Jacobs 2011, 99). Thus, because of the maximization of the human capital, which is substantial for economic development, one should create such possibilities and social conditions that people could become ever more sustainable.

References

Gawor L (2006) Wizja nowej wspólnoty ludzkiej w idei zrównoważonego rozwoju. Problemy Ekorozwoju 1(2):65

Papuziński A (2011) The implementation of sustainable development. Problemy Ekorowoju 6(1):107

Prawo Ochrony Środowiska 2001, Dz. U. 2001.62.627, art. 3 ust. 50

Šlaus I, Jacobs G (2011) Human capital and sustainability. Sustain Open Access J 3(1):97

Sztumski W (2008) Reflection about sustainable development (Is sustainable development fiction, utopia, illusion or swindle?). Problemy Ekorozwoju 3(2):133–134

Umwelltbrief - Hintergrund-Informationen aus Umwelt, Wirtschaft und Finanzen (http://www.umwelt-brief.de/neu/html/Nachhaltig_Leben.html; 10.03.2013)

Zacher L (2008) Trwały rozwój—utopia czy realna możliwość? Problemy Ekorozwoju 3(2):66

Zecha G (2011) The golden rule and sustainable development. Problemy Ekorozwoju 6(1):48

Part VI
Sustainability—Dimensions and Issues

Chapter 21
Embedding Sustainable Development in Organisations Through Leadership: A Conceptual Framework

Emiliya Stoyanova

Abstract The article examines key concepts in relation to sustainable development as they relate to organisations. It provides a conceptual framework covering the most common theoretical models for sustainability policies, namely corporate social responsibility and leadership. Examples from business and social life expounding the importance of these theoretical concepts in a global world economy are presented. Considering the most popular leadership styles today, the article focuses on the uniqueness of a transformative leadership perspective and its potential to enable organisations to improve their performance. The main conclusion of this analysis is that sustainability issues always need to be associated with leadership concepts and good management practices, from corporate social responsibility to community and the environment. Such clarification of the basic definitions, concepts and leadership perspectives within a business context can benefit practitioners who want to embed sustainability models into their organisations.

Keywords Corporate social responsibility · Sustainable development · Leadership · Management · Institution · Business

Introduction

Globalisation and the associated processes of global integration significantly impact all aspects of social, economic and political life. The business world represented by domestic firms and multinational companies is the major driver of global change and thus bears the consequences and criticism from the rest of

E. Stoyanova (✉)
Economics and Management Department, Varna University of Management,
13 Oborishte Street, 9000 Varna, Bulgaria
e-mail: emiliya.stoyanova@vumk.eu

L.W. Zacher (ed.), *Technology, Society and Sustainability*,
DOI 10.1007/978-3-319-47164-8_21

society. Common topics, such as pollution, global warming, violation of human rights and corrupt practices, are just a few of the critical issues which influence and/or derive from unsound business practice. Sustainable development emerged as an important issue in the 1980s, but the lack of drastic changes in the business environment indicates that such a shift has never been more topical than it is now. Setting up a conceptual framework for sustainable development related models could facilitate the actions of entrepreneurs, innovation hunters, leaders and CEOs in a pursuit for improving their business practices. This article examines the main concepts that inform such a transition, namely corporate social responsibility and leadership for sustainable development and their capacity to create a new conceptual framework for business.

Corporate Social Responsibility in a Global World

The concept of corporate social responsibility (CRS) refers to "the obligations of firms to society or more especially to those affected by corporate policies and practices" (Smith 2003). It emerged in the 1960s as a reaction to adverse practices, such as corruption, large-scale pollution, labour discriminative, misuse of child labour and safety negligence. Often performed by multinationals, such actions raised numerous environmental, social and human rights concerns demanding urgent reorientation of these companies' policies (Lasserre 2012). They were pushed by governments and public opinion to reconsider their business operation and assume responsibility for their actions in relation to all stakeholders. This was in strong contrast to Friedman's concept of social responsibility according to which business must correspond strictly to its initial purpose of existence, namely the generation and increase in profit (Friedman 1970). In many ways, this attitude continues to dominate the business world based on the following assumptions:

1. Accountability can be traced only on a personal level and
2. Shareholders invest their money for profit, not for social or environmental purposes.

Indeed, many managers presently share this point of view and reject the idea of being socially responsible which they perceive only as a way to create a better image and improve the marketing of their products and services. More interestingly, concerns about the natural environment, ethical behaviour and social accountability and any imposed measures on companies are seen as a call for hypocritical behaviour that does not align with that companies' true business model. For example, young managers report the fact that in the name of profit they were asked to "do things that they believe are sleazy, unethical and sometimes illegal" (Badaracco and Webb 1995, p. 8).

Challenged by rising terrorism, pollution problems and climate change, however, global and national companies are forced to adapt (either willingly or unwillingly) CSR-related practices for their businesses. Unprecedentedly, present day global firms are required to meet the expectations and cover the criteria of numerous stakeholders. As a matter of fact, in most cases particularly in post-communist countries, employees and top managers are compelled to promulgate socially responsible organisational demeanour (although they have to play "a double cross") regardless of their convictions.

The particular parties pushing the firm to sustainable development through CSR procedures and practices can be precisely grouped as follows (classification adapted and amended according to Lasserre 2012):

1. *Local governments*—either as a business partner of the host country or as an investor, a company has to adhere to the particular legislation and sustainable development policy of the particular country and its various municipalities;
2. *International institutions*—such as the European Union, World Bank, International Monetary Fund, United Nations, etc. periodically re-evaluate and update their requirements and documentation on environmental, social and economic matters. Disregarding their restrictions or neglecting the new opportunities these institutions give to companies through sustainable development incentives is more likely to result in business retardation;
3. *Industry associations* (e.g. CAUX)—fight for social justice, defend the rights of disadvantaged groups and encourage companies to stay in pace with the demands of society, thus making a good presentation of global companies relevant to market success. One of the focuses of the CAUX foundation for the year 2013, for instance, was to work towards a sustainable world, including presenting practical solutions for action at corporate level. CAUX aims to equip people with practical tools, grounded on moral and ethical framework, for use in their places of work (Bhagwandas 2013). While conducting conferences such as TIGE (Trust and Integrity in the Global Economy), CAUX also targets organisational change in business and economic life, expressed in preventing corruption and abuse of power practices;
4. *Non-governmental organisations*—the most popular organisations in this group are Greenpeace, Amnesty International, Transparency International, Corpwatch, etc. These organisations and others like them advocate for the transparency and accountability of multinational corporations in terms of human rights (including business context), ecology, social injustice, etc.;
5. *Public opinions*—this includes media groups, educational and religious institutions. The ongoing process of digitalisation, increasing impact of forums, blogs, social media and online professional networks has a profound influence on economic and social life, becoming platforms for sustainable development discussions, which are relevant to businesses as well as the individual;
6. *Financial markets*—innovations help companies to survive global economic crises. To compete successfully, industry leaders fund and then utilise the research of non-profit organisations such as SRI International, allowing them to apply the principles of innovation and positive change into their enterprises (SRI 2013).

From CSR Towards Sustainable Development

Undoubtedly, the "final destination" of CSR is the well-being of all stakeholders and the overall sustainable development of society. Presently, sustainability as a concept and practice is more topical than at any other time in history. A clear, business-oriented formulation of the concept has been given by Procter & Gamble, presenting it as "an activity that ensures a better quality of life for everyone now and for generations to come" (P&G 2011a, b). The classical framework for sustainable development was set in 1987 by the World Commission of Environment and Development: "Development that meets the needs of the present without compromising the ability of future generations to meet their own needs" (WCED 1987). Industries such as forestry, mining, oil, agro business, biotechnology, etc. usually bear a larger accountability with regard to sustainable development.

However, when considering a much broader context, each individual is responsible for the development of sustainable culture that would consequently lead to the development of relevant and valuable contents for the above-mentioned framework. As stated by the World Business Council for Sustainable Development, the perfect contents should correspond to the following definition: "Development means the building of societies in which people are able to enjoy security, sufficient food, good health, decent housing, clean water and modern power supplies" (wbcsd 2010). The Council also emphasises on corporate social responsibility as a path to sustainability: "Corporate social responsibility is the continuing commitment by business to behave ethically and contribute to economic development while improving the quality of life of the workforce and their families as well as of the local community and society at large" (Holme and Watts 2000, p. 8). The implementation of sustainable principles in the day-to-day life of employees is achieved through motivation and training. However, practicing CSR across all corporate levels might be a great challenge for CEOs. In the present day world of social instability, economic crisis and ethics contradictions, workers need to see their leaders as role models. Sustainable models of conduct could be clearly demonstrated through stewardship behaviour. Stewardship in business is often associated with the obligations of individuals to protect the welfare of the company they work for. Therefore, the notion can be defined as "the extent to which an individual willingly subjugates his or her personal interests to act in protection of others' long-term welfare" (Hernandez 2012). Related to sustainability, the stewardship behaviour is pro-social action intended to positively affect other people (Penner et al. 2005). This type of behaviour is not restricted to particular levels of organisation, because it is not related to a certain position, possessing institutional power. Thus, through leadership, stewardship theory could be enacted across all levels of the companies' structure.

Leadership for Sustainable Development

But how does the manager guide subordinates towards stewardship and sustainable development behaviour? What is the key to successful change management in organisations as far as the interests of all stakeholders and society are concerned? Undoubtedly a crucial factor for organisational transformation is the leadership style, performed across all business unit levels. Leadership is commonly defined as "the process of inspiring others to work hard to accomplish important tasks" (Bird 2010). The concept encompasses encouraging beneficial efforts and enthusiasm for positive development, as well as motivation for hard work and organisational commitment towards achieving a particular goal.

Leadership is important also for the establishment of a sustainable corporate culture, which is believed to be a system of shared assumptions that strongly influences the beliefs and behaviours of its followers (Schein 2010). Research conducted by Schaubroeck et al. (2012) encourages leaders to engage all members of the organisation in ethical leadership development and thus embed their expectations concerning ethical conduct. The essence of ethical leadership is the socially responsible use of power (De Hoogh and Den Hartog 2009), which assumes that it concentrates predominantly on social issues, concerning sustainability. A much more recent study, however, reveals a current critical side of the leadership business phenomenon. According to its findings, only seven per cent of subordinates accept the behaviour of the senior management as absolutely consistent with their words (Maritz Research 2010). The majority of the organisational leaders apply traditional leadership models, and yet they seem incapable of earning the trust of their employees or the support of society at large (Perucci 2009). The loss of trust towards leaders and organisations (Bandsuch et al. 2008) turns out to be indicative of the need for a new approach to leadership in a world that seemingly lacks moral compass (Paine 2003). The new model of transformative leadership proposed by Caldwell et al. (2012) appears to be an adequate response to the necessities of the corporate world. It integrates key features of six highly respected leadership styles that would enable leaders to incorporate its principles to earn the trust of employees and society. These are:

1. *Transformational leadership style*—enabling leaders to honour synergistic duties owed to both individuals and organisations (Burns 1978). It is grounded on moral foundations concentrating on motivation through inspiration, power of influence, stimulation of intellect and an individualised approach to situations (Bass and Steidlmeier 1999). Transformational leadership triggers organisational excellence while motivating employees to seek out improvements to their company's welfare on an individual, business unit and society level. Followers are inspired to believe that they are able to create trust and a high-performance work culture which produces increased profitability and long-term sustainability (Pfeffer 1998). Apparently, the principles of transformational leadership set grounds for an integrated commitment to (1) individuals, (2) organisations and (3) community (Manville and Ober 2003).

2. *Charismatic leadership style*—stimulates a strong personal connection between leaders and followers, where the specific relationship is based on the conviction that the leader's highest asset is their "extraordinary character" which inspires followers to achieve unprecedented results (Conger et al. 2000, p. 748). Charismatic leaders challenge people's hearts and minds to change so that they become passionately committed to a great ideal—thereby helping their organisations to also fulfil their potential (Anding 2005).
3. *Level 5 leadership style*—integrates the personal humility of the leader with the need to achieve previously unrealised organisational outcomes (Collins 2001). Leaders from this category are not aware of their contribution to the development of their organisation; moreover, they prefer to acknowledge the efforts of others for organisational recognition (Collins 2008). The personality of the Level 5 leader is illustrated as one who "looks in the mirror" when problems occur, and "looks out the window" to give credit to others for success (Singh 2008, p. 740).
4. *Principle-centred leadership style*—based on values and principles, it focuses on governing oneself and honouring relationships with others that present leadership as a highly ethical obligation to honour implicit duties owed to others (Covey 1999). This leadership perspective integrates a pursuit of high ideals for becoming a better person with an obligation to create a more productive and moral society. Covey (2004), Lennick and Kiel (2008) argue that moral leadership aims to add value today, to do no harm, and to contribute to the welfare of individuals and society in the future. The ethical obligations of the principle-centred leadership style facilitate organisational change, serving as a foundation on which the justification and clarification of sustainable development can be built upon and explained to individuals not aware of its benefits.
5. *Servant leadership style* offers a broader view of the leadership concept, concentrating on the individuals' and stakeholders' interests. Defined as "providing leadership that focuses on the good of those who are being led and those whom the organisation serves" (Hamilton and Nord 2005, p. 875), servant leadership is a good starting point for Corporate Social Responsibility training within the firm. The common description of servant leadership implies that the leader sets the needs, desires, interests and welfare of others above his or her personal self-interest (Ludema and Cox 2007). Acting as a responsible steward, the servant leader helps others to achieve their own goals and at the same time honours duties owed to individuals, the organisation and society (Savage-Austin and Honeycutt 2011).
6. *Covenantal leadership style* focuses on the ability of the leader to act as a role model or an exemplar and at the same time as a contributor to partnerships development within the organisation (Pava 2003). This assumes generating a learning culture within organisations (Senge 2006) and is a key to the development of sustainability practices. Constantly seeking new truths, covenant leaders possess the power "to unleash the great human potential which is often dormant and silent" (Pava 2003, pp. 25–27). Sustainability could be accomplished through innovation opportunities coming from culture of learning and

creativity stimulation, where modern organisations create "disruptive innovations", which are the key to economic growth (Christensen 2011).

According to the research team of Caldwell et al. (2012), transformative leadership is an ethically based leadership model that integrates a commitment to values and outcomes by optimising the long-term interests of stakeholders and society and honouring the moral duties owed by organisations to their stakeholders. Transformative leadership focuses on sustainable development issues such as need for long-term wealth accumulation, achievement of terminal priorities through application of instrumental values, respect for business ethics and moral standards as a deontological (or duty-based) requirement etc. Although it can be challenging to apply the ethical standards and commitment to excellence of the transformative leadership concept, it can have a huge impact on the sustainable development of organisations and society.

Embedding Sustainable Development in Organisations: A Conceptual Framework

Based on the above-presented concepts of CSR, stewardship, sustainability and various leadership styles, the author suggests an integrated approach for embedding sustainable development within organisations. Figure 22.1 puts forward a conceptual framework, combining all sustainable development theories and ideas examined in this chapter. Called "Conceptual Framework for Organisational Sustainability", this illustrative figure could serve as a guideline for managers and leaders, undertaking training at all levels of the organisation. Practitioners who would like to embed sustainability models and need clarification of the basic definitions, concepts and leadership perspectives within a business context could also benefit from it.

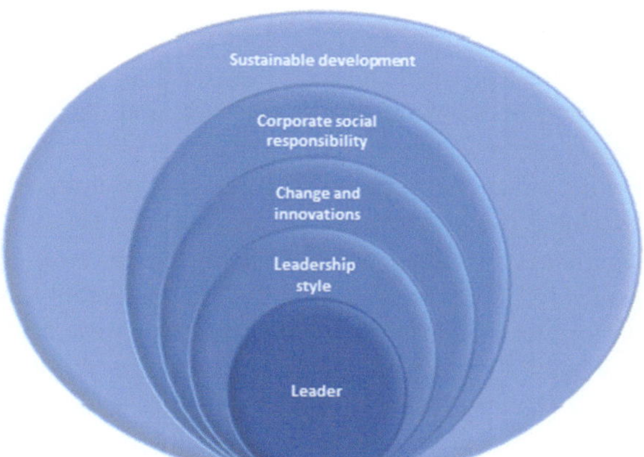

Fig. 22.1 Conceptual framework for organisational sustainability. *Source* Own Graphic

The basic idea of the framework is to unite the sustainability-related theories and ideas in a single model. Unleashing the potential of a leader to influence followers through a proper leadership style would facilitate change management, and thus, the company would excel in its innovation policy. Undoubtedly, organisational transformation would help it adapt to the pace of the modern business world, the expectations of the stakeholders and the society at large. Hence gradually, the personal skills and characteristics of a single leader could lead to sustainable development of the local economy which, consequently, would redound upon the world economic, social and environmental welfare.

Recommendations for Future Research

The presented framework focuses on the key concepts related to sustainable development within companies. However, it does not examine the personality of the leader. Further research could be conducted on relevant characteristics of those working in business and the ways in which their behaviour could be influenced through relevant leadership practices. Another relevant object for further research would be the correlation between leadership and organisational change management processes as well as the role innovation plays towards the goal of achieving sustainable development within companies and wider society. Furthermore, exploration of the interrelatedness between transformative leadership style and organisational sustainability, along with research on the hazards involved in the process of employing leadership techniques on problem people, might also prove beneficial.

Conclusion

The high moral features and commitments to excellence of the various leadership perspectives (particularly the transformative leadership style) could have a huge impact on the sustainable development of organisations and society. Despite the collision between the world economy and environmental dilemmas, the globalisation processes during the last twenty years have gradually set a new standard for organisational success. The findings of a recent survey and report conducted by global management consulting company Hay Group (2012) identifies the most significant characteristics that distinguish the world's most admired business companies from the rest. Amongst the critical factors enabling these enterprises to outperform is placing a high value on leadership, achieving success through people and building processes to sustain long-term performance. Evidently, sustainability issues should always be associated with leadership concepts and good practices in management of community and environment corporate social responsibility. Companies that concentrate on values and moral standards attract and retain the most talented people which in turn contribute to their excellent presentation and prosperity in the long run. Global sustainable development definitely demands more and more companies capable of outperforming through implementation of innovative practices and application of evidently successful leadership approaches.

References

Anding JM (2005) An interview with Robert E. Quinn. Entering the fundamental state of leadership: Reflections on the path to transformational teaching. Acad Manag Learn Educ 4(4):487–495

Badaracco JL Jr, Webb A (1995) Business ethics: a view from the trenches. Calif Manag Rev 37(2):8–21

Bandsuch M, Pate L, Thies J (2008) Rebuilding stakeholder trust in business: an examination of principle-centred leadership and organisational transparency in corporate governance. Bus Soc Rev 113(1):99–127

Bass BM, Steidlmeier P (1999). Ethics, character and authentic transformational leadership behaviour. LQ 10:181–218

Bhagwandas M (2013) Trust and Integrity in the global economy: towards economic justice and environmental sustainability. http://www.caux.iofc.org/en/trust-and-integrity-global-economy-towards-economic-justice-and-environmental-sustainability. Cited 14 Jan 2014

Bird A (2010) Leading through values: an interview with Paul Polman of Unilever. McKinsey Quarterly 2010(1):97–101

Burns J (1978) Leadership. Harper and Row, New York

Caldwell CR, Dixon RD, Floyd LA, Chaudoin J, Post J, Cheokas G (2012) Transformative leadership: achieving unparalleled excellence. J Bus Ethics 109(2):175–187

Christensen CM (2011) The innovator's dilemma. The revolutionary book that will change the way that you do business. Harper, New York

Collins J (2001) Good to great: why some companies make the leap…and others don't. Harper and Collins, New York

Collins J (2008) Celebrity leadership. Leadersh Excel 25(1):20

Conger J, Kanungo J, Menon S (2000) Charismatic leadership and follower effects. Int J Organis Behav 21(7):747–767

Covey SR (1999) Principle-centered leadership. Simon and Schuster, New York

Covey SR (2004) The 8th habitat: from effectiveness to greatness. The Free Press, New York

De Hoogh A, Den Hartog D (2009) Ethical leadership: the socially responsible use of power. In: Tjosvold D, Wisse B (eds) Power and independence of organizations: Cambridge companion to management. Cambridge University Press, Cambridge, pp 338–354

Fortune Magazine and Hay Group (2012) Lighting the path to success: Hay Group survey. http://www.haygroup.com/fortune/downloads/2012-FORTUNE-Lighting-the-path-to-success.pdf. Accessed 29 Oct 2015

Friedman M (1970) The social responsibility of business is to increase its profits, The New York Times Magazine, Sep 13. http://www.colorado.edu/studentgroups/libertarians/issues/friedman-soc-resp-business.html. Accessed 29 Oct 2015

Hamilton F, Nord WR (2005) Practicing servant-leadership: succeeding through trust, bravery and forgiveness. Acad Manag Rev 30(4):875–877

Hernandez M (2012) Toward an understanding of the psychology of stewardship. Acad Manag Rev 37(2):172–193

Holme R, Watts P (2000) Corporate social responsibility: making good business sense. WBCSD Report. http://research.dnv.com/csr/PW_Tools/PWD/1/00/L/1-00-L-2001-01-0/lib2001/WBCSD_Making_Good_Business_Sense.pdf. Accessed 14 Jan 2014

Lasserre P (2012) Global strategic management, 3rd edn. Palgrave Macmillan, London
Lennick D, Kiel F (2008) Moral intelligence: enhancing business performance and leadership success. Wharton School Publishing, Upper Saddle River
Ludema JD, Cox CK (2007) Leadership for world benefit: new horizons for research and practice. In: Piderit SK, Fry RE, Cooperrider DL (eds) Handbook of transformative cooperation: new designs and dynamics. Stanford Business Books, Palo Alto, pp 333–373
Manville B, Ober J (2003) A company of citizens: what the world's first democracy teaches leaders about creating great organisations. Harvard Business School Publishing, Boston
Maritz Research (2010) Managing in an era of distrust: Maritz Poll reveals employees lack trust in their Workplace. http://www.maritz.com/Maritz-Poll/2010/Maritz-Poll-Reveals-Employees-Lack-Trust-in-their-Workplace.aspx. Accessed 14 Jan 2014
Paine S (2003) Value shift: why companies must merge social and financial imperatives to achieve superior performance. McGraw-Hill, New York
Pava L (2003) Leading with meaning: using covenantal leadership to build a better organisation. Palgrave Macmillian, New York
P&G Suppliers (2011a) Supply chain environmental sustainability scorecard. http://www.pgsupplier.com/en/current-suppliers/environmental-sustainability-scorecard.shtml. Accessed 14 Jan 2014
P&G Suppliers (2011b) Sustainability guidelines for suppliers. http://www.pgsupplier.com/en/pg-values/sustainability.shtml. Accessed 14 Jan 2014
Penner LA, Dovidio JF, Piliavin JA, Schroeder DA (2005) Pro-social behaviour: multilevel perspectives. Annu Rev Psychol 56:365–392
Perucci R (2009) America at risk: the crisis of hope, trust, and caring. Rowman and Littlefield, Lanham
Pfeffer J (1998) The human equation: building profits by putting people first. Harvard Business School Press, Cambridge
Savage-Austin R, Honeycutt A (2011) Servant leadership: a phenomenological study of practices, experiences, organisational effectiveness, and barriers. JBER 9:49–54
Schaubroeck JM, Hannah ST, Avolio BJ, Kozlowski SWJ, Lord RG, Treviño LK, Dimotakis N, Peng AC (2012) Embedding ethical leadership within and across organisational levels. Acad Manag J 55(5):1053–1078
Schein H (2010) Organisational culture and leadership, 4th edn. Wiley, Hoboken
Senge PM (2006) The fifth discipline: the art and practice of the learning organisation. Crown Publishing, New York
Singh J (2008) Imposters masquerading as leaders: can the contagion be contained? J Bus Ethics 82:733–745
Smith C (2003) Corporate social responsibility: whether or how? CMR 45:1–25
SRI International (2013) Company Website. http://www.sri.com/. Accessed 14 Jan 2014
World Business Council for Sustainable Development (wbcsd) (2010) Current context – human development. http://www.wbcsd.org/changing-pace/current-context/currentcontexthumandevelopment.aspx. Accessed 14 Jan 2014
World Commission on Environment and Development (WCED) (1987) Our common future report. Oxford University Press, Oxford

Part VII
Sustainability—Practical Orientations

Chapter 22
Can the EU Sustainable Consumption and Production Action Plan Realize the Sustainable Development Principle?

Joanna Kielin-Maziarz

Abstract The paper describes the idea of Sustainable Development in the European Union Policy as well as in the EU environmental law. The discussion in based on the Strategy of Sustainable Development on the basis of which the concept is realized. The EU Strategy concerning sustainable development contains seven main challenges which are necessary to be fulfilled in order to achieve the sustainable development or the path leading to the sustainable development. These challenges are: Climate change and clean energy, Sustainable transport, Sustainable production and consumption, Better management of natural resources, Public health threats, Social inclusion, demography and migration and Fighting global poverty. The one which is the most interesting from the point of view of this article is the challenge concerning Sustainable production and consumption. That is why the text is based on the problem of realization of the Sustainable production and consumption—in order to answer the question if the legal measures which are adopted on the basis of this challenge can lead to that point. The legal measures which are examined are: Ecolabel Regulation, Ecodesign Directive, Energy Labelling Directive.

Keywords Sustainable development · Sustainable consumption and production action plan · Environmental law · Sustainable development strategy

The sustainable development principle occurred in the primary law in the Treaty of Amsterdam.[1] The Preamble of the EU Treaty and its article 2 as well as in articles 2 and 6 of the EC Treaty contained the notion of sustainable development. The EU

[1]Treaty of Amsterdam (1997) O.J. (C340).

J. Kielin-Maziarz (✉)
Kozminski University, Jagiellońska 57/59, 03-301 Warsaw, Poland
e-mail: jkielin@alk.edu.pl

L.W. Zacher (ed.), *Technology, Society and Sustainability*,
DOI 10.1007/978-3-319-47164-8_22

Treaty and the EC Treaty understood sustainable development in the same way as the "Our Common Future" Report.[2]

During the Göteborg Summit in 2001, a decision was made to create the Sustainable Development Strategy and its main stipulations treated it as the fundamental aim of the treaty. The European Council in Göteborg (2001) adopted the first EU Sustainable Development Strategy (SDS).[3] This was complemented by an external dimension in 2002 by the European Council in Barcelona in view of the World Summit on Sustainable Development in Johannesburg (2002).

According to the Strategy, the sustainable development principle and the question of environmental protection should be included in each of the European Policies. It can be stated that because of that, the sustainable development concept changed the whole EU Policy. EU Policies should sustain each other in the three dimensions—economic, social and environmental.

In June 2006, the Sustainable Development Strategy was changed. The New Strategy[4] indicated that the sustainable development is the major aim of the EU. According to the new Strategy, SD is the rule which should be reflected in each action of the European Union. As the Strategy states, it is possible to achieve the state of sustainable development when the societies are based on the rule of sustainability. Strategy stresses that "EU SDS recognises that investments in human, social and environmental capital as well as technological innovation are the prerequisites for long-term competitiveness and economic prosperity, social cohesion, quality employment and better environmental protection".[5]

The Renewed Strategy from 2006 notices that despite the efforts made since the first Strategy adopted in Göteborg, the aim of the Strategy, which is achievement of the sustainability, was not done as the Strategy stresses; EU was not even on the path to sustainability. According to the Strategy, since 2006 the unsustainable trends are still observed. It is noticeable, especially in the fields of climate change and energy use, threats to public health, poverty and social exclusion, demographic pressure and ageing, management of natural resources, biodiversity loss, land use and transport.[6]

Having in mind the obstacles in achieving sustainable development, the Renewed Strategy (2006) indicated that the main problem is to change the existing models of consumption and production. The main challenge is to gradually change our current unsustainable consumption and production patterns and the nonintegrated approach to policy-making.

---

[2]Report of the World Commission on Environment and Development: Our Common Future Transmitted to the General Assembly as an Annex to document A/42/427—Development and International Co-operation: Environment.

[3]COM (2001).

[4]Renewed EU Sustainable Development Strategy, Brussels, 9 June 2006, 10117/06.

[5]Renewed EU Sustainable Development Strategy.

[6]See footnote 5.

The EU Strategy concerning sustainable development contains seven main challenges which are necessary to be fulfilled in order to achieve the sustainable development or the path leading to the sustainable development. These challenges are: Climate change and clean energy, Sustainable transport, Sustainable production and consumption, Better management of natural resources, Public health threats, Social inclusion, demography and migration and Fighting global poverty. The one which is the most interesting from the point of view of this article is the challenge concerning Sustainable production and consumption.

The question of realization of one of the challenges, which is sustainable consumption and production, is enclosed in The Sustainable Consumption and Production Action Plan included in the communication from the Commission to the European Parliament, the Council, the European Economic and Social Committee and the Committee of Regions from (2008).[7]

The Action Plan states that the challenges are directly linked to our way of life. This is because of our consumption and production methods we contribute to global warming, pollution, material use and natural resource depletion.[8] That is why the Commission stresses the urgent need of more sustainable patterns of consumption and production.[9]

As it is stressed in the document, the aim of the Action Plan is to create measures which will improve the environmental and energy performance of products. This also should be popularized among consumers. In order to achieve this, the ambitious standards should be established throughout the Internal Market. It can be achieved by a more coherent and easier labelling system of the energy efficiency products. Such change will raise the level of awareness among consumers about the environmentally friendly products. As the Action Plan stresses, the approach is to address mainly the products that have a large potential for reducing its environmental negative influence.[10] The measures, which are predicted in the Action Plan, concern the improvement of the environmental performance of the products during their life cycle. The most attention in the product policy is devoted to its energy efficiency, which is one of the aspects of the climate change policy. That is why the Action Plan pays attention to the promotion and stimulation of the demand for better products and production technologies and "helping consumers to make better choices through a more coherent and simplified labeling".[11]

---

[7]The Sustainable Consumption and Production Action Plan included in the Communication from the Commission to the European Parliament, the Council, the European Economic and Social Committee and the Committee of Regions from (2008), Brussels, 16.7.2008, COM(2008) 397.

[8]Environmental Impact of Products (EIPRO), Joint Research Centre—IPTS; National Accounting.

Matrix with Environmental Accounts (NAMEA), EEA. These studies show that in the EU, the consumption areas of eating and drinking, housing (including: heating, water, domestic appliances), and travelling bear between 70 and 80 % of all environmental impacts.

[9]The Sustainable Consumption and Production Action Plan.

[10]See footnote 9.

[11]See footnote 9.

The Action Plan indicates the legal measures which should be implemented in order to improve the environmental performance of the product. As mentioned, the aim of the product policy is also to improve the energy performance of the product. That, according to the plan, can be achieved by the amendment or by the implementation of the new legal measures concerning the product's energy efficiency.

One such measure is the Ecodesign (EuP) Directive.[12] The Directive establishes a framework for setting Ecodesign requirements for energy-using products. Measures predicted by the Action Plan include also the schemes concerning product labelling. EU environmental law predicts such labelling schemes: Energy Labelling Directive[13] or the Ecolabel Regulation.[14] As the Action Plan mentions, there are also other schemes that have been developed by Member States, retailers and other economic operators which provide consumers with information on the energy and environmental performance of products.[15]

In order to achieve the model of sustainable production and consumption, the Action Plan predicted the amendment of the norm mentioned above. In this paper, these norms of the Action Plan which are devoted to the product, especially product energy efficiency, will be the only norms presented.

The first norm mentioned in the Action Plan which, according to the document, should realize the concept of the sustainable consumption and production is the Directive on the ecodesign. The aim of the Directive is to reduce the environmental impact of products, including the energy consumption throughout their entire life cycle. According to the Action Plan, the scope of the Directive on the ecodesign of energy-using products should be extended in order to cover all energy-related products.

The new Directive 2009/125/WE of ecodesign was adopted by the EU on 21 October 2009.[16] This Directive replaced the Directive 2005/32/EC of the European Parliament and of the Council of 6 July 2005, establishing a framework for the setting of ecodesign requirements for energy-using products.[17]

The aim of the Directive is to achieve a high level of protection for the environment. It can be done by decreasing the negative environmental impact of the products which are energy related. As the Directive states, it will be beneficial to consumers and other end-users.

[12]Directive (EC) (2009).

[13]Directive (EU) (2010).

[14]Regulation (EU) (2010).

[15]See footnote 9.

[16]OJ L 285/10, 31.10.2009.

[17]OJ L 191, 22.7.2005, p. 29.

According to the amendment Directive,[18] the ecodesign of products is a crucial factor in the Community Strategy on Integrated Product Policy. It realizes the prevention principle, one of the European Environmental Law principles, which is stipulated in the Treaty.[19] It is designed in order to optimize the environmental performance of products while maintaining their functional qualities.[20]

The concept of the ecodesign implies taking action on the energy-related products already at the stage of design. As the Directive stresses, pollution caused during a product's life cycle is determined at that stage, and most of the costs involved are committed then.[21] To aim at continuous improvement of products, implementing measures will also indicate a date for the review of minimum requirements and benchmarks on the basis of the foreseeable pace of technological change for the product group concerned.[22]

As the Action Plan stipulated, the scope of the Directive was widened. According to the new Directive, it covers all energy-related products. As the Directive states, the extension of the scope to all energy-related products ensures that ecodesign requirements for all significant energy-related products can be harmonized at the Community level.[23]

The Directive enables the Commission, assisted by a Committee, to enact implementing measures on specific products and their environmental aspects (such as energy consumption, waste generation, water consumption, extension of life time) following impact assessment and consultation of interested parties.

Is it then possible to state that the amendment of the Directive, which is the result of the realization of the measures predicted by the Action Plan on the sustainable consumption and production, can realize the concept of sustainable development? As the amendment Directive states "Sustainable development also requires proper consideration of the health, social and economic impact of the measures envisaged. Improving the energy and resource efficiency of products (which is the aim of the Directive—JKM.) contributes to the security of the energy supply and to the reduction of the demand on natural resources, which are preconditions of sound economic activity and therefore of sustainable development".[24]

The Directive also stresses that "in the interest of sustainable development, continuous improvement in the overall environmental impact of those products should be encouraged, notably by identifying the major sources of negative environmental impacts and avoiding transfer of pollution, when this improvement does not entail excessive costs".[25]

---

[18]Directive 2009/125/WE, OJ L 285/10, 31.10.2009.

[19]Art. 192 of Treaty.

[20]See footnote 18.

[21]See footnote 18.

[22]See footnote 9.

[23]See footnote 18.

[24]See footnote 18.

[25]See footnote 18.

It then can be summarized that the aim of the Directive, as well as the scope of its amendment, intends to realize the concept of sustainable development. In addition, having in mind that the scope of the Directive is now widened, it can be concluded that the measure included in the Action Plan can contribute to the realization of the concept. Products which use energy, as the Directive notes, take a large share of the market being the subjects of a broad consumption and have an important impact on the environment. Increasing the impact of the Directive by enhancing products in relation to which they are applied, ecodesign causes a positive impact on the environment through this change. Because of that, more products achieve more suitable character. Thus, it can be stated that the amendment of the Directive contributes to the realization of the concept of sustainable development. Besides, that Directive states that its aim, which is the enhancement of the energy efficiency of the product, is to contribute to the program of the sustainable production which is a part of the 10-year framework of programs on sustainable production and consumption adopted during the World Summit on Sustainable Development in Johannesburg in (2002).[26]

The second norm, which should be examined in light of the realization of the concept of the sustainable consumption and production, is product labelling under the Energy Labelling Directive and Ecolabel Regulation.

Directive 2010/30/EU on energy labels was adopted by the EU on 19 May 2010.[27] The aim of the Energy labels is to help consumers in choosing energy-saving products, which at the same time are cost effective. Simultaneously, they provide incentives for the industry to develop and invest in energy efficient product design.[28]

Directive 2010/30/EU replaced the Directive 92/75/EEC,[29] the scope of which was very narrow. It referred only to household appliances. The Action Plan[30] mentioned above also indicated the necessity of the extension of the scope of Directive 92/75/EEC. According to the Action Plan, the Directive should also be applicable to the energy-related products which have a significant direct or indirect impact on energy consumption during use. Directive reinforce potential synergies between existing legislative measures, in particular Directive 2009/125/EC of the European Parliament and of the Council of 21 October 2009 establishing a framework for the setting of ecodesign requirements for energy-related products[31] (which was the subject of the previous paragraph—JKM.).

---

[26]See footnote 18.

[27]See footnote 13.

[28]Available at: http://ec.europa.eu/energy/efficiency/labelling/labelling_en.htm, 23.08.2012.

[29]Council Directive (1992).

[30]See footnote 9.

[31]Directive 2010/30/EU OJ L 153/1, 18.6.2010.

As indicated in the Action Plan, the scope of the Directive had been widened. The Energy Labelling Directive is extended in order to cover a wider range of products, including energy-using and other energy-related products.

The Directive[32] states that there is a need for a compulsory scheme; the voluntary one would not be efficient, as only some products will fulfil the requirements and thereby only some of them would be labelled or supplied only with standard product information. This could lead to the misinformation for some of the end-users. The amendment Directive stipulates that the labelling, which informs about the consumption of energy and other essential resources, is addressed to all the products concerned.

In order to be able to determine the extent to which the revised Directive could contribute to the concept of sustainable consumption and production, it seems necessary to refer to the content of the Directive. This is because the Directive states that the answer to this question will provide an overview of progress and report on the implementation of the Action Plan on Sustainable Consumption and Production and Sustainable Industrial Policy, which will be carried out by the Commission in 2012.[33] In the overview, the Commission will, as the Directive states in particular, check if there is a necessity for further action in order to improve the energy efficiency of the products. The Commission will also review if it is possible to inform consumers about the carbon footprint of products or the products' environmental impact during their life cycle.

It also should be mentioned that this Directive should not prejudice the application of Directive 2009/125/EC. According to the norm, this Directive and other Union instruments form part of a broader legal framework and, in the context of a holistic approach, bring about additional energy savings and environmental gains.[34] This statement could create the argument that the mentioned norms together, being complementary to each other, contribute to the realization of the concept of the sustainable consumption and production.

The third norm which should be presented is the EU Regulation concerning Ecolabel. The amendment Regulation of the European Parliament and of the Council on the EU Ecolabel 66/2010/EC was adopted on the 25 of November 2009.[35]

The Ecolabel is a voluntary scheme. That means it is possible for the producers, importers and retailers to apply for the label, which could be added to their products. The aim of the previous regulation concerning the Ecolabel scheme, which was the Regulation (EC) 1980/2000 of the European Parliament and of the Council of 17 July 2000 on a revised Community ecolabel award scheme,[36] was to create a voluntary ecolabel award scheme that aimed to promote products with

[32]See footnote 31.

[33]See footnote 31.

[34]See footnote 31.

[35]Regulation 66/2010/EC OJ L 27/1, 30.01.2010.

[36]Regulation (EC) No. 1980/2000, OJ L 237, 21.9.2000, p. 1.

a reduced negative environmental impact which is measured during its whole life cycle. Science-based information on the characteristics of the environmentally friendly products should be delivered to the consumers in the most reliable way by the ecolabel.[37] According to the Action Plan, the ecolabel will complement the information provided to consumers as a voluntary label. "It will act as a 'label of excellence' to signal to consumers those products that perform at such a level when many environmental criteria are considered over the whole life-cycle".[38]

The current Regulation stresses that the Ecolabel scheme is part of the sustainable consumption and production policy of the European Union. The aim of the scheme is to promote products in an environmentally friendly method through the use of the EU Ecolabel.

To this effect, it is appropriate to require that the criteria with which products must comply in order to bear the EU Ecolabel be based on the best environmental performance achieved by products on the Community market. Those criteria should be simple to understand and to use and should be based on scientific evidence, taking into consideration the latest technological developments. Those criteria should be market oriented and limited to the most significant environmental impacts of products during their whole life cycle.[39]

The way the environmental criteria are built was predicted in art. 6 of the Regulation. According to the article, Ecolabel criteria should be based on the environmental performance of products, taking into account the latest strategic objectives of the EU in the field of the environment.[40] The criteria are necessary to be fulfilled in order to receive the product label. The criteria are based on the scientific basis, which concerns the whole life cycle of the product. These are determined by: their impact on climate change, on nature and biodiversity, energy and resource consumption, generation of waste, emissions to all environmental media, pollution through physical effects and use and release of hazardous substances, the substitution of hazardous substances with safer substances, as such or via the use of alternative materials or designs, wherever it is technically feasible; the potential to reduce environmental impacts due to durability and reusability of products; the net environmental balance between the environmental benefits and burdens, including health and safety aspects, at the various life stages of the products; where appropriate, social and ethical aspects.[41]

The Ecolabel Regulation is to be revised to simplify and streamline the process of obtaining an ecolabel and extend the product coverage.

[37]See footnote 35.

[38]See footnote 9.

[39]See footnote 35.

[40]See footnote 35.

[41]See footnote 35.

According to the Action Plan, the norms mentioned above are to be further developed and, following a review of the Ecodesign Directive in 2012, complemented as appropriate by an Ecodesign Labelling Directive to provide consumers with information about the energy and/or environmental performance of products.[42]

In 2009, the Commission prepared communication including the Review of the European Union Strategy for Sustainable Development.[43] According to the second Progress Report, some development in the scope of Sustainable Consumption and Production has been achieved since the previous revision. However, these changes, as stated in the report, showed a rather mixed picture.[44] The better results were more noticeable in the sphere of sustainable production than consumption. According to the Commission, some progress was achieved in the production sphere. The Commission noted that the positive trends could be observed in isolation of economic growth from natural resource consumption but still in the consumption area there were unfavourable signs.[45]

The Action Plan stipulated that the review of the progress of its implementation would be conducted in 2012. The report, which would then be prepared by the Commission, should analyse whether further action to improve the energy and environmental performance of products was needed, in particular by extending the Ecodesign and Labelling Directives to cover nonenergy-related products.[46] However, no report had been prepared. Ahead of this 2012 review, the Commission was reviewing the progress of the Action Plan by means of this mid-term evaluation. A new report was prepared for the Commission by ECORYS in 2011 titled: Mid-term Evaluation of the Sustainable Consumption and Production and Sustainable Industrial Policy Action Plan.[47]

The evaluation stated that the Action Plan includes varied instruments which aim to create more sustainable consumption and production. At the same time, these instruments provided competitiveness in the European economy.[48] In the evaluation, there were recommendations which include actions necessary to implement the concept of the sustainable consumption and production.

It also should be mentioned that the Action Plan noticed the obstacles in its realization.

---

[42]See footnote 9.

[43]Brussels, 24.7.2009, COM (2009).

[44]See footnote 43.

[45]See footnote 43.

[46]See footnote 9.

[47]Available at: http://ec.europa.eu/environment/eussd/pdf/14.%20SCP-SIP%20AP%20Mid%20Term%20-%20Final%20Report.pdf, 24.08.2012.

[48]Mid-term Evaluation of the Sustainable Consumption and Production and Sustainable Industrial Policy Action Plan.

As the Action Plan stressed, the realization of the sustainable consumption and production might be weakened as the measures concerning the product policy addressed only the specific aspects of the product's life cycle. The other question, which did not influence positively the realization of the concept, was that the measures of the obligatory and voluntary character were not connected, which was why the potential effect of synergies between them was not used.[49] The potential effect of the Internal Market, which could help the effectiveness of the concept, was also not used. It happened because as the Action Plan stressed, the requirements built on the national and regional level were not coherent. This caused the misunderstandings between the producers, and because of that, these measures rested ineffective.[50]

According to the recommendations in the area of sustainable consumption and production, there is still a need for more ambitious and effective policy. The problem of the source scarcity still exists, and there are still growing negative effects of the excessive consumption; these effects are especially noticeable in climate change. That is why there should be undertaken the measures which will change the European economy. This could be achieved by introducing the eco-innovation measures. However, as the estimating document stated, "the high level indicators on resource efficiency are proving difficult to move in the correct direction".[51] The recommendation states that the high level policy statement should include the long-term objective of reducing consumption, having in mind that it is not a fast process. Recommendations stressed that its achievement was difficult and required further researches. This was because it aimed for hardly achievable targets for resource efficiency and linked them with the targets such as energy efficiency, waste generation, competitiveness, integrated product policy, sustainable development and social inclusion which are also high level aims.[52]

As the recommendation suggested, in order for the Action Plan to achieve its goals, it should include the aspects as follows: the level of the awareness between the consumers should be widened; this would have influence on the more environmentally friendly behaviour. Besides that, according to the recommendation, in order to achieve greater participation in the Action Plan by the society, it should take advantage of the existing activity like the Competitiveness and Innovation Programme, Life+ and Interreg and new activity, which aim to change the consumers' behaviour by being more conscious of their environmental impact.

[49]See footnote 9.

[50]See footnote 9.

[51]See footnote 48.

[52]Ibidem.

Despite the necessary action which aimed to influence consumers' behaviour, the recommendation noticed the low impact of Ecolabel. Authors of the report stressed that the scheme suffered from low awareness and low uptake. Such evaluation did not allow the conclusion that the current system contributes to the concept of sustainable consumption and production. The authors of the report then stated that there are two options for its future appearance: "to be either to stay as a small, niche [sic] label or to radically alter its criteria and nature possibly by seeking to set principles to which sector and MS specific labelling schemes could adhere".

In relation to the Ecodesign process, the recommendation summarized that the Action Plan had not done much in order to develop it. The reason for the low development of the Ecodesign scheme was the complexity of the modification of the scope of products, from energy using to energy related had a role. The other obstacle, which might be the reason for the weak popularity of the scheme, was the inherent technical complexity, resistance by some stakeholders and lack of political drive.[53]

Summarizing the scope and realization of the Action Plan, the recommendation stressed that the wide scope of issues that the Action Plan sought to address "makes it difficult to achieve a balance that is seen as ideal in the eyes of all stakeholders".[54] It happened, as the report observed, because each stakeholder tended to have a focus on a particular issue/objective and in their eyes, this objective was to be the most important.

Assessing the Action Plan for implementation of the sustainable development concept in light of the three criteria of sustainability, economic, environmental and social, the authors of the report noticed the environmental issues were covered in the most comprehensive way. Also, economic issues received, as the authors stated, reasonable coverage. It happened by the measures including the promotion of eco-industry, or by encouraging it to produce goods with a lower environmental impact. The report stressed that the third dimension of the sustainable development concept was not clearly defined by the Action Plan.

Despite that, the report stated that having in mind the fact that the low budget was associated with the Action Plan, which made its realization harder and gave the plan as a whole a relatively short life, it could be regarded as a satisfactory performance.

The estimation of the norms, which are the subject of the paper, shows that the mandatory instruments, Energy Labelling and Ecodesign, have been achieving their goal, so it then could be stated that they contribute to the realization of the concept of sustainable consumption and production. They play a greater role compared to the voluntary schemes. But the overall effect is not as satisfactory as the

[53]Ibidem.

[54]Ibidem.

authors of the Action Plan wished to achieve. According to the report, the improvement is noticeable but still too small; at the same time, it is very hard to prove that the changes are the result of the instruments included in the Action Plan.[55]

On the other hand, the report states that the instruments predicted in the Action Plan can be seen as "a good conceptual match with the objectives". That means that the report estimates that the measures of the plan can realize its aim. The report states that the instruments adopted in the Action Plan might influence production and consumption.[56] But, it also stresses that the nature of these instruments is too soft and because of that, they are not able to achieve the expected scale of change.[57] The other weakness of the Action Plan is the lack of high level targets. Such a situation does not help in estimating if they are fulfilled. These refer also to each of the instruments predicted in the Action Plan. They also do not have clear targets. Because of that, the answer to the question the instruments predicted in the Sustainable Consumption and Production Action Plan can realize that the sustainable concept is not clear. As the report stated, the main idea of the norms, which were included in the Action Plan as those which contributed to the realization of the concept, was correct. However, lack of their clear targets and problems with their estimation made it so it was not possible to give a straight answer to this question. Despite the troubles concerning the clear estimation of the measures predicted by the Plan, the report indicted some data of the effects of its realization. The Action Plan and the following reports and policy measures were valuable experiences paving the way to sustainable consumption and production in the EU. Works and actions still are in progress, and they are increasingly more ambitious.

References

Communication from the Commission (2001) A sustainable Europe for a better world: a European union strategy for sustainable development. Brussels, 15 May 2001, COM (2001) 264

Communication from the Commission to the European Parliament (2009) The Council, The European Economic Social and Social Committee and the Committee of the Regions: Mainstreaming sustainable development into EU policies: 2009 Review of the European Union Strategy for Sustainable Development, Brussels, 24.7.2009 COM (2009) 400 final

Council Directive 92/75/EEC of 22 September 1992 on the indication by labelling and standard product information of the consumption of energy and other resources by household appliances OJ L 297, 13 Oct 1992

Directive 2009/125/EC of 21 October 2009 establishing a framework for the setting of ecodesign requirements for energy-related products OJ L 285/10, 31.10.2009

[55] Ibidem.

[56] Ibidem.

[57] Ibidem.

Directive 2010/30/EU of 19 May 2010 on the indication by labelling and standard product information of the consumption of energy and other resources by energy-related products, OJ L 153/1. 18 June 2010
Environmental Impact of Products (EIPRO), Joint Research Centre -IPTS; National Accounting Matrix with Environmental Accounts (NAMEA), EEA
Mid-term Evaluation of the Sustainable Consumption and Production and Sustainable Industrial Policy Action Plan (2012)
Regulation 66/2010/EC of 25 November 2009 on EU Ecolabel, OJ L 27/1, 30 Jan 2010
Regulation (EC) No. 1980/2000, OJ L 237, 21 Sept 2000, p. 1
Renewed Sustainable Development Strategy, Brussels, 9 June 2006, 10117/06
Report of the World Commission on Environment and Development: Our Common Future Transmitted to the General Assembly as an Annex to document A/42/427—Development and International Co-operation: Environment
The Sustainable Consumption and Production Action Plan included in the Communication from the Commission to the European Parliament, the Council, the European Economic and Social Committee and the Committee of Regions from 2008, Brussels, 16 July 2008, COM (2008) 397
Treaty of Amsterdam (1997) O.J. (C340)

Chapter 23
Implication for China's Resource Demand on Sustainability in Australia

Jin Hong, Wentao Yu, Dora Marinova, Xiumei Guo and Margaret Gollagher

Abstract The recent rapid growth of the Chinese economy combined with urbanization and changes in consumption patterns are increasing Chinese demands for iron, oil, gas and coal. These changes have greatly influenced China's public policies, especially energy and resource policy and foreign policy. For instance, China has moved to strengthen cooperative relations with foreign resource-intensive countries. Australia, and Western Australia (WA) in particular, is an ideal target for China in this regard. This chapter first studies the existing relationship between China's demand for resources and supply from Western Australia. It then analyses different effects produced as a result of this resource trade. China's demand for resources has been a key driver of economic growth in Western Australia in recent decades, and this is likely to continue for the foreseeable future. However, this resource-driven economic growth is inconsistent with a sustainable approach to development in WA. The chapter discusses some potential conflicts and predicts some trends in the Sino-Australian resource trade. In conclusion, it presents suggestions for policy-makers.

Keywords Resource supply and demand · China · Australia · Sustainability · Western Australia · Economic growth · Trade

---

J. Hong (✉) · W. Yu
School of Management, University of Science and Technology of China,
No. 96 Jinzhai Road, Hefei 230026, Anhui, China
e-mail: hongjin@ustc.edu.cn

D. Marinova · X. Guo · M. Gollagher
Curtin University Sustainability Policy Institute, Curtin University, GPO Box U1987,
Perth, WA 6845, Australia

L.W. Zacher (ed.), *Technology, Society and Sustainability*,
DOI 10.1007/978-3-319-47164-8_23

Introduction

China has been experiencing uninterrupted economic growth for over the last three decades. China's GDP has increased by 22 times since 1978 and grew at an average growth rate of 9.9 % between 1979 and 2011 (National Bureau of Statistics of China 2012). In 1978, the total primary energy consumption in China was 0.57 billion tons of standard coal equivalent (SCE), and in 2007 it increased to 2.22 billion tons of SCE (Guerrero, n.d.). China is currently the global second largest oil consumer, and the largest energy consumer in the world. Correspondingly, resource and energy demand in China have increased tremendously. In 2011, China's oil consumption growth accounted for 50 % of the global oil consumption growth. Although annual growth rates in China have recently slowed down to around 7 %, the country's economy is likely to go strong for another two decades, with this continuous expansion escalating resource demands (US Energy Information Administration 2013).

China's growing demands for energy and other resources have greatly influenced China's public policies, especially energy and resource policy, and foreign policy (Kristen 2005). For instance, China has moved to strengthen cooperative relations with foreign resource-intensive countries. Australia, and Western Australia (WA) in particular, is an ideal target for China in this regard. China can secure a stable, quality supply of resources from WA, which in turn has experienced a commodity boom associated with significant economic benefit. However, the short-term and long-term effects of this supply–demand relationship are very complex and have significant implications for sustainable development in WA (Government of Western Australia 2003).

China's Resource and Energy Demands

China has experienced rapid economic growth over the last decade (see Fig. 23.1). In 2001, China's gross domestic product (GDP) was 1096 billion RMB, while in 2011 GDP grew to four times that amount. The average economic growth rate was about 10 % per annum during the period of 2001 to 2011 (see Fig. 23.1). However, this high growth rate has been achieved, to a great extent, at the expense of natural resources and the ecological environment.

Structural Analysis of China's Resource and Energy Demand

China's economic growth has been associated with dramatic increases in energy and resource use (Wu 2004). The country is the world's largest consumer of coal,

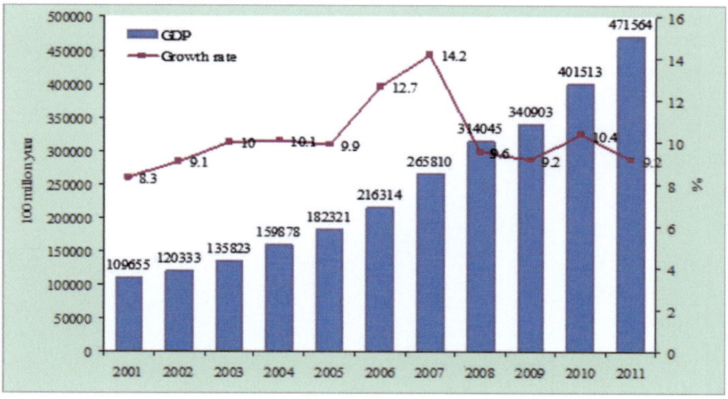

Fig. 23.1 China's economic growth from 2001 to 2011. *Source of data* National Bureau of Statistics (2013)

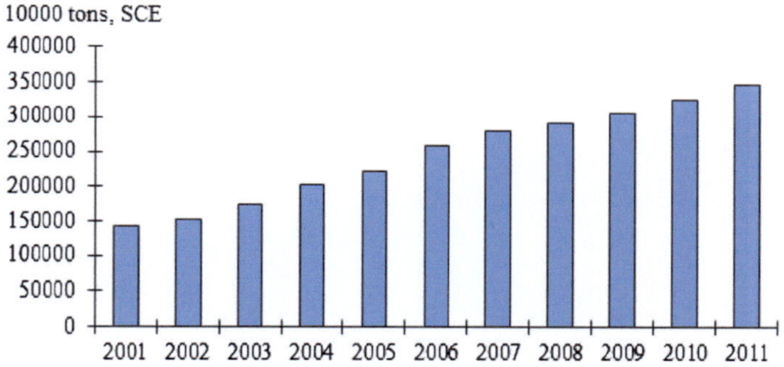

Fig. 23.2 Total primary energy consumption during 10th and 11th Five-Year Plan. *Source of data* National Bureau of Statistics (2013)

iron and steel, and copper, and the second largest consumer of oil and electricity (Wang 2009, p. 125). Its demand for energy and resources impacts significantly on the global market. Recently production and consumption in China have been increasing rapidly. Figure 23.2 shows the change of total primary energy consumption, measured by standard coal equivalent (SCE), in the 10th and 11th Five-year Plan (FYP). According to Fig. 23.2, total primary energy consumption has increased steadily over the last decade.

China consumes more energy and resources than it produces, as shown in Table 23.1, particularly in crude oil and iron ore.

A breakdown of China's energy consumption by source is shown in Fig. 23.3, coal is the dominant source of energy in China, 68.4 %, followed by oil, 18.6 %; hydroelectric power, 8 %; and natural gas, 5 %. Clearly, coal and oil remain the dominant sources of energy in China, while natural gas and hydroelectric power together account for only 13 % of energy consumption.

Table 23.1 Production and consumption of energy and resources in 2010

	Product	Growth rate (%)	Consumption	Growth rate (%)
Total primary energy (SCE, 100 million tons)	31.8	7.0	34.8	7.0
Coal (100 million tons)	35.2	8.7	34.9	9.7
Crude oil (100 million tons)	2.04	0.3	4.4	2.7
Natural gas (100 million m^3)	1030.6	8.7	1174.2	12.0
Hydroelectricity, wind electricity, and nuclear electricity (SCE, 10,000 tons)	27,982.9	0.26	27,840.2	−0.37
Iron ore (100 million tons)	1.3	27.2	9.2	–
Copper (10,000 tons)	206.1	14.88	330	11.9

Source of data National Bureau of Statistic of China (2013)

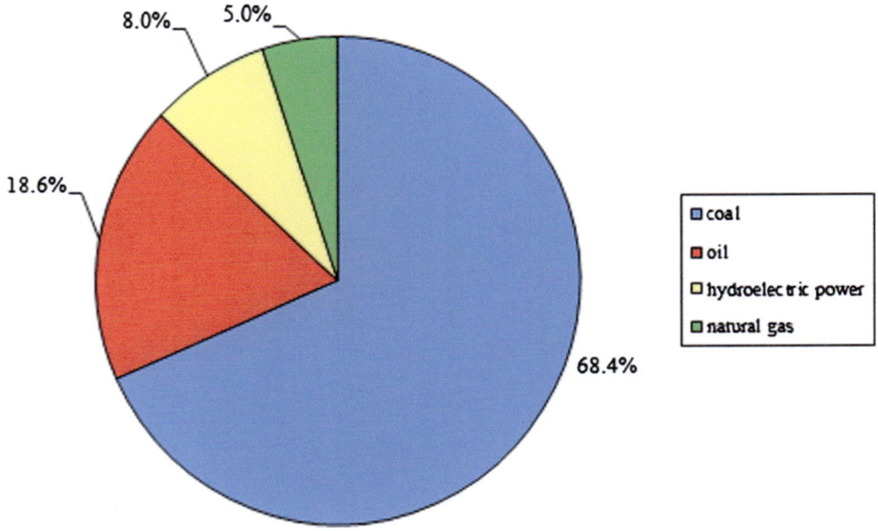

Fig. 23.3 China's energy consumption in 2011. *Source of data* National Bureau of Statistic of China (2013)

Causes of the Rapid Increases in China's Resources and Energy Use

For the last three decades, China' intensifying industrialization and fast increasing urbanization have driven China's energy demand which is expected to continue to rise in the coming decades (WorldWatch Institute 2010). A number of factors drive this increase in energy demand. These include: First, in developing countries, as elsewhere, economic growth is typically closely correlated with increased demand for energy. As China's enormous and robust economy has grown, energy use has increased correspondingly. Secondly, energy consumption for non-production purposes has also increased dramatically. As living standards rise in China, people increase their spending on home appliances such as air conditioning, refrigeration, cooking, and space and water heating, as well as on automobiles and other energy-intensive products. Increasing urbanization has amplified this process. A third major factor contributing to high levels of energy consumption in China is inefficiency.

Future Development Trends

It is difficult, and somewhat risky, to accurately predict what China's future resource demands are likely to be. According to National Bureau of Statistics of China, China's annual energy consumption in 2010 totalled 3.25 billion tons of standard coal, up 5.9 % from 2009, with its natural gas consumption jumping 18.2 % year on year in 2010, while coal, crude oil and electricity consumption also increased 5.3, 12.9 and 13.1 %, respectively, year on year (China Daily 2011). Generally, China's resource and energy demands will continue to grow significantly for the foreseeable future, despite recent downturns in economic growth. With China's strong and continued economic growth, the country became a net oil importer and its demand for oil has increased, from 4.6 mb/d in 2000 to over 8 mb/d.

In 2011, China imported more than 5 mb/d of crude oil, accounting for 54 % of its total demand (IEA 2012). In addition, it was reported that within 4 years, China will consume more oil than the USA. It is also projected that China's primary oil demand will increase to 12.2 mb/d and that the country will spend $500 billion a year on crude oil imports by 2020 (IEA 2012; Forbes 2013). In any case, China is likely to continue to be a big consumer in the world resource market.

In addition, energy consumption structure will be improved gradually, for instance, the consumption of renewable energy and natural gas has been increasing rapidly. But because of price and supply, the basic composition of energy consumption will not be greatly affected. Coal is expected to continue to play a dominant role in China's energy mix for the foreseeable future, although there is a serious coal pollution in China.

Australia's and Western Australia's Resource and Energy Supply to China

As a member of many organizations such as APEC, the G20, WTO, and OECD, Australia has multiple trade flows with numerous countries, such as China, the USA, and Japan (see Table 23.2). However, Australia's recent economic progress has been heavily reliant on trade with China. In 2009, China became Australia's largest export market, surpassing Japan (see Table 23.2). Resources continue to underpin

Australia's exports to China. Australia is China's 11th largest merchandise trade partner, 11th largest import source, and 13th largest export destination. Australia exported 266.2 million tonnes of iron ore to China in 2009, an increase of 45.2 % over the same period. China is also Australia's largest source of imports. In 2011, China became Australia's largest import market, surpassing the European Union. China exports many different major products to Australia such as clothing, communication and sporting equipment, electronic devices, goods for children, and furniture.

As the world's largest exporter, China's share of worldwide value-add manufacturing has doubled since 2002 (Australian Government 2012). Western Australia, in particular, has a strong trade relationship with China. WA is booming on a trade and investment relationship based on the export of minerals, energy, agricultural goods and education services and Chinese investment in the WA resources sector. In 2011, more than 70 % of Australia's total merchandise exports to China are from WA which also accounts for 80 % of China's investment into Australia over the last 5 years (Australia China Business Council 2013).

Australia's Resource and Energy Supply to China

More importantly, the two economies are complementary and have great potential for trade. China was Australia's largest export market, accounting for 26.1 % ($82.5 billion) of total exports in 2011–2012 (see Table 23.2). Iron ore and concentrates, coal, education-related travel services, and crude petroleum are Australia's largest exports (see Table 23.3). As showed in Table 23.3, most of the natural resource and services exports to China increased rapidly in year 2009–2011. These trades are underpinned by sizable long-term contracts, for example for iron ore and the sale of liquefied natural gas into the Guangdong market, and will remain the basis of Australia's export trade for many years to come.

Table 23.2 Australian merchandise trade shares, by selected countries and country groups

Period	South East Asian regions	China	Europea n Union	Hong Kong	Japan	New Zealand	Republic of Korea	United States of America
Annual exports (%)								
2008–2009	9.2	17.0	10.4	1.4	22.9	3.7	8.3	5.0
2009–2010	9.9	23.2	7.9	1.4	18.5	4.0	8.2	4.8
2010–2011	10.1	26.4	7.3	1.3	19.1	3.1	9.2	3.7
Annual imports (%)								
2008–2009	19.9	16.9	20.7	0.6	8.1	3.2	3.0	11.5
2009–2010	20.2	17.9	18.9	0.6	8.7	3.4	3.5	10.7
2010–2011	18.4	19.2	18.0	0.5	7.8	3.4	3.3	10.8

Source of data ABS (2013)

Table 23.3 Australia's top 10 goods and services exports to China (A $ million)

						% growth	
	Rank commodity	Sector	2009	2010	2011	2010–2011	5 year trend growth
1	Iron ore and concentrates	Minerals	21/790	34,685	43,960	26.7	44.9
2	Coal	Fuels	5651	5191	4543	−12.5	78.9
3	Education-related travel services	Services	4040	4343	4091	−5.8	13.1
4	Crude petroleum	Fuels	796	1668	2902	74.0	64.5
5	Wool and other animal hair (incl tops)	Agriculture	1382	1621	2022	24.7	4.9
6	Copper ores and concentrates	Minerals	1050	1314	1500	14.2	7.0
7	Cotton	Agriculture	180	426	1470	244.8	31.2
8	Gold	Other goods	1	220	1284	484.2	449.6
9	Copper	Manufactures (STM)	890	1066	1211	13.6	46.6
10	Nickel ores and concentrates	Minerals	610	1121	1094	−2.3	−0.4
Total	–	–	36,390	51,655	64,077	862	–

Source of data Australian Government (2013)

As a resource-rich state, WA is supplying China with a large amount of natural resources, and its top 10 exports to China in 2011 include iron ore, crude petroleum, nickel ores and so on.

Effect of China Rise on Australian Sustainable Development

Generally, the effects of China's resource demands on sustainable development in Australia include positive and negative dimensions. However, it is difficult to distinguish between these two effects in real situations. What is positive or negative often depends on what measures are adopted to deal with these challenges. The impacts of China's resource demands are complicated and manifold, such as economic, ecological, political, social, and cultural.

Economic Effects

China's huge and continuing demands have produced and will produce some positive results. First it propels the Australian economy. Second it is an important opportunity to restructure the Australian economy and industry, but this opportunity must be exploited in a timely fashion. In economic development, natural resources should be treated as a potential source of income, some of which is saved and converted into capital to support increases in future output levels. For example, resource rents may be used for transport development, modernization of telecommunication systems, health and educational programs, science, and technology activities.

On the other hand, this boom results from the advantage of Australia's abundant natural resources, and will continuously strengthen resource industries in the Australian economy. But this assumption is risky. Overdeveloped resource industries may restrain development of other industries, especially manufacturing and technology-intensive industries. At the same time, it is difficult to keep the advantages brought by natural resources for a long time, and high dependence on it will result in losing sustainable dynamics of economic growth.

According to research conducted by economists Sachs and Warner (1995), countries which base their economies on natural resources, such as Russia, Nigeria and Venezuela, tend to be examples of development failures over the longer term. In contrast, countries which have had only limited access to natural resources, such as Japan, Korea, Singapore, and Switzerland, have experienced extremely high economic growth rates. Some economists have further studied these issues and produced some interesting conclusions.

At the discovery of a natural resource, the sudden income increase may lead to sloth and less need for sound economic management and for institutional quality. The boom may also create a false sense of security and weaken the perceived need for investment and growth-promoting strategies. Naturally resource abundant economies benefit less from the technology spillovers that are typical in manufacturing industries because the exports of these industries are harmed by an appreciation of the local currency, e.g., through the inflationary pressure resulting from increased domestic demand. Finally, as the natural resource sector expands relative to other sectors, the returns to human capital decrease and investments in education decline (Sachs and Warner 1995, 2001; Papyrakis and Gerlagh 2004).

Ecological Effects

Rapid growth of resource industries has resulted in some ecological and environmental issues in China. Non-renewable resources are decreasing; some landscapes have been destroyed; water and soil are polluted; some residential areas are damaged. As China's demand for resources will continue for a long time, it is enhancing the expansion of WA's resource industries and its relevant departments, such as minerals and oil production, transport construction, population, and labor growth, which will bring pressure to ecological system in Australia and Western Australia, creating additional problems. In addition, economic prosperity does not play an absolutely negative role in environmental protection. In fact, the recent resource boom has strengthened WA's financial ability, making challenging tasks such as ecological restoration and environmental infrastructure development much more economically feasible.

Political Effects

As a new economic power, China will have a definite influence on the Australian political system. Some politicians and parties, who keep close relation with resource industries, will play a more important role in political and administrative activities on both a national and state levels. At the same time, political powerful interest groups, which are associated with natural resource abundance, are emerging and attempting to influence politicians and to adopt policies that may not favor environmental protection. Some policies will be challenged. If parties do not adjust their policies, it will produce some adverse consequences. Furthermore, natural resources provide rents so that they promote rent-seeking competition rather than productive activities. In addition, rents induce economic agents to bribe the administration in order to gain access to them (Krueger 1974; Sachs and Warner 1995).

Social and Cultural Effects

With the growth of two-way trade and personnel exchange, diverse factors of Chinese culture influence the social and cultural life of WA. Although Chinese have been living in WA, Chinese elements brought by them are thin and a little ostensible. In other words, there has been little or no mainstream Chinese cultural influence on WA. However, Chinese culture will play a more important role in the cultural diversity of WA in the future, as more Chinese people are immigrating into the state. Essentially, Chinese culture is rooted in Confucianism and Australian mainstream culture is rooted in European culture whose core values are originated from a Christian culture. So a clash of civilizations, to some extent, may take place in WA.

Suggestions to Policy-Makers as Well as for Educating the Broader Community

In order to achieve the sustainability goals of social advancement, economic prosperity and environmental protection, governments, industries, and social communities respectively need to take on important responsibilities. The first involves making sure that each understands their social, economic and environmental responsibilities and that each group, appreciates the importance of sustainability by developing an awareness of how to implement it within their sector scope of influence.

Promoting Communication and Diffusion of Ideas and Knowledge of Sustainability

Theoretically, sustainability knowledge and education should be provided to everybody within in a certain area. However, from the view of efficiency and practice, it should focus on some special individuals and organizations, which could be called *key actors*. These will be critical stakeholders that provide an important impact on sustainability strategy. Such stakeholders might include some special persons, administrative institutes, economic agents, NGOs and local communities. More importantly, these key stakeholders should be integrated into a key stakeholders network, which will become the main target for sustainability education and thus a hub of knowledge diffusion and strategy implementation. Another issue that needs special attention is that lots of Chinese enterprises lack strong awareness of environmental protection and sustainability. So Australian economic agents and governments should popularize ideas and knowledge of sustainable development and social responsibility. This not only improves economic cooperation but also helps Chinese business accept an more sustainable development model.

Promoting Industrial Reconstruction and Adjusting Economic Strategy

To help ensure continued development of the Australian economy, the most import factor to consider is that Australia should base its economy and industries on technological progress and innovation, not on natural resources. Resource booms only provides a strategic opportunity by which certain parts of the national and state economy (e.g., WA Resources Industry) might adjust their industrial structure and surpass previous performance by a great leap. The development of Australia's future business trade with China is unlikely to be only limited in the resources sector. Australia has strong capacities of the provision of technology and expertise that would help Asia, including China, in safeguarding food and energy supplies (Australian Government 2012). Enhancing technology-intensive manufacturing industry is also a critical strategy, as it will bring more talents, more capital and more knowledge. Otherwise, overreliance on a resource boom may prove to be more of a curse than a boon.

In summary, Australia continues to put more effort into enhancing sustainability and communities are also embracing a low-carbon and environmentally friendly society. The strong trade relationship between Australia and China will also bring more positive Chinese cultural elements such as harmony between humans and nature, Confucius and other oriental wisdom. This has been strongly encouraged in Australia's strategies developed in the White Paper of Australia in the Asian Century (Australian Government 2012). Therefore, Australia would become a green and innovative country with quality environment, economic prosperity, cultural diversity, and social inclusion.

Acknowledgments The first two authors acknowledge the financial support from National Natural Science Foundation of China (71172213), China's Department of Education (09YJA630153), the large bid project from Social Sciences Fund (08&ZD043). The last two authors acknowledge the financial support of the Australian Research Council.

References

Australian Bureau of Statistics (ABS) (2013) International trade in goods and services. Cat No. 5368.0. http://www.abs.gov.au/ausstats/abs@.nsf/mf/5368.0. Cited 28 Jan 2014

Australia China Business Council (2013) Western Australia: overview. http://www.acbc.com.au/default.asp?id=1,26,40,39. Accessed 28 Jan 2014

Australian Government (2012) Australia in the Asian century: executive summary of White Paper. http://pandora.nla.gov.au/pan/133850/20130914-0122/asiancentury.dpmc.gov.au/white-paper.html. Accessed 28 Jan 2014

Australian Government (2013) Australia's exports to China 2001 to 2011. http://www.dfat.gov.au/publications/stats-pubs/australias-exports-to-china-2001-2011.pdf. Accessed 28 Jan 2014

China Daily (2011) Energy consumption per unit of GDP drops 4 %. http://www.chinadaily.com.cn/china/2011-02/28/content_12090709.htm. Accessed 28 Jan 2014

Forbes (2013) Within four years, China to consume more oil than U.S. http://www.forbes.com/sites/kenrapoza/2013/08/25/within-four-years-china-to-consume-more-oil-than-u-s/. Accessed 28 Jan 2014

Guerrero D-G (n.d.) China's ecological time bomb. Focus of the Global South. http://focusweb.org/node/1355. Accessed 28 Jan 2014

Government of Western Australia (2003) Hope for the future: the Western Australia state sustainability strategy. Department of Premier and Cabinet, Perth, Australia

International Energy Agency (IEA) (2012) Oil and gas security, emergency response of IEA countries. http://www.iea.org/publications/freepublications/publication/name,3755,en.html. Accessed 28 Jan 2014

Kristen AD (2005) China's environment and the challenge of sustainable development. M.E Sharpe, Armonk

Krueger AO (1974) The political economy of the rent-seeking society. Am Econ Rev 64(3):291–303

National Bureau of Statistics of People's Republic of China (National Bureau of Statistics) (2012, 2013) http://www.stats.gov.cn. Accessed 28 Jan 2014

Papyrakis E, Gerlagh R (2004) The resource curse hypothesis and its transmission channels. J Comp Econ 32:181–193

Sachs JD, Warner AM (1995) Natural resource abundance and economic growth. NBER Working Paper No. 5398. National Bureau of Economic Research, Cambridge, MA

Sachs JD, Warner AM (2001) The curse of natural resources. Eur Econ Rev 45:827–838

US Energy Information Administration (2013) International energy outlook 2013. http://www.eia.gov/forecasts/ieo/. Accessed 28 Jan 2014

Wang T-F (2009) China's economic development and its challenges. In: Hsiao H-HM, Lin CY (eds) Rise of China: Beijing's strategies and implications for the Asia-Pacific. Routledge, Oxon, pp 107–130

WorldWatch Institute (2010) Renewable energy and energy efficiency in China: current status and prospects for 2020. WorldWatch Report 182. WorldWatch Institute, Washington, DC

Wu Y (2004) China's economic growth. Routledge Curzon, London

Chapter 24
Integrated and Sustainable Approaches to Address City Inundation in China

Baohui Zhao, Dora Marinova and Xiumei Guo

Abstract Each year, many Chinese cities are inundated with water and this results in substantial losses (people, infrastructure, etc.) and environmental challenges to the state and its residents. The subsequent reaction to this problem largely focuses on the inefficiencies of the cities' sewer systems, which is depicted as 'the conscience of the city' in Hugo's (1862) novel 'Les Miserables.' This idea of blaming inefficient sewer systems for a city's inundation troubles is exploited by many critics in China. However, Hugo's emphasis on poor sewer management as the principle cause of inundation and his subsequent advocating for the construction of a 'Paris-style' sewer are poles apart from addressing the problem of city inundation. This chapter contributes to the debate about wastewater management through the case study of the City of Jiaonan in China. We find that given the exponential urbanization experienced by China, when making efforts to address city inundation, sustainable approaches in urban stormwater management should be taken, instead of relying solely on engineering technologies. With China's rapid urbanization as a result of its population and economic growth over the decades, urban infrastructure such as water distribution, wastewater treatment and sewer systems needs to be much more adequate and efficient, particularly stormwater management infrastructure. In many Chinese cities, the stormwater management infrastructure has been constructed many years ago and is deteriorating fast. Therefore, there is an urgent need for government strategies in support of redesigning or upgrading current facilities. This needs massive capital investment and is also time-consuming. To keep the ever-expanding cities running, it has become common practice in China to simply add new segments of infrastructure to the already existing system. This study outlines a sustainability strategy model for urban stormwater management, and its findings encourage policy makers to adopt sustainable and integrated approaches toward urban stormwater management while quality economic growth is sustained.

B. Zhao (✉) · D. Marinova · X. Guo
Curtin University Sustainability Policy Institute, Curtin University, GPO Box U1987, Perth, WA 6845, Australia
e-mail: baohui.zhao@postgrad.curtin.edu.au

L.W. Zacher (ed.), *Technology, Society and Sustainability*,
DOI 10.1007/978-3-319-47164-8_24

Keywords Sustainable · Inundation · Sewer · Stormwater · Urbanization · Water resources, policy

Introduction

Urban stormwater management is an important component in urban water resources management. Preventing cities from being inundated has always been a priority in urban planning and governance. In essence, the traditional urban stormwater management approaches are engineering-based within a linear system rather than an integrated system (Rudlin and Falken 1999). It is characterized by 'big pump out' (Newman and Mouritz 1996). City stormwater is pumped out through the engineering facilities such as the sewer, channelized creeks and pump stations. The engineering achievements such as the magnificent sewer system in European cities highlight the significance of stormwater management. However, these growth-oriented, 'supply-side' approaches are unsustainable and have reached their social, economic and environmental limits (McManus 2005). Sustainable approaches in urban stormwater management cover concepts such as water sensitive urban design (WSUD), water management circular system at local level (Rudlin and Falken 1999; McManus 2005; McManus and Haughton 2006) and integrated land–water approaches (FAO 1993; Falkenmark 2008). These approaches involve on-site stormwater capturing or harvesting, recycling and reuse within a city and will substantially reduce stormwater runoff.

In line with China's exponential urbanization over the decades, hundreds of Chinese cities are annually inundated during the rainy seasons resulting in substantial losses. A Web site review finds that most of the subsequent uproar arising as a result of city inundation largely focuses on the inefficiencies of the city's sewers, which is somehow described in Victor Hugo's novel as 'the conscience of the city' (Hugo 2013). This idea is capitalized by many critics in China to blame the government for its failure in stormwater management and city inundation prevention. It is further optimized to criticize the widely existing disjuncture between the governments' promises and performances. However, Hugo's interpretation where the blame is placed on the sewer system and a Paris-style system is advocated for, risks being preyed upon by 'construction-happy' local officials. This chapter contributes to the debate through the case study of the City of Jiaonan (COJ) of China. It finds that integrated approaches rather than the 'pump out' technologies work better when addressing city inundation.

Context Analysis

The COJ is a county-level city in Shandong Province China and is situated on the west coast of the Yellow Sea. To its north, west and southwest are vast farming hinterlands. The northwestern part of the city is hilly, while the southeastern

Table 25.1 Urban area and population of Jiaonan, 2002–2009

Years	Population (thousand)	Area (m^2)
2009	54	365
2008	53.2	356
2007	52.7	332.4
2006	51.2	N/A
2005	48.6	321
2004	45.2	303
2003	N/A	N/A
2002	33.8	221

Source Quingdao Government (2012)

part is coastal plain. Jiaonan's topography is characterized by its disproportion in elevation, high in the west and north, but low in the east and south (Li et al. 2005). The annual average rainfall ranges between 560 and 990 mm and is in decline. Precipitation mainly takes place from June to September (Bureau of Water Resources Jiaonan City 2010).

As an agriculture-based town, Jiaonan city's rapid urbanization is a showcase of China, expanding quickly both in land area and in population. It covers an area of 54 km^2 with a population of 335,100 as of December 2009 (Quingdao Government 2012). Within eight years (2002–2009), its population has increased over 65 %. Table 25.1 shows the urban growth in Jiaonan. With the city's quick development, a variety of issues have arisen and concerns have been growing among all actors of the city. One of them is the city's water resource management which falls into a vicious circle (see Fig. 7.1). On the one hand, COJ is experiencing a drastic scarcity of available water supply. The annual rainfall in 2009 was 562.4 mm, much lower than the average rainfall (751.7 mm) of the previous years, representing a drastic drop of 43.3 % from 992.4 mm in 2008 (Quingdao Government 2012). On the other hand, the city is prone to inundation in summer. In August 2001, COJ was stricken with a severe flooding. The affected area exceeded 10 km^2 with average flood water as deep as 1 m (Li et al. 2005). In August 2002, COJ was again inundated with flood water over 1.5 m deep in the city's CBD (Quingdao Government 2012) (Fig. 25.1).

Having suffered from water scarcity as well as inundation, the COJ has been engaged on two fronts in its water resources management. To combat water scarcity, traditional approaches have been adopted. Securing more water resources and increasing supply are among the countermeasures. Five medium-sized dams have been constructed in the city's hinterlands, mainly in cropping areas. Water is piped to COJ to meet its increasing demand as a result of population growth and improvements in living standards.

Unfortunately, water scarcity is far from being eased in Jiaonan city. It experiences critical water shortages during more than half of the year. Unless water demand is addressed, efforts and technologies will only focus on approaches geared toward increasing supply (McManus 2005). The COJ is yet to win the battle on the first front. To address city inundation, as in many other Chinese cities,

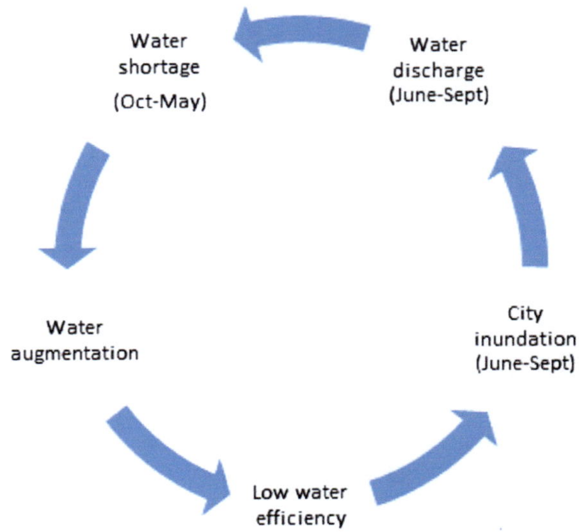

Fig. 25.1 Water scarcity and city inundation. *Source* Own graphic

expensive engineering operations have been taken and millions of dollars have been pumped into water resource management projects or 'hydrological projects' over the years. About $20 million was spent in 2002 alone. A reservoir was built in the upper catchment of the Xiaoxin River that flows across the city. The river was re-channeled and its dykes consolidated. Pumping stations were built in the lower stream in order to pump water out in case of too much stormwater. These 'seeable' 'individual career-enhancing' engineering hydrological projects are preferred by city policy-makers across China. Such practices are typical of nineteenth-century management solutions, where stormwater is first collected by expensive sewer systems and then discharged into receiving water bodies, with any engineering work focused on nearby rivers and applicable drainage systems (Newman and Mouritz 1996). As a result, city inundation has remained an annual event in Jiaonan for many years. The fact that COJ keeps being flooded indicates those approaches are neither cost-effective nor cost-efficient.

Integrated Sustainable Approaches

The City of Jiaonan used to be a cozy coastal county town where one could cycle to most destinations along bicycle paths lined with native trees without fear of being knocked down any time by a crazy road user. Several clean rivers and creeks flowed across the city ending in the sea in its southeast, where the beautiful beach and seawater were crystal clear and clean. Residents had never experienced city inundation as frequent and severe as in the 1990s. Jiaonan was a city with a character of its own, which is an invaluable asset. Urbanization and industrialization have ripped off the city's character, degrading it to just another identical Chinese city. Water scarcity, city inundation and other environmental woes became factors that might undermine the goals of building a well-off (xiao kan) and harmonious (he xie) society. In recent years, COJ has adopted integrated sustainable approaches rather than piecemeal solutions to meet this challenge.

To regain its character and build a sustainable city, COJ intends to develop clean industries such as services, education, commerce and tourism industries (Han 2006). It will shift to a responsible sustainable development model rather than the previous 'drain the pond for fish' approach, which was robbing natural resources from future generations. The needs of current and future generations, as well as local natural ecosystems, will now be considered. Farming lands will be protected for social harmony, future food safety, biodiversity and the preservation of the local ecosystems. It will maintain a proper urban scale rather than expanding to 384 km^2 (half of Singapore) with a projected population of only 700,000 as planned (Han 2006). Residents will benefit from the city's development, and their welfare will be looked after regardless of their household registration (hu kou) categories.

The COJ will now manage its urban inundation in a systematic way. City inundation is not caused by rainfall within the urban area itself. The aim is to address the problem from not only within, but also outside the city at the same time. In the city's hinterlands and in its upper catchment, large-scale re-vegetation projects have been undertaken. Closer to the city, buffer zones (greenbelt in the north and west) were built. This greenspace with selected species of plants plays an important part in the city's stormwater management. Stormwater from hinterlands catchments, from the northern and western hills is now retained when running down into the city.

Wetlands have been constructed in the city. Some of them are in close proximity to urban residential compounds. They have become attractive local community features (Jiaonan Government 2011; Qindaonews 2011). Stormwater is directed into these wetlands for natural filtering and purification before it reaches waterways. Within the urban area, COJ discarded the traditional method of 'channeling or straightening up the rivers' and has chosen to maximize the benefits of meandering rivers in arresting and retaining stormwater, where it would slowly filtrate into the ground, thus recharging aquifers (Li et al. 2005). The creeks and ponds which had been covered up in the process of urban development over years are now recovered to trap stormwater runoff before releasing it into the water courses. Swales are also used along some of the city's major roads to slow and reduce stormwater runoff and filter it before it goes to the waterways. Stromwater formed within the city will either be harvested, stored by government-invested facilities, run into ponds, creeks and wetlands (re-surfaced or newly built) or flow into water aquifers through off-road swales, broken kerbings and pervious pavers. Only limited stormwater now runs into the Xiaoxin River after flowing through the re-vegetated spaces on the river banks and into the wetlands in the city's south. This complete stormwater arresting and harvesting system coupled with natural drainage is more reliable than expensive sewer systems, and more importantly, it works when there is heavy rainfall (McManus 2005).

Integrated approaches as highlighted above help to solve problems, save money and enhance sustainability. The COJ's integrated city inundation system is now well prepared for heavy rainfall conditions in the summer. Having reviewed the city's practices to manage stormwater and address city inundation, this chapter proposes several recommendations to be included in the COJ's stormwater management strategies. From a sustainable development policy perspective, Jiaonan city needs to take effective measures to control urban expansion and stabilize its population, so that farming lands and vegetation would not be destroyed in the surrounding catchments. This is vital because they retain stormwater and most importantly provide primary supplies for the residents of the city in addition to many other important functions.

As to the technical side, the COJ needs to take the following steps:

- Establishing multiple-purpose corridors that will integrate drainage and conservation functions to detain stormwater runoffs in the city's northwest and northeast where most of the runoff forms and runs into urban areas. Native plant and vegetation species are preferable for these corridors.
- Rehabilitating Xiaoxin River to regain its original nature, manage it in a more environmentally friendly way and restore this major urban creek into a meandering watercourse. Xiaoxin River is the scapegoat for all the city inundation over the years. This rehabilitation process involves but is not limited to:

    1. Clearing all buildings built on top of parts of the river;
    2. Changing the two-way roads along the west bank of the river into one way, re-claiming all land within 50–100 m from the east bank and re-vegetating the reclaimed space with native species, turning the spaces along the river into natural parklands; and
    3. Within the city, bringing back to the surface the ponds and creeks which used to connect with Xiaoxin River; outside the city, in the lower discharge areas, restoring buried natural wetlands or building artificial wetlands and linking them together and further connecting them with the Xiaoxin River system.

- Applying more water sensitive urban design practices into urban design planning as one of the components of the city's systems management of stormwater. This can be aided by using more local instead of expensive exotic species in urban road greening works, building more roadside swales, adopting broken kerbings, reducing paved areas within the city and introducing more pervious paving materials.

Discussion and Conclusion

With China's rapid urbanization as a result of its sustained population and economic growth over the decades, urban infrastructure such as water distribution, wastewater treatment and sewer systems tends to become insufficient. In many cities, the stormwater management infrastructure has been constructed over the years; the redesign and upgrading of the facilities need large capital investment and are time-consuming. Due to widespread financial constraints, the stormwater management infrastructure in many cities is deteriorating. To keep the expanding city running, it has become a common practice in Chinese cities to simply add infrastructure to the existing system (FAO 1993). In addition, many cities are yet to adopt sustainable and integrated approaches for urban stormwater management. Many cities take piecemeal solutions in stormwater management. Negative consequences resulting from outdated infrastructure and unsustainable approaches to stormwater management, in an ever-expanding city, are therefore unavoidable. City inundation under heavy rainfall conditions can be particularly severe and costly.

In his ‘Les Miserables,’ Victor Hugo dedicated one chapter to explore the ancient history of the sewer. He wrote: ‘The sewer is the conscience of the city. Everything there converges and confronts everything else. In that livid spot there are shades, but there are no longer any secrets... A sewer is a cynic. It tells everything' (Hugo 2013: 1258). Firstly, it is obvious that Hugo’s ‘conscience’ from his famous piece meant nothing in relation to *liang xin*, the Chinese version of conscience, and individual integrity. Secondly, in today’s cities, the traditional engineering-based stormwater management approaches represented by expensive sewer systems are no longer the preferred solutions to city inundation. Given the history of urbanization in China, it is no longer appropriate to cling onto the outdated nineteenth-century stormwater management practices.

Stormwater has always been seen as a flood-triggering factor in China and elsewhere. Huge capital is invested in stormwater management and city inundation prevention, with the cities’ sewers sucking away a big proportion of the investment. As the cities keep expanding or sprawling, this public-money-sucking machine will never stop. Jiaonan is a showcase of China’s rapid urbanization as well as transition from traditional to sustainable and integrated approaches in addressing city inundation. Stormwater is taken as a natural resource in COJ (Geng et al. 2007). Its integrated approaches have yielded the expected results. For example, the city recorded the heaviest rainfall in August 2007. However, the deepest flood water in the city’s CBD was only 0.35 m and it receded within 2 h (Dazhong Daily 2007). The COJ’s case reveals that a city’s sewer, somehow literally understood by some critics to represent a city’s *liang xin*, is just one of the areas that need to be given priority when addressing city inundation. Quoting an individual sentence from Hugo’s novel and overstating its real-life implications when advocating a Paris-type sewer context is misleading in the first place. Secondly, it does not contribute to the cities’ endeavors to explore sustainable approaches to address the problems. Furthermore, it risks being optimized by ‘construction-happy’ city officials. In today’s cities, inundation has not much to do with Hugo’s ‘conscience of the city.’

Sustainable and integrated approaches should be adopted in addressing Chinese cities’ inundation. They involve the integration of sustainable development, urban design and stormwater management (Mouritz 2000). Stormwater is a resource rather than an inundation source that should be harvested, recycled, reused and harnessed in sustainable cities. In order to achieve these, as suggested by UNEP (2012), the combination of new and retrofit network-based technologies and infrastructure is essential in the integrated and systematic approaches to the construction of sustainable cities, including the City of Jiaonan.

References

Bureau of Water Resources Jiaonan City (2010) Annual hydrological report. Bureau of Hydrology, Jiaonan City, China. http://www.jnsswj.cn/Show-news.ASP?ID=1025. Accessed 10 Apr 2010

Dazhong Daily (2007) Three approaches taken to address the inner city flooding in the City of Jiaonan: three consecutive flooding wakes up the dreamer. http://dzrb.dzwwww.com/pdf/2007/08/15/db15.pdf. Accessed 20 Sep 2010

Food and Agricultural Organization (FAO) (1993) Land and water integration and river basin management. In: Proceedings of a FAO internal workshop. 31 Jan–2 Feb 1993. Retrieved from http://www.fao.org/publications/card/en/c/747649f4-1cd8-5b03-8c05-8a3e1e5039f3/. Accessed 31 Oct 2015

Falkenmark M (2008) Water and sustainability: a reappraisal. Environment 50(2):4–16

Geng R, Yang SZ, Yan WX (2007) Practices and explorations on optimizing stormwater in the City of Jiaonan. Water Resources Forum, Shandong (in Chinese)

Han JY (2006) Connecting with Qingdao City. http://www.sina.com.cn/c/2006_04_27/05258800095s.shtml. Accessed 10 Apr 2010

Hugo V (1862) Les miserable. http://www.online-literature.com/victor_hugo/les_Miserable/324/. Accessed 12 Nov 2011

Hugo V (2013) Les Misérables. Signet Classics, New York

Jiaonan Government (2011) Wetland park planning 2009. http://ghj.jiaonan.gov.con/xiangxi.asp?id=53517. Accessed 28 May 2011

Li FH, Liu ZP, Ma BG (2005) Tackling the flooding dilemma of the 'old city' by adopting a scientific development concept, Shandong. Hydrology 10:22 (in Chinese)

McManus P (2005) Vortex cities to sustainable cities: Australia's urban challenge. University of New South Wales Press, Sydney

McManus P, Haughton G (2006) Planning with ecological footprints: a sympathetic critique of theory and practice. E&U 18(1):113–127

Mouritz M (2000) Water sensitive urban design-where to now? A paper presented at water sensitive urban design workshop. 30–31 August 2000. Melbourne, Australia. Retrieved from http://www.ewater.org.au/archive/crcch/archive/pubs/pdfs/technical200107.pdf. Accessed 31 Oct 2015

Newman P, Mouritz M (1996) Principles and planning opportunities for community scale systems of water and waste management. Desalination 106(1–3):339–354

Qindaonews (2011) A harmonious chapter between the hills and the sea—a sketch of urban development in the City of Jiaonan 2005. http://www.qingdaonews.com/content/2005-07/04/content_4998033.htm. Accessed 19 May 2011

Quingdao Government (2012) Qingdao economic brief for the year 2002–2003. http://qdsq-ls.qingdao.gov.cn/shizhi/laoshanshizhi.ssf/8441323bd3cf67fc48256dl. Accessed 10 Apr 2012

Rudlin D, Falken N (1999) Building the 21st century home: the sustainable urban neighborhood, 2nd edn. Architectural Press, Oxford

United Nations Environment Programme (2012) Sustainable, resource and efficient cities—making it happen! http://www.unep.org/urban_environment/PDFs/SustainableResourceEfficientCities.pdf. Accessed 12 July 2012

Chapter 25
Nano-biotechnology for Water Sustainability: Bibliometric Analysis

Li Xu and Dora Marinova

Abstract Nano-biotechnology is regarded as having high potential for solving challenges related to water, food and biodiversity. Of particular interest to sustainability is its promising ability to enhance the supply and security of water resources for human use. The chapter applies bibliometric analysis to describe the trends in the development of nano-biotechnology for issues related to water supply, contamination prevention and treatment. A co-occurrence analysis is used to identify the types of technologies emerging in this area, namely related to bioengineering, chemical engineering, microbiology and material sciences. The majority of the new knowledge comes from the USA, but researchers from China, South Korea, the Netherlands, India and Australia are also making their mark.

Keywords Technological innovation · Water supply · Water contamination · Water treatment · Research trends · Nano-biotechnology · Bibliometric

Introduction

Water is vital for life on our planet, but expanding population, industrialisation and industrial agriculture put immense pressure on this important resource. Taken for granted in developed countries, nations in the developing world are still struggling to provide clean and safe water. Security and scarcity are two major challenges. While the problems across the globe may vary, many regions face multiple changes in water supply and water quality that undermine its accessibility and safety, thereby destabilising human well-being.

L. Xu (✉) · D. Marinova
Curtin University Sustainability Policy Institute, Curtin University, GPO Box U1987, Perth, WA 6845, Australia
e-mail: li.xu2@postgrad.curtin.edu.au

L.W. Zacher (ed.), *Technology, Society and Sustainability*,
DOI 10.1007/978-3-319-47164-8_25

Access to clean and safe drinking water and sanitation is a human right and a prerequisite for adequate standards of living (UNESCO 2011). Yet 884 million people have limited drinking water and 2.6 billion people live without proper sanitation (UNESCO 2011). Although the quantity of water on Earth has been stable for millions of years, its quality has deteriorated progressively due to a combination of human demographic and activity factors (PWC 2012). Climate change coupled with degradation of the quality of surface and groundwater reserves exacerbates the already serious global and region-specific issues.

Technological solutions have always been compelling in assisting with water challenges, ranging from construction of dams, piping and drainage to desalination, filtering devices and wastewater treatment. According to PricewaterCooper (PWC 2012), R&D and innovation can generate solutions related to water production, including alternative sources from sea water and marginal quality water, reuse, intelligent consumption and optimised sanitation. In recent years, nano-biotechnology, a technological field combining nanotechnology and biotechnology solutions, is seen as a significant potential to solve water challenges across the globe and enhance the supply of clean water for human use (Diallo and Brinker 2011). Nano-biotechnology is the branch of nanotechnology that deals with biological and biochemical applications and uses (Venkatesh 2009), or in other words, it represents "the use of nano-science for specific biological applications" (Gazit 2007, p. 13). Research in nano-biotechnology (also referred to as nanobiology) developed in embryonic stage in the mid-2000s and is still in its infancy. A search of published papers in Scopus up to year 2013 generates only 9 entries pertaining uniquely to water sustainability and nano-biotechnology. This small number indicates that although nano-biotechnology is promising its overall impact is still negligible. The two areas from which nano-biotechnology emerged, namely nanotechnology and biotechnology, however have been going strongly for decades. Immediate and fast solutions related to water sustainability need to rely on the progress and advancement of knowledge made in these two individual classes of technologies.

How can these new technologies contribute to dealing with water security and scarcity? What is the scientific evidence that they can offer useful and workable solutions? Diallo and Brinker (2011) point out the potential of nanotechnology solutions for safe environment and water resources in efficiently supplying potable water for human use and clean water for agricultural and industrial applications. Similarly, many argue the potential biotechnology holds for biotreatment and bioremediation to control water quality, decontaminate wastewaters, and monitor and prevent pollution (e.g. Zechendorf 1999; Gommen and Verstraete 2002). Both nanotechnology and biotechnology can be used in water treatment, for example nanofiltration membranes for producing potable water from brackish groundwater (Hillie and Hlophe 2007) or biofilm bioreactors for wastewater treatment (Van Loosdrecht and Heijnen 1993).

Using a bibliometric approach, we analyse in this chapter how active the nanotechnology and biotechnology research field is in relation to water. We describe this as nano-biotechnology. While several bibliometric studies have examined nanotechnology and biotechnology through patent or publication activities (Meyer 2001; Marinova and McAleer 2003; Schummer 2004; Leydesdorff and Rafols 2009; Rafols and Meyer 2010; Thursby and Thursby 2011), not much is known about issues related to water sustainability. Despite their promising potential in relation to water, the links between nano- and biotechnology are yet to be firmly established. In order to address water sustainability priorities in the time being, we need to understand the individual trends within the two individual technology groups, and this is where a bibliometric analysis of nano-biotechnology can be very informative. It is also important to see how researchers connect with each other in collaborative efforts to address the water challenges.

Hence, we conduct a study of nano-biotechnology research on water sustainability issues with the aim to present the global trends in the area and give direction for future quantitative studies. We firstly screen publications indexed by the Web of Science (Thompson Reuters ISI) and Scopus to describe the general trends in nano-biotechnology research in the past decades. This allows for the more appropriate database for further analysis to be selected. The methods, data used and scope of the analyses are presented in "Methodology and Data" section. "Results and Discussion" section presents the results from the bibliometric investigation based on co-occurrence analyses. We conclude the study with a discussion and further research directions in "Conclusion" section

Methodology and Data

Bibliometrics aims to quantitatively analyse the publications covered in scientific sources (De Bellis 2009). Described as the foundation for the science of science (De Solla Price 1963), it applies the methods of science to examine scientific output. The two most commonly used bibliographic databases of academic work are Web of Science established by the Institute for Scientific Information (ISI) and more recently acquired by Thompson Reuters, and Scopus owned and operated by Elsevier.

Data Collection

Choice of Bibliographic Database

In order to determine which of the two databases has a better coverage and is better suited for this study, we firstly conduct queries, using nanotechnology and biotechnology as search words in the total number of publications in Scopus and Web of Science. We search for publications which contain these two words separately in either their titles or lists of keywords. The specific journal databases that we use for the search are Science Citation Index and Social Science Citation Index in Web of Science and Physical Sciences and Social Sciences and Humanities in Scopus (see Fig. 25.1).

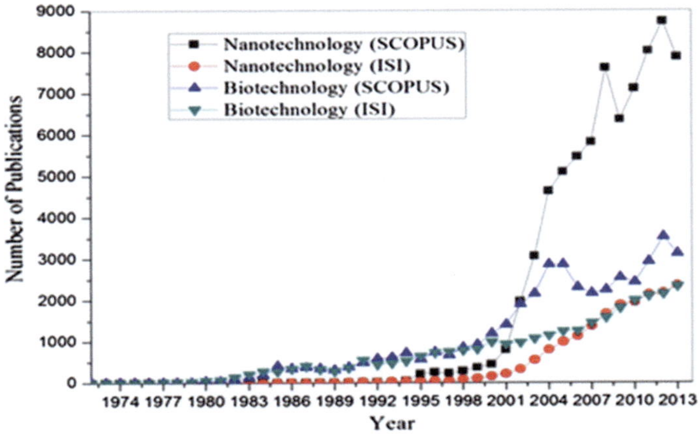

(data accessed July 2014)

Fig. 25.1 Number of publications per year with "nanotechnology" and "biotechnology" in the title or keywords

The total numbers of publications in Scopus and Web of Science are 74,361 and 42,375 for nanotechnology and 18,038 and 29,482 for biotechnology. The peak numbers for nanotechnology publications are 8719 for Scopus in 2012 and 2351 for ISI in 2013, while for biotechnology they are 3529 for Scopus in 2012 and 2333 for ISI in 2013. The historical trends in the two databases show that biotechnology research preceded nanotechnology—the first two biotechnology publications appeared in 1972 in Web of Science, while the first nanotechnology publication was registered in 1978 in Scopus. Although there might be earlier publications in both areas in other bibliographic databases, overall research in biotechnology started prior to that in nanotechnology.

Given the broader coverage of publications by Scopus on both technologies, this bibliographic database is chosen for further analysis and we believe it represents well the progress made in the respective fields. We are particularly interested in research and advancement of knowledge through these two types of technologies in relation to water sustainability. This requires narrowing down the publication fields using specific keywords.

Keywords and Dataset

Growing demands for water and increasing agricultural and industrial pollution continuously intensify the stress on surface and groundwater resources and on supply systems. The three main issues related to water sustainability are supply, contamination and treatment (Gray 2010). These are the three retrieval keywords that we used to examine the research progress made in nano-biotechnology in relation to the sustainability of water resources. Hence, the keyword combinations

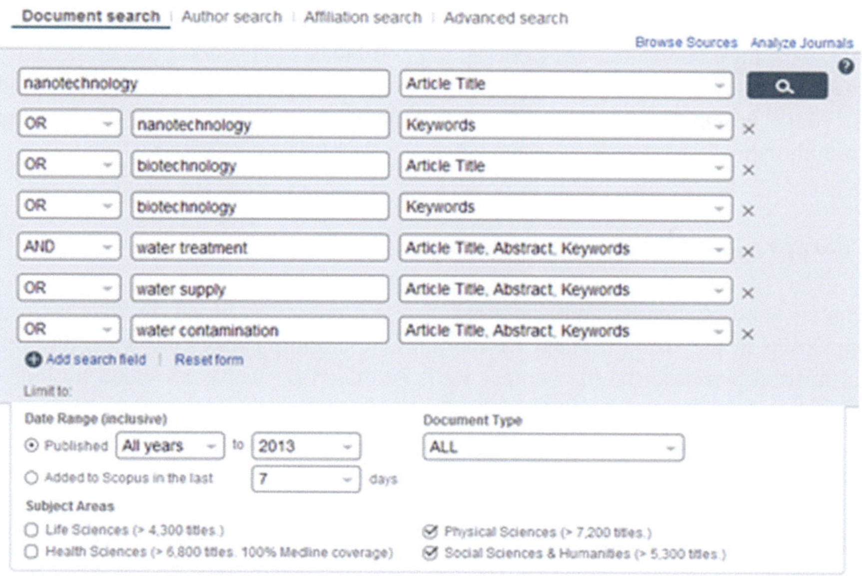

Fig. 25.2 Keywords combinations used in Scopus

are: nanotechnology and water supply, nanotechnology and water contamination, nanotechnology and water treatment, biotechnology and water supply, biotechnology and water contamination and biotechnology and water treatment (see Fig. 25.2). In order to ensure data integrity, we manually cleaned the obtained data set by deleting any repeated publications to avoid double counting and removing all documents for which author(s) names were not available.

Data Analysis

Co-occurrence analysis is applied to identify research hot points and connections in the field of nano-biotechnology for water sustainability. The linguistic term "co-occurrence" refers to analysis of related words (Kroeger 2005), and in this case, the mutual occurrence of two units in the same metadata field, e.g. "climate change" or "Newman and Kenworthy". This type of bibliometric analysis has been previously used to describe research development, including to explore concept networks and reveal research themes (Courtial 1994; Ding et al. 2001; Ronda-Pupo and Guerras-Martin 2010; Hu et al. 2013). One of the first studies applying a co-word analysis was conducted by Rip and Courtial (1984) who mapped developments in biotechnology.

This study is the first to specifically examine water-related issues. The co-occurrence method here comprises two sections, namely co-word and co-authorship analyses, and is employed to reveal on which topics nano-biotechnology researchers focussed their studies on water sustainability and how well they connected with one another. Bibexcel is used for the bibliometric analysis, and Pajek (a software for large networks representation) combined with ArcMap is applied to visualise the networks of dominant topics and co-authorship.[1]

Co-word Analysis

The co-word analysis reveals the research hot points. It is based on the keywords appearing in the lists provided by the authors assuming that they were properly scrutinised and describe the content well. Extended keywords provided by indexers are excluded as they do not always indicate exactly what the authors have done and may lead to double counts. The higher the co-occurrence frequency of two words, the closer relationship they have.

The co-word data set is built by cleaning meaningless publications according to two criteria: (1) when the extracted keywords from the list overlap with the ones used as searching criteria to identify the publication (see Fig. 25.2); and (2) when the keywords from the list are not representatives of a well-defined topic (e.g. being too general, such as “water” or “modelling”). For example, the frequency distribution extracted 114 “nanotechnology” and 16 “modelling” from the keywords lists of all publications. These publications are excluded from the sub-data set because “nanotechnology” is a keyword we used in searching for publications and “modelling” does not reveal any specific subject domain. We use words which appear more than ten times in the co-occurrence matrix and co-words which appeared more than five times in the matrix to build the descriptive networks. This avoids having too many vectors and arcs in the graphic representation which might make the visualisation unclear.

Co-authorship Analysis

The co-authorship analytical method is a good way to discover social networks, scholarly collaborations, scientific evolutions and the research performance of various fields (Barabasi et al. 2002; Liu et al. 2005; Abbasi et al. 2011). We constructed global co-authorship networks for nano-biotechnology research in water sustainability to reveal the scholarly connections among leading authors around the world. To visually present the global connections through the authors’ networks, they are graphed onto a world map according to the places with which the researchers affiliate.

[1]Original data and routine files are available from the authors.

Similar rules as for the co-word analysis are applied for the data cleaning in this part. That is, all authors are firstly counted up from the data set of the names appearing in the publications. Then, we conduct frequency distribution analyses in the order of the list. We use five as the minimum number of times an author's name has to appear in the documents to establish a co-occurrence matrix. However, all co-authors are included in the co-occurrence matrix because of the higher complexity in authorships than in keyword use. According to the rules we adopted, the matrix of co-authors is smaller than that of co-words, which is easier and clearer to visualise in the networks. The nodes in the networks represent authors, and the line between two nodes (authors) denotes that they co-authored a publication. The more lines a node has, the more collaborations that author has with others.

Although this way presents well the most productive authors and collaborations between researchers, some links may be missing because of the emphasis on larger number of publications. For example, some authors may have strong relationships with others who are not included in the networks because of their publications' number not matching the rules, e.g. author A and author B are both in the networks and they have two collaborations; author A, however, has four collaborations with author C who is not in the networks because of having only four documents in the defined contexts. This limitation similarly occurs in the co-word networks; however, it does not create significant effects on revealing the relationships between leading scientists as all authors in the networks have more than 5 publications.

Results and Discussion

A general description of the trends is first provided. Following this, we present the results from the two co-occurrence analyses.

Research Trends in Nano-biotechnology for Water Sustainability

A total of 4126 publications in nano-biotechnology research were found in Scopus in the contexts of water supply, contamination and treatment, which account for 36 % of all nano-biotechnology research on broader water issues (11,469) and 3.5 % of all nano-biotechnological studies (Fig. 25.3).

The earliest study on nano-biotechnology addressing water issues was published in 1971. Since then, the publication rate grew steadily until the first rise in 1985, following which it continued to increase gradually and later on witnessed a dramatic jump in 2000. A similar trend happened in research on water

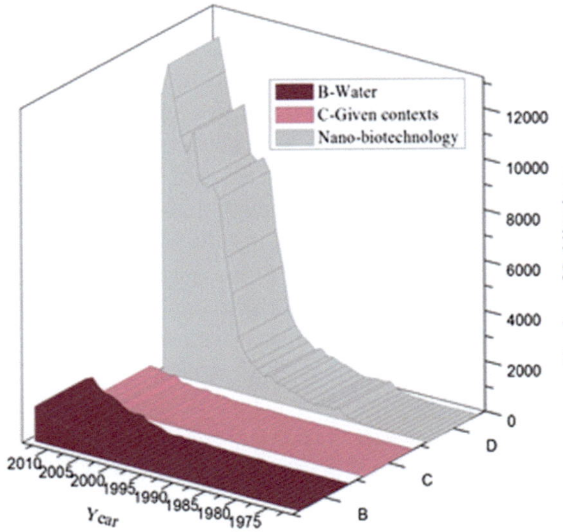

Fig. 25.3 Comparison between "nano-biotechnology" in total water research and within sustainability context

sustainability problems (namely supply, treatment and contamination). After the first study on water sustainability problems in 1981, there was a steady growth reaching a peak of 548 in 2013. The number of publications which apply nano-biotechnology to water supply, contamination and treatment rose dramatically after 2001. Afterwards, it experienced a period of seven years (2002–2008) of relatively stable development prior to the next soar between 2009 and 2013.

Co-occurrence Analysis

After 43 documents were removed from the data set during data clearing, the final number of publications for the co-occurrence analysis is 4083. The results from the two co-occurrence analyses based on these publications are presented below.

Co-word Analysis

Overall, 131 key vertices and 1186 lines constitute the network of co-words for all publications. Lines with a value of less than two were removed from the data set to make the visualisation of the networks clearer. Using a manual semantic analysis, the keywords were firstly classified into three groups, namely water treatment—yellow nodes, water supply—red nodes, and water contamination—green nodes (see Fig. 25.4). This is in line with the keywords used originally to identify the publications combined with the keywords in the titles and abstracts of the articles in which they appeared.

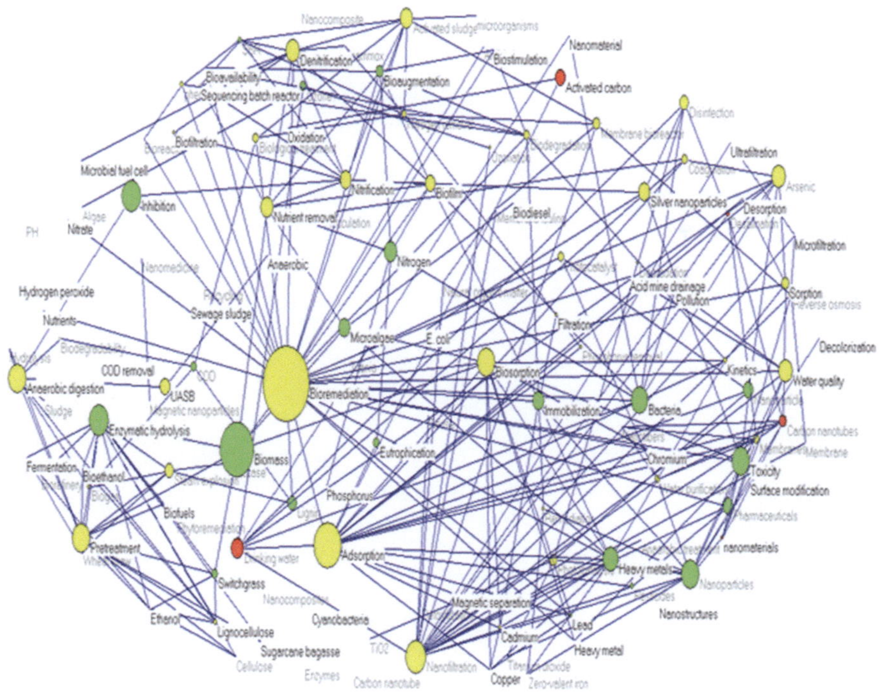

Fig. 25.4 Co-word networks by topics relevant to water sustainability research

Figure 25.4 shows that bioremediation, biomass and drinking water are the three main topics attracting research attention in water treatment, water contamination and water supply, respectively. Water treatment studies dominate the domain of nano-biotechnology research on water sustainability. This indicates that bioremediation is the main way researchers approach the treatment of wastewater. Biomass is the most popular keyword in water contamination studies. This is easy to understand as water treatments are mainly directed towards controlling and dealing with water contaminations, while research on pollution is geared towards managing biomass and other pollutants such as heavy metals, enzymatic hydrolysis, bacteria and toxicity surface modification. Research in nano-biotechnology for water supply, however, is less focused than the other two areas as far as the domain defined by the keywords of this study is concerned. The central topics of water supply are on drinking water and activated carbon related to material sciences.

In order to compare the topics and observe research distribution by main subjects, we use the following classification of the publications' keywords (see Fig. 25.5): biophysics—blue nodes, bioengineering—orange nodes, biomedicine—white nodes, biochemistry—pink nodes, microbiology—red nodes, material sciences—green nodes, and chemical engineering—yellow nodes. This is also diagrammed in the pie chart on Fig. 25.6.

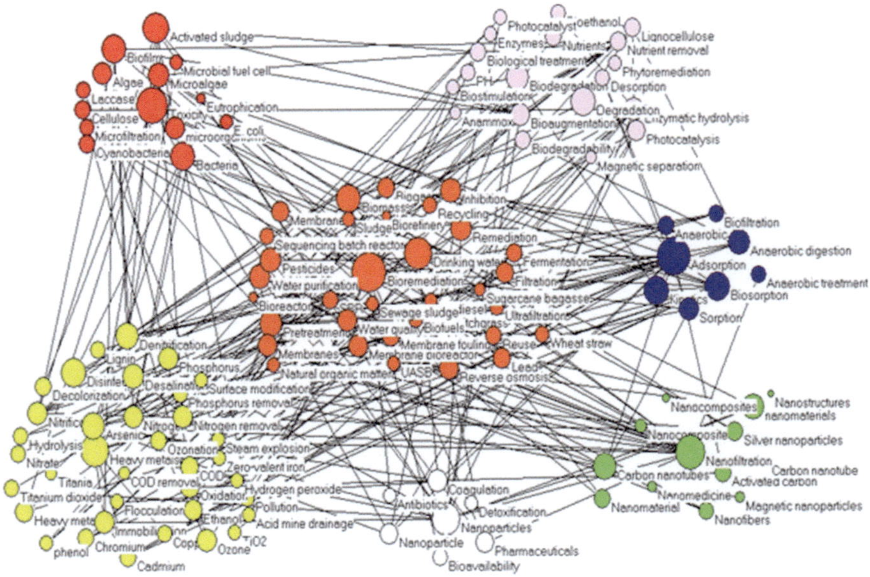

Fig. 25.5 Co-word networks by keyword subjects in water sustainability research

Fig. 25.6 Composition of studies of nano-biotechnology on water sustainability by subject

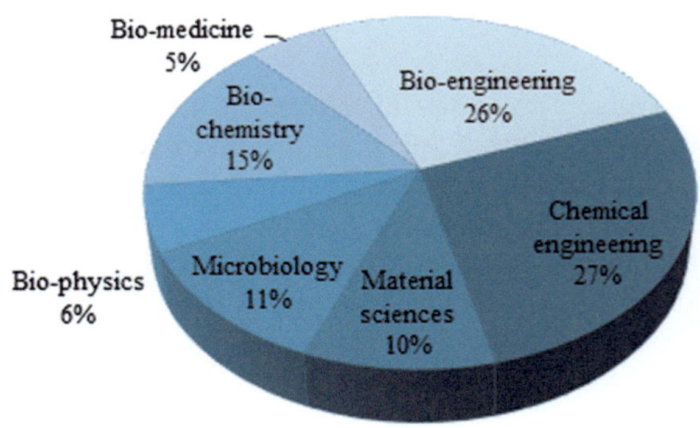

Similar numbers of studies have been published in bioengineering and chemical engineering in the field of nano-biotechnology for water treatment, supply and contamination. They also have strong connections with the other topics. Microbiology and material sciences have again similar but smaller contributions, while limited research has been conducted in biomedicine and biophysics. On the other hand, biophysics and biomedicine have weak connections in the whole network.

Co-authorship Analysis

Based on the 4083 articles, 16,496 authors in total conducted research on either nanotechnology or biotechnology related to water-specific issues. After data cleaning for publications without stated authors and elimination of authors whose names appear less than five times in the documents, the 97 most productive authors were identified for inclusion in the co-authorship analysis. The networks are shown in Fig. 25.7 and are classified in 16 groups according to the degree of the connections between the authors. They are further organised into two classes (see Fig. 25.7): high collaborations (Class One) and low collaborations (Class Two).

The collaborations in Class One are more than those in Class Two meaning that the relationships between authors are closer in Class One. There is a gap between the two classes. Choi, H. is the only author from Class Two linked to Class One by publications (collaboration between Choi, H. and Kim, Y.). In Class One, the connections between authors are closer than those in Class Two. To further assess the relationships between the most active authors, we removed those whose names occurred together with others in the documents less than 6 times and generated new networks (see Fig. 25.8).

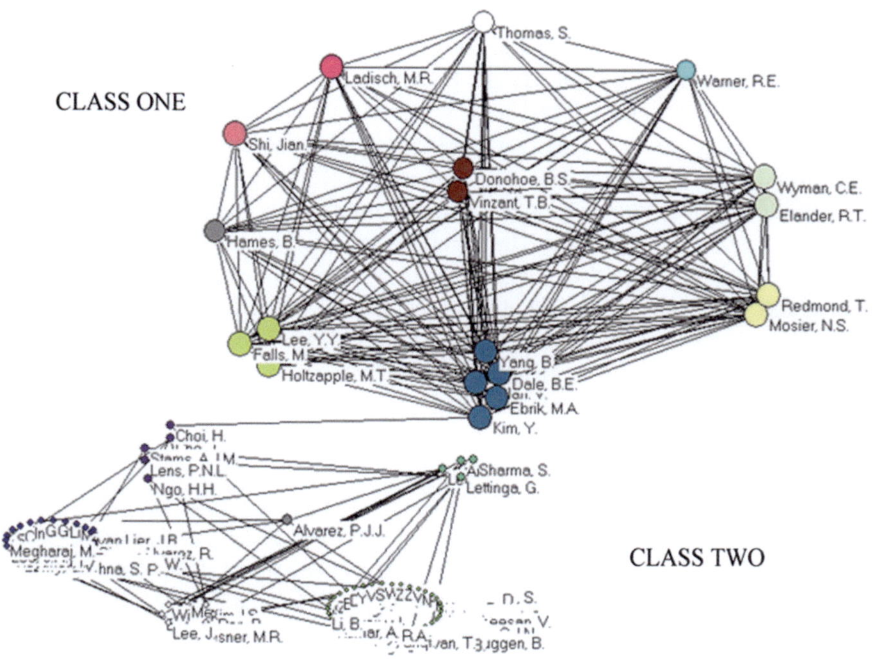

Fig. 25.7 Co-authorships grouped by degrees of connections

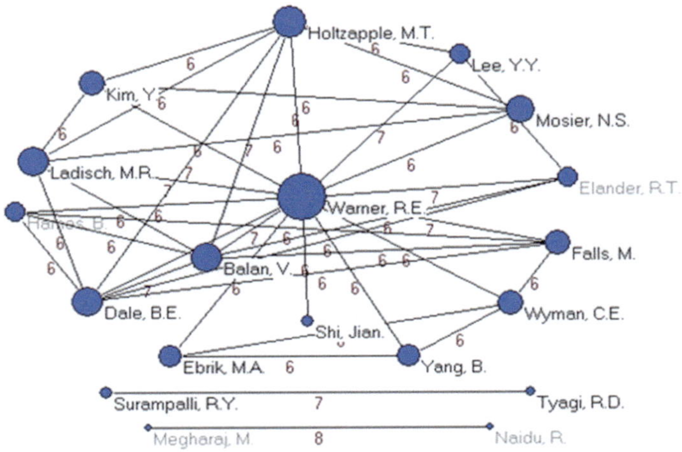

Fig. 25.8 Closely connected authors' networks

Figure 25.8 shows the close relationships between the active authors with nodes representing authors and a line between two nodes meaning that the respective two authors have joint publications. The values on the lines denote the number of publications the authors have co-authored. Warner, R.E. has the most relations with others than any other active author in the networks. Although Megharaj, M. and Naidu, R. have co-authored the most, neither of them has collaborations with other researchers. Similarly, Surampalli, R.Y. and Tyagi, R.D have high numbers of publications (11 and 10, respectively) and are, respectively, the second and the third most productive authors in the data set; however, they have less collaborations with others.

Within the co-authorship networks, it is also interesting how authors are linked to each other among the various countries as this reveals the global collaboration picture in nano-biotechnology. The locations of the 97 authors were determined from the affiliations listed in their publications. For authors with multiple affiliations, the telephone number and mailing address (if applicable) were used to determine the country in which they are based or, if these do not provide a clearly indication, we used the first affiliation as the location base. The spatial distribution of the linked authors is presented in Fig. 25.9. It shows the various productive authors and their closer relations with others who have a similar status in terms of publication numbers and with whom they have collaborated.

At a regional scale, the USA is the leading country in nano-biotechnology studies on water sustainability. The US authors have the majority of collaborations in this field, followed by China, South Korea, the Netherlands, India and Australia. Notwithstanding this, the collaborations conducted by the US researchers were mostly within their own country (see Fig. 25.10).

The main collaborating organisations are Michigan State University, National Renewable Energy Laboratory, University of California (Riverside), Texas A&M University, Auburn University, Ceres Inc. in Thousand Oaks and Purdue

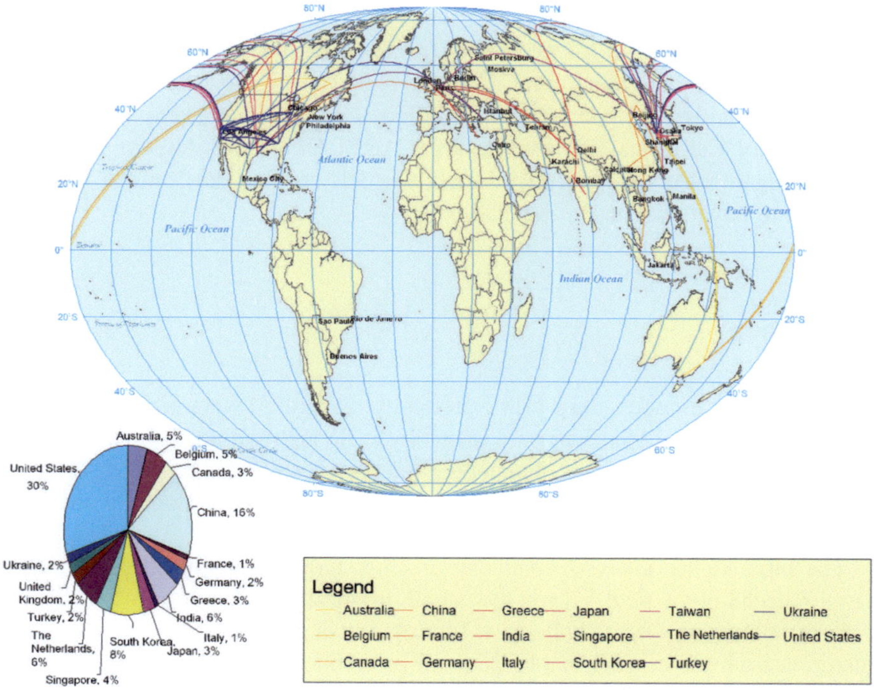

Fig. 25.9 The global spatial networks of linked authors

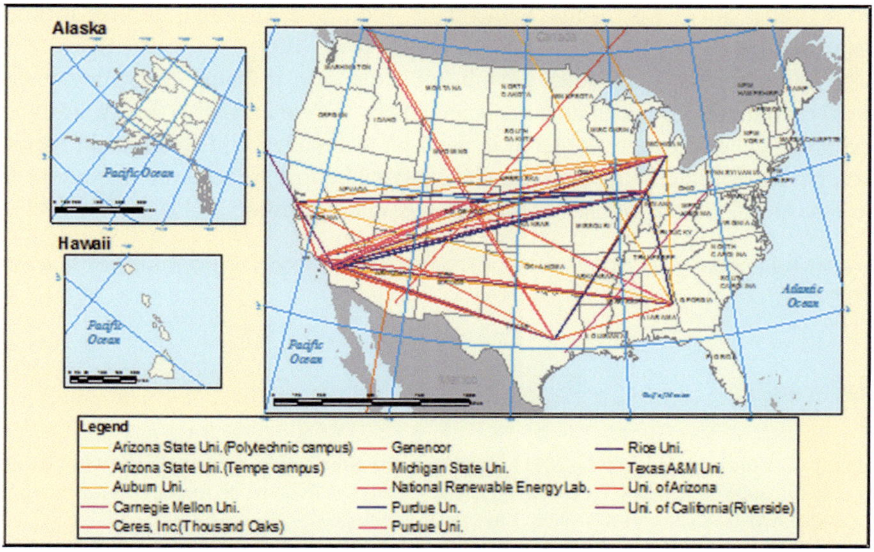

Fig. 25.10 Co-authorship network distributions in the USA

University in the USA, with the University of California (Riverside) being the most involved organisation (accounting for 18 % of the total US authors in the data set). On an individual level, Warner, R.E. from Genencor participated in the largest number of collaborations. External connections were built by the University of California (Riverside), Texas A&M University, Auburn University and Michigan State University (the stretching left purple lines, middle red lines and right light yellow lines in Fig. 25.10).

Conclusion

In the last three decades, the world community recognised water as a global issue which needs to be approached on a planetary scale and that negligence and ignorance could lead to problems threatening human survival (PWC 2012). Progress made in research related to nano-biotechnology represents humanity's ability to respond to these global challenges.

The bibliometric analysis conducted in this study shows that the specialised nano-biotechnology field is yet too small to make a meaningful contribution towards addressing water challenges. More promising technological solutions are emerging from the broader combined nano-biotechnology field whose publication output is consistently growing. The most active publication areas relate to bioengineering, chemical engineering, microbiology and material sciences confirming that water sustainability is a truly interdisciplinary and transdisciplinary research field. Despite the relatively large number of very active researchers, the bulk of the new knowledge is generated within the USA. The contributions by researchers from China, South Korea, the Netherlands, India and Australia are also making their mark.

With continuing population growth and expansion in human activities, water supply, contamination and treatment are likely to remain highly active areas of technological endeavour. Research in nano-biotechnology will also remain important, but further internationalisation and concerted effort are required for society to be able to address the global water challenges and priorities.

Acknowledgments The second author acknowledges the support of the Australian Research Council.

References

Abbasi A, Altmann J, Hossain L (2011) Identifying the effects of co-authorship networks on the performance of scholars: a correlation and regression analysis of performance measures and social networks analysis measures. J Informetr 5:594–607

Barabasi AL, Jeong H, Neda Z, Ravasz E, Schubert A, Vicsek T (2002) Evolution of the social networks of scientific collaborations. Phys A 311:590–614

Courtial JP (1994) A coword analysis of scientometrics. Scientometrics 31(3):251–260

De Bellis N (2009) Bibliometrics and citation analysis: from the science citation index to cybermetrics. Scarecrow Press, Lanham, MD

De Solla Price D (1963) Little science, big science. Columbia University Press, New York, NY

Diallo M, Brinker CJ (2011) Nanotechnology for sustainability: environment, water, food, minerals, and climate. In: Roco MC, Mirkin CA, Hersam MC (eds) Nanotechnology research directions for societal needs in 2020. Springer, Boston, pp 221–259

Ding Y, Chowdhury GG, Foo S (2001) Bibliometric cartography of information retrieval research by using co-word analysis. IPM 37:817–842

Gazit E (2007) Plenty of room for biology at the bottom: an introduction to bionanotechnology. Imperial College Press, London

Gommen R, Verstraete W (2002) Environmental biotechnology: the ongoing quest. J Biotechnol 98(1):113–123

Gray NF (2010) Water technology: an introduction for environmental scientists and engineers, 3rd edn. Elsevier, Oxford

Hillie T, Hlophe M (2007) Nanotechnology and the challenge of clean water. Nat Nanotechnol 2:663–664

Hu CP, Hu JM, Deng SL, Liu Y (2013) A co-word analysis of library and information science in China. Scientometrics 97(2):369–382

Kroeger P (2005) Analyzing grammar: an introduction. Cambridge University Press, Cambridge

Leydesdorff L, Rafols I (2009) A global map of science based on the ISI subject categories. JASIST 60(2):348–362

Liu X, Bollen J, Nelson ML, Van de Sompel H (2005) Co-authorship networks in the digital library research community. IPM 41:1462–1480

Marinova D, McAleer M (2003) Nanotechnology strength indicators: international rankings based on US patents. Nanotechnology 14:R1–R7

Meyer MS (2001) Patent citation analysis in a novel field of technology: an exploration of nanoscience and nano-technology. Scientometrics 51(1):163–183

PricewaterhouseCoopers (PWC) (2012) Water: challenges, drivers and solutions. Retrieved from http://www.pwc.com/en_GX/gx/sustainability/publications/assets/pwc-water-challenges-drivers-and-solutions.pdf. Cited 30 Oct 2015

Rafols I, Meyer M (2010) Diversity and network coherence as indicators of interdisciplinarity: case studies in bionanoscience. Scientometrics 82:263–287

Rip A, Courtial JP (1984) Co-word maps of biotechnology: an example of cognitive scientometrics. Scientometrics 6(6):381–400

Ronda-Pupo GA, Guerras-Martin LA (2010) Dynamics of the evolution of the strategy concept 1962-2008: a co-word analysis. SMJ 33(2):162–188

Schummer J (2004) Multidisciplinary, interdisciplinarity, and patterns of research collaboration in nanoscience and nanotechnology. Scientometrics 59(3):425–465

Thursby J, Thursby M (2011) University-industry linkages in nanotechnology and biotechnology: evidence on collaborative patterns for new methods of inventing. J Technol Transf 36(6):605–623

United Nations Educational, Scientific and Cultural Organisation (UNESCO) (2011) The right to water. Retrieved from http://www.unesco.org/new/en/media-services/single-view/news/the_right_to_water/#. VcjU6iig Tzg. Accessed 30 Oct 2015

Van Loosdrecht MCM, Heijnen SJ (1993) Biofilm bioreactors for waste-water treatment. Trends Biotechnol 11(4):117–121

Venkatesh MN (2009) Nanobiotechnology, foundation for biotechnology awareness and education. Retrieved from http://fbae.org/2009/FBAE/website/our-position-nanobiotechnology.html. Accessed 30 Oct 2015

Zechendorf B (1999) Sustainable development: how can biotechnology contribute? Trends Biotechnol 17:219–225

Part VIII
Sustainability and Business

Chapter 26
Stakeholders' Strategic Thinking for Sustainable Development

Anna Czarczyńska

Abstract This article analyzes interdependencies of main stakeholders of sustainable development process based on innovative improvement and collective action. I start from the meaning of sustainable development in the cross-sector relations, examining to what extent lack of interdisciplinary strategic thinking and innovative attitude can be main obstacles in sustainable development of innovative economy. Technological gap in the sense of new ways of construction, distribution, and utilization of collaborative knowledge supporting sustainable development placed the whole economy below the potential sustainable growth level.

Keywords Sustainable development · Stakeholder analysis · Innovative improvement · Collective action · Innovative economy

Introduction

Improving an economy's sustainability is an important challenge, especially for the post-transition countries like Poland. Following the creation of the post-communist state, Poland has had to face competition from knowledge-based economies. Because the basic cost advantage is not working anymore, further development depends mostly on the ability to produce a creative and sustainable economy. Increasingly apparent limitations of traditional economic growth models, based on resources and production factors, show that a balanced approach, based on integrated values, may be the only alternative in building a new variety of the welfare state model.

A. Czarczyńska (✉)
Kozminski University, Jagiellońska 57/59, 03-301 Warsaw, Poland
e-mail: czarczynska@kozminski.edu.pl

L.W. Zacher (ed.), *Technology, Society and Sustainability*,
DOI 10.1007/978-3-319-47164-8_26

However, in the case of Poland, the main obstacle to achieving this goal is the lack of cooperation and strategic attitude between main the stakeholders: inefficient public sector, society with growing but still low awareness of sustainability, and underdeveloped education and research industries. The emerging grassroots social movements increasingly emphasize the need for sustainable development with transparency, value, and trust, taking into account the common good of society, which seems to be optimistic.

This article analyzes the interdependencies of main stakeholders in the sustainable development process based on innovative improvement. I start with the meaning of sustainable development in cross-sector relations, examining to what extent a lack of interdisciplinary strategic thinking and innovative attitude can be an obstacle to the sustainable development of an innovative economy. The technological gap in the sense of new ways of construction, distribution, and utilization of collaborative knowledge supporting sustainable development has placed the whole economy below the potential sustainable growth level.

Experience over the past years shows that there is no discretion to treat imbalances as the driving force of the economy. The ability to develop sustainably with cross-sector cooperation is quite substantial. However, this potential is used in line with the institutional order on the rational behavior base, instead of the mindfulness-developing process. The main reason for this is the existence of systemic barriers and the lack of cooperation and strategic attitude. However, the successive generations that have been bred in the Internet age seem to be increasingly aware of the environmental consequences of economic activities and the strength of their impact on the global level—sustainability also means a balance among generations (environment, labor market, and public debt).

Systemic change toward a pro-sustainable approach can help to better organize cross-sector cooperation, improving strategic attitudes. Limited public resources can be better located thanks to civil society's involvement and active role in the R&D sector. These goals can be achieved by balancing a creative sustainability approach with the interdisciplinary strategic thinking. The sustainable solutions on a macroscale can be implemented only with awareness and cooperation of stakeholders.

What's Wrong with the Global Economy?

The financial crisis of 2008 and the subsequent tsunami waves in the real economy are unprecedented and only to some extent can be explained using the classical theory of economics. The world of the twenty-first century no longer behaves in accordance with the principle of general equilibrium in particular countries and reacts with unpredictable barriers generated at the global level, where even open-economies analysis is not adequate.

The question is whether we really know what happened to the so perfectly self-perpetuating machine of the global economy, and whether it is indeed a perpetual-motion machine, in which we so dearly wanted to believe (e.g., financial markets seemed to believe in cheap financing for ever). This painful economic collapse is a symptom of the disease of civilization (generally having a chronic course), which we earned in either an active or passive way. So what is the difference between this wave of recession cycles and the turbulence of twentieth century?

The clearest consequence of the economic and financial crisis is the change in the balance of power in the world's major economies. Although the USA maintained its leadership position, its prevalence is melting quickly, and even in this decade, the main force of the global economy as envisaged will be in Asia (Czarczyńska 2012a). The crisis of recent years was, in this case, a kind of catalyst for change, which can be observed from the end of the last century. In addition, this catalyst also speeds up the pace of change in the geopolitical world.

This strong turbulence resulted in the strengthening of asymmetric costs and benefits, as financial markets mourned the loss of millions of price-takers' funds. The feasibility of a quick profit for a small group of price-setters directly affected the condition of the world market, and transitional difficulties in dominant countries became unbalanced at the expense of small countries (in terms of impact on the global economy).

Large countries have a much larger arsenal for a rapid response and a much larger portfolio of assets that can run (including the printing of money and influence due to seniority, which is carefully used in the USA). More than in times of recession, we hunt for "black swans" and broaden its sphere of influence, realizing the long-term strategy of taking control over natural resources in different parts of the world (especially in the field of rare elements), as does China (Moran 2010).

In sum, crisis accelerates change, showing the natural tendencies of individual countries (including the over-indebtedness as in the case of Greece), as well as the voracious and predatory nature of both the financiers and the great powers that compete for the palm (Friedman 2005). The only question after such turmoil is what is under the table and whether the prosperity of global competition still helps to solve the problems of humanity.

Globalization and What Next? Manuel Castells is identifying and analyzing the determinants of the emergence of the global economy pointed in its political origins (Castell 2010, s.109–170). Building a network society was based primarily on the spread of information and communication technologies, but the process was largely based on political support of the interaction between markets and governments and financial institutions. Universally dominant "only correct" neoliberal ideology has been reinforced by policy-makers interested in personally pursuing profit opportunities, both during and after their time in office, through a network of contacts (e.g., ex-politicians advise the Nord Stream Company), and to control access to particular markets (various forms of bribes and corrupt lobbying from Russia to the European Parliament).

According to the Maslow's law, if we are holding an instrument in the shape of a hammer, we treat every problem like a nail. From a liberal global economy, supported by technological advances, we expect a miraculous solution to human problems at both the economic and institutional, which is in itself a contradiction. However, while it was relatively easy to promote the progress of globalization on the political-institutional level, it is difficult at the same level to reduce the avalanche nature of the process once it is started.

The current global economy is a network of interrelated market segments, key national actors playing the roles of relay nodes. The exclusion of such a node within a global network causes the power to be cut in the form of information, technology, and production factors, which is a de facto power cut to development. For the rest of the system, such exclusion is not relevant for all countries outside the USA and China. This means that in the new global system, all countries, except for the two largest countries, are small in terms of international economic relations, which does not affect the functioning of the system as a whole and also works perfectly when a node is bypassed. However, note that the separation of the economy from a single system at a fast pace can result in necrosis.

Globalization in this form leaves no alternative to the ideology of efficiency at all costs, which becomes the core of the new economy (Beck 2001). A chance to change the direction of the avalanche process of commercialization is paradoxically a shock caused by the collapse of financial markets and to realize the separation of the real economy at the local level. The changing geopolitical balance of power and the emergence of centripetal force acts on individual economies as a particular form of response to globalization and the crisis in its broadest sense, including financial markets, environmental risks, and uncertainties in energy markets (Sachs and Warner 2007).

Gradually, however, there are an increasing number of trends and initiatives that promote collective action to gain independence from global flows based on entirely different values and local potential, both socially and economically. We are on the way from an information society to a stakeholder society, where members of a society have both rights from it, and duties or responsibilities to it. In this new era, we have to treat society not only as a rule taker, but also as a rule setter by the direct creation of the reality (Czarczyńska 2012b).

The common denominator in the crisis of recent years, appearing on many interdependent levels, is a crisis of confidence in the institutions and authorities responsible for economic and social order of the modern world (Yunus 2007). The reasons for departing from a sustainable development path can be traced not only to the economic factors of production, but also to the environmental factors, funding current needs at the expense of future generations, and the ethical aspects of human economic activity. The global crisis in the area of value not only led to the financial losses of millions of people, but also increased the sense of danger from the financial markets, geopolitical stability, and natural hazards due, to a large extent, to the lack of ecological balance, which led to a growing sense of dissatisfaction. Money no longer guarantees the safety, and the redistribution of public resources in many countries is no longer based on the principle of equivalence.

The record of economic relations, both extremely selfish behavior (particularly common in post-communist countries, where the system of internal moral scheme has largely been destroyed) and altruistic behavior, does describe the typical way of using economic instruments, nor the economic consequences. At the macrolevel, too much inequality leads to social unrest, where we are ready to accept the loss of one's own in the name of the struggle for social justice.

The effect of economic activities depends not only on economic calculations, but also on the value we choose. We make a choice when spending our money on goods and services produced by the company on terms consistent with our beliefs, and also by engaging in activities, which give us a sense of meaning in the labor market ("selling" your work to some extent the buyer can choose). Behavior consistent with our values takes into account economic relations as the basis for successful transactions in the future, where mutual trust allows us to reduce the cost of protection against the risk. Trust lowers transaction costs, and the observation of this fact makes the pursuit of values based on direct contacts and trust (Walsh 2011).

The concept of fair plays an increasingly important role in the distribution of social goods, especially in the sphere of relations producer–consumer and the level of government–society. Moreover, not only is the result of the process is important (e.g., blocking Acta), but also the process (consultation, treatment partners). Contemporary politics, and thus the economy, must be managed by the appropriate narrative. People pay attention to the final decision announced officially, but in its assessment of itself even more strongly emphasize the integrity of the process of generating a given position. Classical models of production based on factors such as raw materials, labor, and capital do not depend so much on the condition of the human mind and the knowledge economy. The basis for such an economy is to create an intellectual resource and share it, which is not easily achieved without the voluntary cooperation (Chan and Mauborgne 2002). This approach goes far beyond simply improving the efficiency of the economy and optimal resource allocation through the mechanisms of supervision and control, and rewards the desired behavior of the premium system.

Wealth creation (which is much more than the production process), even according to classical economics, is a social process, rather than an individual process. Adam Smith recognized the division of labor as the foundation of value creation, so that you can achieve higher profits at both the scale of production and international trade (Kennedy 2005). Each individual enrichment strategy is also based on social interaction and exchange (Fukuyama 1996). So, our prosperity depends on our relationship with the environment, which in turn is a reflection of the level of maturity and emotional intelligence unit. Paradoxically, both the individual and society gain more if we are focused on economic cooperation with partners and do not seek to profit at the expense of operating and social costs (Eisler 2007).

An integrated approach enhances the welfare of all and restricts activities harmful to society (similarly as it did in traditional economic theory, except that the spectrum of relevant factors is much higher, because it includes the effects of tangible and intangible). Contribution to social development is assessed on the basis of taking into account the net effect of external costs in the bill pulled, which shows whether the positive economic growth is offset by an excessive and harmful to the environment consumption. Such perceptions of the economy are based on the paradigm of integral perception of reality, which is also part of the economic activity of man.

Power Diffusion

Social transformation changing the balance of power is true not only in the international sphere, but also in social relations. And in this respect, the crisis has become a catalyst of change initiated by the IT revolution and reinforced by the processes of globalization. The main direction of change offsets the impact of government for social actors and individual actors by obtaining access to the same information as the group already monopolizing information. Joseph Nye has called this phenomenon the diffusion of power (Nye 2011), due to the impact of technology, especially information on creating opportunities for participation in international events by de facto eliminating the costs of such participation. Rapid technological progress and the dramatic fall in prices of telecommunications services virtually abolished the barriers to entry on the global market for goods and services, but also enabled access to global data, which weakened the government position based on information asymmetry. Information as a commodity is highly valued by the market and was previously reserved for large organizations, governments, and transnational organizations (often cooperating with the government), because of the high cost of obtaining it.

Changing the balance of power has a huge impact not only on the global political relations of governments, but also, and above all, on the nations and the awareness of open economies. The worldwide wealth of information is not considered sacred, which is clearly shown in the WikiLeaks example. Almost every day in the media we can find examples of authorities in the international arena, which in reality are addressed to national audiences. This shows the growing awareness of the interdependence of the authorities and a number of small social and economic actors causing the only appropriate ad hoc coalitions and merging.

Existing policy-makers are losing the race in the monopolization of information. Increasingly, their success depends on the skills of dialogue and establishing effective and transparent cooperation (one false note is quickly intercepted) with distributed external entities, which have no institutional leverage. Hitherto regarded as fundamental, values and mechanisms of interaction used by the ruling no longer work. Entering into adulthood is generation Y, subjected to the dictates of stick and carrot, which see many other ways to meet their needs. The dominant need for freedom is fulfilled mainly on the Internet, which is regarded as an essential part of human rights (EU funds supporting broadband infrastructure, legally guaranteed access in some countries). The new generation carefully reviews the mechanisms of power and quickly obtains an information advantage.

A decline in the importance of governance is a universal phenomenon, regardless of the political system, wherever we are not dealing with an isolated society.

Even so, it is evident that the value of democracy has ceased to be an asset to developed countries. Democracy is the most attractive system if given the chance for a fairer redistribution of income, which has improved the quality of life. An example of the development path of China's economy shows that democracy itself ceases to be a value if the public has access to a growing number of distribution of goods, among which is the dissemination of ICT technology. Its social impact strengthens social capital (*guanxi* networking) as the foundation of the release Chinese capitalism.

Processes occurring in many countries, both internal and external, are very similar in nature. The common external threats force us to find a formula of cooperation independent of ideology. The sphere of the real economy through successive crises atomizes movement toward decision-making centers and multi-dimensional society *homo agens* (Chmielewski 2011, s.317–343). The elements of this society are less and less the subjects, and operators are increasingly changing their environment, including institutional environments. Strongly individualized social actors (both physical and legal), on the one hand, are an essential component of the community and, on the other hand, are becoming, more and more, creators—thanks to a better utilized ability to create a social organism, which puts them in a dual role in the system. The activities of these new social forces manifest themselves in many new forms, including not only civil and political impact, but also the new pro-environmental initiatives, regional, and self-help in the form of organizations, associations, and movements for purposes of ad hoc one issue (e.g., anti-Acta).

Knowledge Environment

The beginning of twenty-first century has been characterized by a radical transformation of different sectors of economy, which, so far, have operated on the national and global area in a traditional manner, whereby the search for possibilities of creating and developing high-value innovative operations is clearly marked. Currently, the research orientation is taking into account the latest trends related to the change of paradigm in social sciences, as a consequence of the development of global network organizations, importance of intangible assets, and the growing role of new information and communication technologies (Fig. 26.1).

The most interesting field of research in this context is posed by innovative, flexible, high-value societies, as they are able to most efficiently adjust to the challenges of a changing and increasingly uncertain environment. For transformational change in knowledge management based on value policy, it is necessary to identify key stakeholder-related sectors that influence stakeholders' participation and the domains where the integration takes place in innovation diffusion.

The recipe for a knowledge environment has three main components: social capital in the sense of network quality and diversity, human capital as a sense making resources, and absorptive capacity in the meaning of systems, structures,

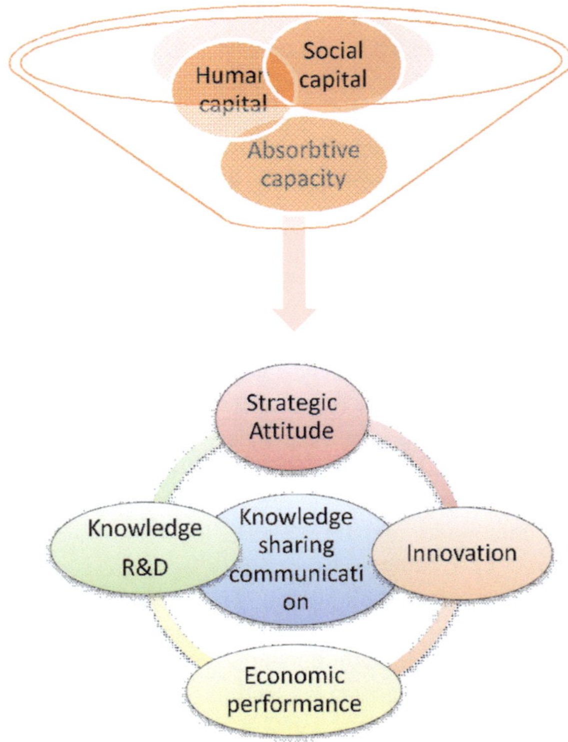

Fig. 26.1 Knowledge environment scheme

and routines. With such a background, the society is able to create the potential for effective economic performance based on R&D knowledge and innovation. However, a high level of innovation and strong research system is not enough when there is a lack of strategic attitudes and effective knowledge-sharing communication channels based on competencies (the combined skills acquired during the course of training which condition the efficient performance of tasks in a job, based on the knowledge, experience, and predisposition of an individual, and displayed in a sequence of behavior) (Baczynska 2015).

An inclusive and innovated global society is under construction, but the entire process is based on the new evolution of knowledge with three pillars:

- *Human Capital* ability to communicate ideas to a range of people, both internal and external, in a language they understand;
- *Social Capital* ability to access learning and knowledge from others through the number and quality of relationships in both formalized networks and informal contacts;
- *Absorptive Capacity* ability to embed routines for knowledge exploration, reshaping, and retention of knowledge.

The role of open technologies in encouraging the sharing and utilizing of innovations more effectively is crucial. The question is how institutional or governmental policies should support this movement. Ways in which public institutions and private sector are effectively working together around new ideas and technologies depend on the channels of collaboration.

Innovation impact is totally different on personal development, public-sector improvement, and business-sector profitability. The measures of innovation impact are well developed in the last group; however, the interaction among these three kinds of stakeholders and the spillover level determining the innovation culture at the national level is not easily measureable (Huntington 2003).

Various state management models have the potential to reduce barriers and produce sustainable outcomes for innovative performance (Wojtowicz and Olejniczak 2015). However, with the positive socioeconomic consequences of innovation, there is a growing recognition of the need for sustainable approaches to strategic development. While innovation has been acknowledged as a tool for growth and for creating a knowledge-sharing culture, it is presumably that conflicting interests of multiple stakeholder groups can hinder innovation in equally achieving social and economic objectives. The integrated approach to a innovation environment combines institutional background, business management tools, and personal attitude to provide a holistic and collaborative innovation system.

Summary

The recent crisis is a signal transition toward a new order of interdependence. It showed the significant limitations of money as a measure of value, detached from the real size, which, through its virtual nature, has value in itself. Moving away from shareholders to stakeholders is the moment of transition from value management in the interest of the owner to the integration of virtual management value by including all stakeholders to the process of creating added value.

A key element of transformation is a return to fundamental values, such as trust, direct contact, and the natural environment, and is not detrimental to the character of goods and services in the economy. Creativity is triggered by the values of the integrated unit activity in not only the professional area, but also mental, emotional, physical, spiritual, cultural, social, and environmental sustainability (Wilber 2001). The value in this approach is not so much the material level (although this is mostly true, or for an individual at a higher level of transformation ceases to be so important), but is widely understood as participation in sustainable development performance, far beyond the classic well-being.

In modern society we have to redefine power, not as good or bad per se; however, too much power can be bad because it can lead to mistaken strategy. That's why so important is to keep a balanced distribution of power and a create win–win situation to deal with challenges that we face. Due to power diffusion, things once restricted to very large organizations, like governments or corporations, are now available to anyone. Information technology is very powerful and important in this process.

Due to globalization, local cooperation is more and more important. Regions become suboptimum, delivering synergy. However, the needs for institutional solutions are still very strong. Regions have an ability to adjust to global competition, but this is a task for politics to determine how to establish rules of cooperation, how to develop potential in the creative sector, and how to have a leading global role in the innovation system for sustainable solutions. There is a time for new ideas based on GET development: governance, education, and trust.

Stakeholders' strategic thinking for sustainable development means building a good community of prosperity in a stakeholder society where members of a society have both rights from it, and duties or responsibilities to it. The background for sustainable development means, in this approach, an attempt to better link together key concepts: knowledge, environment, innovation diffusion, sharing of knowledge and a strong commitment to the implementation of the agreed actions. Innovation, education, or research as the parameters of modern economy topics of the knowledge triangle are not enough to achieve development. The key concept is to create interaction between research, education, and innovation as the key drivers of a knowledge-based society. Stakeholders for sustainability care about sustainable—people centered—technology created on a participatory base of technology-creation process facilitated by state.

References

Baczyńska AK (2015) Analytical, practical and emotional intelligence and line manager competencies. Manag Bus Adm 23(4):34–54

Beck DE (2001) Stages of social development: the cultural dynamics that spark violence, spread prosperity, and shape globalization

Castell M (2010) Społeczeństwo sieci. PWN Str. 109–170

Chan KW, Mauborgne R (2002) Fair process: managing in the knowledge economy. Harvard Business School Publishing Corporation, EBSCO

Chmielewski P (2011) Homo agens Instytucjonalizm w naukach społecznych. Poltext Str., Warsaw 317–343

Czarczyńska A (2012a) Europejskie aspiracje globalnej polityki klimatycznej wobec Chin i ich konsekwencje. w: Europa- Chiny PAN, Warszawa

Czarczyńska A (2012b) Świat w obliczu zagrożeń totalnych a wizja ekonomii integralnej. w: pod red. M.Guzek

Eisler R (2007) The real wealth of nations: creating a caring economics. Berret Kochler, San Francisco

Friedman T (2005) The world is flat: a brief history of the twenty-first century. Farrar, Straus and Giroux, New York

Fukuyama F (1996) Trust: the social virtue and the creation of prosperity. Simon and Schuster, New York

Huntington (2003) Kultura ma znaczenie: Jak wartości wpływają na rozwój społeczeństw, Zysk i spółka, Warszawa

Kennedy G (2005) Adam Smith's lost legacy. Palgrave Macmillan, Basingstoke

Moran D (2010) Theodore, feeding the Dragon. Is China trying to swallow the world's natural resources? The Milken Institute Review, Third Quarter

Nye J (2011) Understanding 21st century power shift. World Financ Rev. Nov 2011

Sachs J, Warner A (2007) Natural resources abundance and economic growth. Harvard University, Center for International Development and Harvard Institute for International Development, Cambridge

Walsh J (2011) Business ethics: the central point. Management. 23 Mar 2011

Wilber K (2001) A theory of everything: an integral vision for business, politics, science and spirituality. Shambhala, Boston

Wojtowicz D, Olejniczak K (2015) Towards Better Regulations—The role of regulatory impact assessment in modern state lawmaking systems Jun 2015. Management and business administration

Yunus M (2007) Creating a world without poverty. Social business and the future of capitalism. BBS Public Affairs, NY

Chapter 27
Generative Dynamics: What Sustains the Creation of Shared Business Value

Massimiliano Monaci and Mauro Magatti

Abstract CSR theories and practices which have been diffusing in the management world until very recently refer mainly to reactive strategies of (re-)legitimation of companies vis-à-vis their stakeholders. However, even in light of the challenges posed by today's economic crisis, the present period can be an extremely favourable moment to move beyond this adaptive approach and to formulate and realize a more advanced view of the social dimension of business as sustainable innovation, i.e. a business model based on a twofold dynamic of 'valorization of the context': on the one hand, the inclusion in enterprises' strategies of social instances and resources oriented to the natural environment and quality of life in and around the workplace; on the other hand, the ability to generate economic value through the creation of social value. Drawing on the findings of a research conducted on a sample of Italian organizations, the paper identifies and discusses three distinctive mechanisms which seem to sustain the production of 'integrated value' in these companies: the balance between cultural tradition and exploration; the tendency to expand in the context and, at the same time, to include it; and human resource practices establishing a direct link between citizenship behaviours in and of the organization. In conclusion, the analysis suggests a wider-ranging perspective on the strategic and competitive implications of CSR practices.

Keywords Shared value · Sustainability · Sustainable innovation · Italian companies

---

M. Monaci (✉) · M. Magatti
Department of Sociology, Catholic University of the Sacred Heart,
Largo Gemelli 1, 20123 Milan, Italy
e-mail: massimiliano.monaci@unicatt.it

M. Magatti
e-mail: mauro.magatti@unicatt.it

L.W. Zacher (ed.), *Technology, Society and Sustainability*,
DOI 10.1007/978-3-319-47164-8_27

Introduction

There is no lack of indicators suggesting that, until recently, companies' engagement in the domain of corporate social responsibility (CSR) was mainly part of adaptive and reactive strategies for (re)legitimization in relation to stakeholders, especially crucial for business performance, such as employees, consumers, the local community, and the public authorities. Current times, however, see a significant change of scenario and prospects. The worldwide economic and financial crisis exploded in 2008 has engendered new interpretations and proposals which tend to reconceptualize the links between market organizations and the contexts in which they operate, and stress a business model based on the valorization of social resources and needs with the aim of creating 'integrated value'—economic, social and environmental—over the long term (e.g. Butera 2009; Esty and Winston 2009; Laszlo 2008; Louche et al. 2010; Lubin and Esty 2010; Nidumolu et al. 2009; Porter and Kramer 2011; Smith and Lenssen 2009).

Put briefly, we find two tenets at the core of this reorientation: (a) the conviction that corporate conducts intended to include the needs of the relevant stakeholders *can constitute a lasting source of competitive advantage*; (b) an emphasis on *the need to incorporate attention to socio-environmental issues directly into corporate strategy*, making it structurally inherent to all the links in the value chain and in the everyday management of the company's core operations, as well as, inevitably, in its support processes (such as human resource management). These most recent views represent a further stage in the evolution of the CSR discourse, in which the drive to shift socially relevant instances from the periphery to the centre of business is accentuated, starting from emerging needs or those intensified by the current crisis (e.g. in the areas of health, services to families and the elderly, the fight against the exploitation of environmental resources). These new contributions invite to a deeper understanding in the business world, as well as among stakeholders in the public and non-profit sectors, of how current social issues and demands can concur to redefining markets and therefore of how the capacity to deal with them might form the basis for an enterprise's long-term success.

The above reference to the current economic crisis is decisive here. This phase represents not only the time frame of the empirical research on which this chapter draws, but also a period during which, more than ever before, there has been an endeavour to reconsider corporate behaviour from the point of view of its mechanisms of ordinary functioning, rather than only in terms of its final effects. In other words, this is a singularly favourable moment in which certain ideas can translate into actions. The terrain is now really ready for the development of advanced concepts of corporate sustainable innovation which emphasize the integration of economic and socio-environmental objectives into the central core of the strategic-productive processes of business activity and their realization as a source of competitive advantage. These are notions which have been around for some time, to a certain extent, although they have been often tacitly considered visionary, or they have been treated through simple categorizations within which sustainability 'is "boxed up" as economic + environmental + social' (Couper et al. 2009: 69). It is our intention here to offer support to this view by referring to the results of a qualitative research on a sample of Italian enterprises. The business model of these organizations seems to revolve around a dual dynamic of 'valorization of the context': by this expression we mean, on the one hand, inclusion in the corporate strategy of social demands and resources aimed at greater concern for the environment, the quality of local community life, and the rights and development of persons within and outside workplaces; and, on the other hand, a tendency to produce economic and social value downstream of market activity, earning profits by virtue of the 'context value' generated.

In the organizational experiences analysed by this study, the creation of *integrated value* is based on innovations undertaken through mechanisms already cited in the literature as able to render companies more effective in economic and social-ecological terms (e.g. Lehni 2000; Shaetegger et al. 2004); this is especially the case of product redesign and process re-engineering. But, more generally, the driver of innovative action by most of these enterprises appears to be their tendency to act in a socially proactive manner by developing—to use an incisive image once proposed by Sethi (1975)—an aptitude for considering the directions of social change, with its needs and emerging problems. It is this rethinking of the market, with a propensity to grasp and frequently anticipate the new needs and expectations of consumers, citizens, or specific stakeholders that enables the competitive differentiation of market offer and therefore the construction or expansion of the business target. Among current interpretations, the one most akin to this indication is probably provided by Porter and Kramer with their *shared value* principle, which implies "creating economic value in a way that *also* creates value for society by addressing its needs and challenges" (2011: 4).

It should be made clear from the outset that the relevance of the organizational situations examined does not derive from the possibility of establishing a direct correlation between crisis pressures and certain choices made by these enterprises in the field of sustainable innovation. This is all the more so because, in a number of our cases, investments in this area originate from corporate history and culture (and, as we shall see, this fact as such warrants attention). Rather, the landscape emerging from the study seems significant because it documents *the feasibility of business models based on a virtuous link between sustainability and competitive innovation*, which are consistent with the notions put forward with increasing insistence by management literature in front of the current uncertainties.

The slice of reality depicted by the research suggests considerable heterogeneity in the processes by which the linkage between economic priorities and socio-environmental concerns can be sought and achieved in corporate performance. Nonetheless, at a general level, and within the limits of the exploratory scope of the study, it has been possible to identify a set of central and recurring elements in the sustainable actions of the enterprises observed. In what follows, after a brief description of the study and by concentrating specifically on two cases examined in the research, we shall delve into three distinctive mechanisms which apparently generate these companies' ability to create and integrate values of different kinds, and which we accordingly call 'generative dynamics'. The final section considers the study's implications from a perspective whereby the main challenge for relaunching capitalism coincides with the possibility of transforming the needs of the social environment from constraints to business opportunities which can fuel the creation of integrated value.

Outline of the Empirical Research

Site and Methodology

The empirical research on which this article is based was part of a broader research project entitled 'New Business Models' conducted in a 10-month period between 2010 and 2011.[1] The purpose of the field study was to analyse *in what way social demands for the valorization of a given context generate sustainable innovation among a sample of ten Italian enterprises which*—as reported by a number of 'privileged witnesses'[2]—*displayed an already existing commitment to the inclusion of social and environmental concerns in business management.*

Besides the requirement of visible traces of a sustainability-oriented strategy,[3] a second guiding criterion for selecting the sample was the heterogeneity of the companies in five respects:

1. Size, so as to examine the situations of both small-to-medium organizations and large ones in terms of number of employees;
2. Sector, whose variety, besides reflecting the fundamental difference between business-to-business and business-to-consumer activities, extends beyond traditional domains (such as the food and construction sectors) to more recently defined markets highly exposed to the changes brought about by technological evolution, which are by definition characterized by a need to pursue innovation as a normal competitive requirement (such as IT service suppliers, cross-media communications, and renewable energy);
3. Geographical location, by selecting organizations with headquarters in different areas of central and northern Italy (from Lombardy to Trentino, and from Piedmont to the Marche);

[1]The project was financed by the Chamber of Commerce of Milan and carried out by a team from the Department of Sociology of the Catholic University of Milan. Apart from the authors, the group consisted of Laura Gherardi, Massimiliano Cossi, and Stefano Santini. The details and final results of the project are included in a book published in Italy: Magatti M. (ed.) (2011), *Verso nuovi modelli di business*. Milan: Bruno Mondadori.

[2]This group was formed by academics, business consultants, and public officials who had previously collaborated with several companies, as well as with the authors, in the area of CSR.

[3]Not necessarily or mainly correlated with conventional formal indicators of CSR (e.g. jobs or positions dedicated to sustainability, codes of ethics, social reports, or certifications).

4. Sustainability approach, by selecting organizations with a vocation to human and social sustainability, organizations with an environmental vocation, and organizations with relevant practices in both these areas;
5. Phase of the organizational life cycle, with start-up dates ranging from the mid-twentieth century to the year before the research.

From a methodological point of view, the case studies were carried out using a qualitative approach (Yin 2009). This choice seemed to be especially suited for grasping not only overt actions and outcomes, but also interpretations and expectations, as well as possible tensions, arising from the practice of sustainability. The assumption underlying the adoption of this approach was therefore that such more intangible factors critically affect the strategies, impacts, and issues entailed by an organizational trajectory in sustainable innovation. The main strength of the qualitative case study lies in the multiplicity of data sources used (Eisenhardt 1989), which allows an all-around analysis of the specific organizational reality observed. Accordingly, field data were produced from documentary materials and semi-structured interviews (8–10 for each case study), as well as informal conversations, with organizational members. The interviewees were representatives of senior management and key informants in the crucial areas of sustainable innovation (such as production, marketing, or human resources management), the latter, especially for medium and large organizations, from different hierarchical levels. In addition, as the opportunity arose, direct observations were made of the 'scene of the site' during visits to the organizations (sometimes held continuously for several days) and by attending specific events (such as internal meetings). The checklist used both in fieldwork and in the interpretation phase related to nine central thematic areas: the history and profile of the company; its organizational and strategic model; its climate and culture; human resource management practices; relationships with stakeholders; the economic-financial dimension of sustainability; the human–social dimension of sustainability; the environmental dimension of sustainability; and future prospects.

The situated nature and the small number of the organizational situations observed, combined with the known limitations to generalize in qualitative case studies, conferred upon the research project an exploratory character. This acknowledged, it was believed that, by virtue of the criteria used in the selection of this set of organizations, the reality studied could be, to a certain degree, indicative—albeit not representative—of some current trends in the country.

The Sample and Two Exemplary Cases

Given that it is impossible here to provide a detailed description of the ten organizations investigated and of their specific trajectories in sustainable innovation, Table 27.1 shows the composition of the sample according to its main internal differentiation factors and sketches the sustainability profile of their business models.

These enterprises manifest a set of features which recent literature has identified as constituting the bases of environmental and/or social sustainable innovation

Table 27.1 Composition of the sample

Company	Sector	Foundation	Region	Employees
Engineering	Computer engineering	1980	Lombardy–Lazio	6300
Sustainability	Created through close connections with the institutional system of Italian Chambers of Commerce, it has consolidated its original drive to create value for the entire national community; today it provides services and products in the computer applications sector, including management of infrastructures for third parties, and in particular combines attention to personnel development and to the environment			
Ferrero	Food	1946	Piedmont	21,500
Sustainability	A world leader in the food and confectionary sectors, besides having concentrated on supplies of renewable energy as a company strategy, it has also implemented a corporate wellness system which is at the cutting edge in Italy and offers a wide range of services to satisfy the work–life balance needs of its employees			
Foppapedretti	Furniture	1945	Lombardy	250
Sustainability	A national leader in the wood industry, it produces furniture for children and the home and has progressively incorporated environmental sustainability criteria into its entire design–procurement–manufacturing–assembly–packaging cycle			
Habitech	Construction	2006	Trentino	21
Sustainability	A consortium sponsored by a local public authority (the Autonomous Province of Trento), to a large extent it operates according to market principles and represents around three hundred companies working in sustainable construction, the production of energy from renewable resources, and technologies for territory management			
Innogest	Finance	2005	Piedmont	(Fund)
Sustainability	A 'seed and early stage' venture capital fund, whose business of financing innovative start-up companies (to which it also provides managerial support and relational networks), is defined by a strategy aimed at promoting synergies among public agencies, companies, and research organizations for the development of the local business system			
Loccioni	Industrial systems	1968	Marche	320
Sustainability	(See below)			
MEG	Environmental remediation	1978	Piedmont	145
Sustainability	Also fuelled by the religious formation of the founder, who previously in the USA had studied as an autodidact the active biotransformation of waste, its mission is directed towards the exploitation—with economic profit and social benefit—of products considered waste (e.g. animal manure); today it is active in the fields of environmental remediation and renewable energy, investing in building synergies between its specific know-how and scientific knowledge from the academic world			

(continued)

Table 27.1 (continued)

Company	Sector	Foundation	Region	Employees
TheBlogTV	Communications	2006	Lombardy	100
Sustainability	Almost entirely composed of young people under the age of 35, it is involved in the management of community-building platforms for the collection of web and television productions created directly by its users; among its most significant autonomous social platforms, there are a satellite TV channel and a social network entirely devoted to the 'new Italians' (second-generation immigrants), which represent a means of entertainment and at the same time have civic and educational purposes			
Vita	Communications	1994	Lombardy	30
Sustainability	(See below)			
WIS	Health services	2009	Lombardy	20
Sustainability	A company consisting of a network of private centres operating a franchise, by offering quality health services at accessible rates it contributes to the construction of a sustainable welfare system based on the involvement of users and private and public partners in the creation of relational capital in local communities			

(e.g. Epstein 2008). We find *a long-term orientation*, with an initial willingness—which is apparent both in the history of the more mature companies and in the initial phases of those in existence for only a few years—to expect, but also to wait for, tangible returns in competitiveness. Then there is *timeliness*, that is, adopting from the outset sustainability as a factor in strategic thinking rather than adapting to contextual pressures. We also note *anticipation*, which is closely connected with timeliness and is the ability to position oneself in the vanguard and—one might add—in a position to be in prior compliance with public regulations which may have a significant impact on practices within the sector (e.g. increasingly stringent European laws on ecological requirements in manufacturing processes).

More broadly, in these enterprises' business activities, we note a propensity to combine several value logics and to generate values of different kinds. It does not seem unplausible to see certain aspects of the idea of 'overall value added' into the strategies conceived and then pursued by a good number of these organizations, an idea which, according to Ricotti (2010), distinguishes the 'well-being company'. This notion is very different from the typical financial one of 'economic value added' (EVA), which emphasizes the value of a company in terms of profit, with the aim of rewarding risk capital. The overall value added constitutes *the synthesis or recomposition of multiple values* which concur to create well-being for an enterprise, within the enterprise, and among those who interact with it (e.g. clients or local communities).

This said, in what follows we shall refer primarily to a couple of specific cases examined in the research. The two cases can be considered somewhat exemplary because the recent trajectories of these organizations tend to exhibit very explicitly such *a capacity to generate* (and, first of all, *acknowledge and mobilize*) *a plurality of wellness values in business activities*, condensing—so to speak—a set of elements that in other case studies we find to a varying extent.

Loccioni

Loccioni is a group involved in the measurement and control of industrial products and processes. By means of a high degree of technological innovation and the centrality given to research, it has developed distinctive expertise in the automotive, environment, health, and energy sectors. Organizational reality and culture are still strongly influenced by the founder, Enrico Loccioni, who comes from a family of sharecroppers originally from the small town of the Marche where the headquarters of the company are located. He started his first business venture in the production of electrical systems, soon providing this activity with three unique elements that would be strengthened as key aspects in the subsequent evolution of the company: (a) the choice of working only with large industrial groups (including, today, important international clients such as Ferrari and Porsche); (b) attention to the search for innovative solutions, that is, offers tailored to the client's needs through 'unique projects' (instead of catalogue solutions); and (c) the consequent care for the relational dimension, through the development of an organizational model based—both internally and externally—on trust relationships and knowledge sharing. Its current functioning by matrix organization and business units allows Loccioni to devote itself to multiple lines of activities, but also to inaugurate new ones grafting them onto those already established, and these conditions do not seem unconnected with the increases in sales recorded since 2008, during one of the most critical phases of the economic crisis.

The company's orientation to integrate people and knowledge is the basis of several sustainability practices. As for human resource management, recruitment does not exist in a traditional sense, since the selection of an employee often precedes a clear definition of job roles to which she will be assigned. One of the primary recruitment channels, also expressing a singular openness to local territory's resources and needs, is the project 'Blue Zone', which is part of a wider initiative of continuing education desired by the founder since the 1970s. Addressed to local students at all school levels in order to stimulate positive synergies between school and the world of work, annually the project results in the selection of about thirty candidates among high school and university graduates, who can then start a career in the group. In addition to the 'before phase', Loccioni manages in a special way also the 'after phase' and often supports former employees who become self-employed, by entrusting them with parts of work orders, offering hospitality in its premises to carry out technical processes, and including them in its business network (which, on the other hand, can be enriched by the opportunities brought about by these new entrepreneurs).

Aside from the support to the launch of this sort of spin-offs, the propensity to network—generating both business and social impacts—emerges through at least a couple of other projects. The first is 'Nexus', a permanent laboratory for promoting entrepreneurial culture in the region which monthly gathers practitioners from various worlds (entrepreneurs, artisans, academics, media and sport professionals) in order to encourage them to share their respective experiences and perspectives.

The second project, named 'Land of values', involves a partnership with some local B&Bs and restaurants where the company directs its visitors, thus creating economic advantages for these stakeholders and obtaining reputational benefits for itself. As for environmental sustainability, a sphere in which Loccioni has actively engaged over the last decade (with the use of lighting control systems, condensing boilers, solar panels, and geothermal heat pumps), the most interesting project is the 'Leaf Community'. This is the first fully integrated sustainable community in Italy, where some Loccioni's employees live in a carbon neutral house, move with electrical and hydrogen cars, bring children to a solar energy school, and work in eco-efficient buildings that employ renewable energy sources (e.g. the reuse of rainwater). The Leaf Community has developed through the involvement of the territory and of some large international companies, which participate with the aim of testing and implementing new eco-compatible devices. Besides being an indirect but effective investment in communication, this sort of open-air laboratory has led to the creation of a new business unit called Loccioni Energy Technologies.

Vita

Vita is a publishing company whose aim is to give voice to the Italian non-profit world and to represent the viewpoint of Italian and international civil society by fostering dialogue between the heterogeneous actors that contribute to its pluralism (profit and non-profit organizations, groups of citizens of different national origin, etc.). It was founded in the early 1990s by a group of journalists formerly involved in a programme of Italy's state television titled the 'Courage to live', in particular through the initiative of its conductor who is currently president and editorial director of the company. For several years, the entrepreneurial activity of Vita consisted essentially in the production of the weekly publication 'Vita non-profit magazine', a low-cost magazine launched with the political purpose of putting at the centre of collective discussion the experiences of societal actors who normally have no access to media and positions of power. This goal was—and is—pursued, on one hand, by representing the interests of the third sector and, on the other, by valorizing the specificity of the various components of this world. Through ups and downs, this model has been consolidated without resorting to public funding and solely by means of proceeds from the sale of the magazine. As a matter of fact, in the early 2000s the publication could count on 25 thousand readers and on the creative contribution of a large editorial board with a varied set of competences and professional backgrounds. In those same years, once achieved a balanced budget, the company expanded participation in the shareholding structure also among non-profit organizations, which previously had not been included to prevent them from sharing the debts of the launch phase.

Over the past decade, Vita has grown significantly and has taken the form of a content company operating in areas of action which, although distinct, are integrated by the basic values and purposes of the organization: to deepen experiences and news neglected by other media; to set up a permanent laboratory for providing new impulses to the non-profit sphere and other civil society actors; and to preserve its own independence in the dialogue with multiple stakeholders (e.g. public institutions as for issues regarding the status of the non-profit sector). Currently, the publishing activity related to the magazine retains its relevance and is strengthened by the enhancement of the marketing and advertising function, an increased emphasis on international trends, and the monthly publication of inserts (e.g. about economy and socially responsible investment, environmental sustainability, second-generation immigrants in Italy). Along with this, today two additional lines of activities are gradually expanding. The first supports companies in launching CSR programmes and non-profit organizations in the implementation of marketing and fund-raising strategies; here, one of the main aims is to encourage mutual knowledge and cooperation between for-profit and third-sector actors (among clients there are corporate actors such as Nike, Novartis, and Unicredit Group). The second new business area deals with Internet-based communication through a web portal consisting of multiple sections, which are devoted to: job offer and search (with free insertions); laws, funding opportunities and calls for tender regarding social projects; charitable donations (through a service that connects donors directly to non-profit recipients); a collection of best practices in the promotion of a shared culture of European civil society; the development of cultural relations between Africa and Europe; and public and private initiatives in the field of CSR. Furthermore, the online channel of the company allows to obtain a significant amount of news which normally are not reported by the Italian and international press.

Basically, Vita's present situation poses several challenges at two levels: the organizational level, with an effort to improve the coordination of the (not always converging) positions of people who contribute in various ways to the organization's offer; and the business level, with the goal of increasing sales of the magazine, the customer portfolio of the company's consulting activity, and traffic to the web portal. In this context, what appears to be definitely significant is the recent decision to go public. This choice is not aimed at paying dividends, but instead entails the deliberate and internally shared intent to raise capital for sustaining future development plans; in its promoters' view, not only is it consistent with the company's traditional policy of being autonomous (through a diffuse shareholder base), but it also demonstrates the viability of a market organization that transforms economic value into social value.

The First Dynamic: Innovative Sustainability Between Cultural Embeddedness and Exploration

Within the frame outlined above, the sample reveals a first key dynamic which seems to have played a crucial role in the development of these organizations' sustainability trajectories and which especially concerns their cultural dimension and their collective learning processes. In this regard, it should be pointed out that, in many of these companies, conceptions and values exist and have become routinized (or are in the process of being institutionalized), which: (a) revolve around instances which can be related to sustainability; (b) are linked with strong convictions held by the founders and/or other internal influential people and at the same time have developed through the history of the enterprise and its organizational traditions; and (c) have come, or are tending, to 'personalize' the organization compared with other ones, giving it some sort of DNA which supports the exercise of certain collective competences, among them openness to experimentation and continuing learning that is crucial for innovation.

The Founder/Entrepreneur as a Catalyst for Sustainable Innovation

The first element to be emphasized is the influence of the founding entrepreneur or the group of persons who have set up and launched the company, all of whom tend to maintain a leading role in the current ownership and organizational structure. Among the corporate trajectories analysed, the role performed by this individual variable is not a component of the 'sustainability formula' but rather *a catalyst for the activation of other organizational drivers of sustainable innovation*. The decisive impact of the entrepreneurs/founders undoubtedly relates to their personality traits. It is not rare, in fact, to find charismatic leadership processes, which are clearly evident in the evolution of Loccioni and Vita. More than personality, however, it is the persona of the founders which has left a clear imprint on these companies, where reference to 'persona' encompasses the entire baggage (which we might justifiably call 'heritage') of experiences gained since their childhoods, and which subsequently led to the set-up and management of business. It is the history of the entrepreneurs/founders which proves to be crucial, and which consists, among other things, of specific events (some of them traumatic, such as the wartime experiences of the founders of Foppapedretti and Ferrero), 'illuminating' experiences (such as the period spent by the founder of MEG in the USA), relationships in civil society (which formed the basis of both the WIS and the Vita projects), family affiliations, and strong ties to specific areas of the country and their socio-cultural traditions (as very visible in some of the ways Loccioni does business as well as in market activities of the Habitech district).

The personal repertoire of the entrepreneur, or of the group of founders, has frequently been a significant incubator of business experience ever since the initial decisions and operations. The principal vehicles whereby all this was assembled and conveyed into the business sphere were the conceptions of reality and the value systems acquired, developed, and shared by these individuals in their broader life courses, which were imported and refined, and above all tested (and usually corroborated), in the management and challenges of organizational activity. In other words, the cultural and identitarian references of these individuals have formed a powerful and long-lasting matrix (even more so in the various cases where they continue to play a managerial role) in the creation process of their companies' cultures, by orienting the collective activation of sense-making codes relative to the common perception—within the organization—of the unique nature of their competences and of their market style.

Not only is worth noting the influence of this kind of entrepreneurs on the development of their companies' cultures, which is consistent with the findings of classic research on the role of the founder in the creation of organizational culture (e.g. Schein 1983). Also to be emphasized is the frequent direct inclusion in the belief and value systems translated by the entrepreneur into the company's activities of principles and models of action which presuppose a commitment to sustainability. The role of these 'premises of meaning' is evident in the creation and management of WIS, Vita and Habitech, but it is also apparent in the cases of MEG (through the incidence of the Catholic upbringing of the founder) and Loccioni, where work relationships reproduce the responsibility and risk-sharing model typical of the sharecropping system practised in the founder's home area and by his family.[4]

A Participative and Incremental Development of Organizational Culture

In most cases, therefore, the history and vision of the entrepreneur have provided essential coordinates and criteria used from the outset to define organizational goals and strategies. It is mainly through this channel that the enterprise acquired socially and environmentally significant options already in its start-up phase, when business practices and solutions, later subject to market's assessment, were developed and tested. This does not, however, imply a deterministic process of influence along the lines of the reductionist view of 'cultural engineering' (Alvesson and Sveningsson 2008), which maintains that a company's culture can be shaped unilaterally from above, without considering that it must somehow be connected with the actual shared experiences of organizational members. The evolution of the companies studied confirms that the figure itself of the leader—as typified by founders, by major shareholders or even by professional managers—is embedded within the dynamics of the organizational culture and not above them (Hatch 2004).

[4]In addition to being socially oriented, the cognitive and value maps underlying the creation and management of these companies are sometimes characterized by content which might be defined as alternative and, in certain respects, subversive. A very good example of this is provided by the cultural commitment to the 'valorizing of waste' (meaning tangible or intangible waste products, such as news which does not create a stir). This underlies the global business idea behind MEG and Vita's entry into their sectors: an oxymoron from the conventional viewpoint, according to which 'waste' must at best be disposed of under secure conditions.

This more socially constructed dimension of the organizational culture seems to have emerged since the beginnings of many of these companies. So, it is true that during this phase, the founders tend to surround themselves with collaborators possessing features (values and aspirations, as well as professional abilities) which correspond to their own vision, as very clear in the case of Loccioni, and this represents undeniably one of the most effective ways through which their power to steer their businesses in the desired directions is expressed. However, their influence is not exercised by forging meaning codes which give sense to collective action, but rather by the *capacity to support dissemination of such codes by favouring constant experimentation with distinctive 'ways of doing things', on the one hand, and the subsequent collective understanding of the nexus between the vision underlying practices and the results achieved, on the other*. Applying a seminal concept from Pfeffer (1981), one might say that the decisive contribution made by these entrepreneurs to creating corporate cultures imbued with sustainability springs primarily from a symbolic type of leadership action. The success of this action, however, must necessarily pass through the interpretative acts by which other members of the firm, particularly those responsible for managing critical activities (such as research and development, manufacturing and marketing), actively 'contextualize' the principles and messages which come from on high in everyday operations.

It is through dynamics more complex than those evoked by the seductive image of culture shaping, therefore, that our entrepreneurs' visions have penetrated organizational life, materializing in specific activities and orienting the single functions most crucial for the core business. This *dispersion mechanism*—by which the options underlying creation of the business and its market positioning are translated and, possibly, evolve—seems to be closely linked with what Butera (2009) calls the 'integration forces' of a company pursuing high economic and social performance 'which are based on the values of the founders, and which develop by means of a shared method for putting them into everyday practice' (p. 22), and which reveal the circular relationship between 'non-ordinary entrepreneurs and their creations' (ibid.).

These considerations enable us to focus more directly on the innovative capacities of the companies studied. There is no doubt that, in several of the cases, we find the ability to create and offer products excelling in certain areas (or even just one), and to which clients attach real value. As Christianson (1997) has noted, and as exemplified by Vita's strategy, this ability is a constant feature of innovative organizations because it is a decisive factor for differentiating in the market when other performance conditions (such as capitalization resources) are inferior to those of the competitors. But it is especially in the mechanisms by which organizational culture develops that we can identify a key to explaining the link between sustainability and innovation in these organizations' activities. Such mechanisms relate not only to the creation phase of the organizational culture, but also to the adjustments made to it over time. More precisely, we see in these enterprises' trajectories the importance of processes of 'incremental cultural change' (Gagliardi 1986; Swanson 1999) for their current performance in the market. It is true that this type of change involves experimentation with practices and competences more or less innovative compared with the past. But this is achieved, and is successful, to the extent that the value options underlying new types of behaviour are coherently inserted into the web of meanings and values already in effect and available within the organization, expanding or even reinforcing them. Based on this dynamic, the changing needs and opportunities which arise in a company's day-to-day management induce the adoption of new values integrable with the traditional ones, contrary to what happens in two other situations: 'apparent cultural change', which consists in the mere adoption of patterns of behaviour defined entirely by the nucleus of original values; and the more traumatic 'cultural revolution', by which radical strategic transformations based on traits of collective identity in open conflict with the previous situation (for example, from the concept of the environment as a resource to be exploited to that of the environment as a value for stakeholders) are implemented by means of the socially rather costly removal of the old organizational styles of action (e.g. through high personnel turnover).

Here, cultural incrementalism means that *it is the combination of a series of competences and value patterns developed as much through organizational traditions as by later reorientations of business policies which mainly determines the current position of the companies observed.* A relevant example of this is the path followed by Vita, whose current business orientation in three areas (publishing, consultancy, and web-based communication) derives from cross-fertilization between its initial commitment to valorize social grassroots experiences and a later, wider, focus on contributing to a pluralistic development of civil society based on the inclusion and interdependence of its various actors.

The impression given by a number of cases is that the implementation of evolutionary (and not radical) changes, together with internal dissemination processes of priorities and the inclusion of external stakeholders (see the next section), has supported what Walker (2009) has called the 'sense of agency' of individuals (entrepreneurs and managers in core organizational positions) who set up and coordinate corporate activities marked by sustainable innovation. Most of the key figures working, or who have worked, in the companies analysed are both extremely pragmatic and idealistic. The more visionary aspect, so to speak, of their concept of business seems to have been maintained, and even strengthened, by the sharing, extension, and progressive attainment of a certain corporate project, as shown, particularly, by the gradual consolidation of the non-conformist business choices of Vita and MEG.

Organizational Learning Processes in the Emergence of a Sustainable Business Model

The above-cited cultural evolution mechanisms seem essential for the organizational learning processes underpinning the sustainable innovation conducts of the companies studied.

Firstly, although the level of individual competences is still crucial, there is a capacity of the enterprise as a whole to generate and maintain knowledge independently of single persons. What influences the management of business and its innovative development is not knowledge and creative solutions located 'in the heads' of people (or of certain people), but rather the fact that they are transferred and incorporated into collective practices to become institutionalized within forms of knowledge circulating within a network of social as well as formal relationships. The experience of Loccioni in technology-intensive sectors is paradigmatic. In this company headed by an entrepreneur with no specific technological training, shared learning experiences able to impact on production and services delivery are constantly activated within a network of professional and personal relationships involving workers even after they have left the company to set up their own businesses.

Secondly, the know-how which represents an intangible and decisive resource for these companies, one which is difficult to reproduce in other organizational contexts, does not simply lie in a technical repertoire (such as sector-specific knowledge, operational procedures, or systems for monitoring the environment). In line with what contended above about the influence of cultural and identitarian dimensions, these components intertwine with wider meaning codes on the collective mission to generate, support, and adapt the 'knowledge' on which organizational conduct is constructed. To be reiterated is that this type of learning, which is inevitably technical and socio-cultural at the same time (Gherardi 2005), seemingly characterizes also the companies operating in highly technological sectors (such as Loccioni).

Drawing on an expression commonly used at Loccioni—the neologism 'tradinnovation'—we may also suggest a more general, and perhaps more incisive, interpretation of the learning style which supports the sustainable business models emerging from our sample. In particular, a fruitful interpretation is to be found in the notion of organizational learning developed by March (1991) as *the ability to place oneself as consistently as possible at a point of equilibrium between learning as exploitation and learning as exploration.* By this, it is meant that a fundamental, and relatively common, aspect of the processes of learning and skills development within our companies coincides with a tendency to make extensive use of previously discovered and consolidated ways to define and address problems and to derive the maximum benefit from them, while at the same time remaining open to innovation and the unexpected. This equilibrium also seems to be fostered by the direct inclusion of an orientation towards experimentation and the challenges of 'being in the market' in the original nucleus of the value options of these organizations, first of all in the strictly technological ones but also, recognizably, in others (such as Vita).

Consequently, *there is also a role for paradoxical factors in the experiences of the companies analysed*, to the extent that 'order' does not preclude obtaining resources from 'disorder'. In this sense, the above-mentioned processes of incremental change are not necessarily as linear and immediate as one might think. It is harder, however, to grasp the explicitly contradictory elements emphasized by certain radical studies on learning processes which generate innovation, such as Stark's (2009) intriguing theory of the 'sense of dissonance'. This theory has at its heart the idea that in today's scenario of market uncertainty and changeability, the organizational learning and 'reflexivity' which yield business success are based on ambiguity. According to this logic, 'an entrepreneurial rivalry of performance principles [...] exploits the indeterminate situation by keeping open diverse performance criteria rather then by creating consensus about one set of rules' (2009:16): by keeping 'multiple "orders of worth"' active (p. 11)—for example, economic and ethical values—a company succeeds in producing a 'generative friction' (p. 16) which enables it to operate within multiple 'games' and to manage the conflicting expectations of multiple stakeholders. To the extent that this theory sees the possibility to generate resources for innovation in the 'power' of contradiction, through the constant co-presence of alternative approaches, it is difficult to identify points of contact with our study. On the other hand, a series of suggestions offered by Stark are quite significant in the light of the situations investigated. In particular, in these experiences business action is often—to use the author's words (and in a Schumpeterian sense)—'destructive' and 'recombinant', but with an emphasis also on the latter term. Business-as-usual categories are called into question, and the repertoire of criteria and resources available for company performance is redefined by the intermingling of heterogeneous logics. This dual movement is at the basis of the core business of several of our enterprises, from WIS (whose mission is to provide quality health services at low cost) to Habitech (through the original ways in which it combines the aims of public initiatives with managerial strategies similar to venture capital). And in Vita's decision to be publicly traded, the potential of the break-up/reassembly movement appears to be disruptive, with the absence of financial remuneration for investors and the distribution of 'social dividends' linked to the public value of organizational performance. As it is somehow possible to grasp in some passages of Stark's work, the generative dissonance of innovation does not simply require the coexistence of various options which can be activated as alternatives to each other, but that these criteria interact among themselves. This is precisely the process we can find in the combination of multiple value principles implemented by most of the companies studied.

The Second Dynamic: Stakeholder Engagement Between Centripetal and Centrifugal Moves

Besides the dynamic of learning as exploitation and exploration simultaneously, we may identify a second mechanism, this too somewhat paradoxical, at the basis of the investigated companies' capacity for sustainable innovation: *the interplay of centripetal and centrifugal forces in their normal business functioning*. The former concern the creation of a cultural identity connected with the development of these companies' distinctive competences in the market and in the sustainable management of their activities. As noted before, this process of organizational 'individuation', supported by the original value systems of the founding group and fuelled by the internal dissemination of shared meanings, renders the cognitive and value maps underpinning collective action idiosyncratic and, in certain respects, almost self-referential (at least insofar as they are not validated by the market response). At the same time, however, *the expansion mechanism we have already identified 'vertically', through the refining and strengthening of the company's distinctive mission, is re-proposed 'horizontally' in the form of an uninterrupted movement towards the extension and opening of the company's business to the outside*; all this nourished prevalently by the cultural schemes themselves at the core of the organization's identity.

This mechanism of extension is very visible in two situations which can occur together. The first is when organizational evolution is accompanied by substantial size growth, diversification of areas of activity, or even internationalization (e.g. Ferrero, Engineering, Loccioni, TheBlogTV, and Vita). The second significant case is that of the companies operating in sectors with a high rate of technological change (Loccioni, MEG, Engineering, Habitech, and TheBlogTV): these areas naturally favour multi-sector expansion and pressures to cross borders, with cross-fertilization among different domains of activity (especially between ideation and production processes) and the emergence of forms of lateral collaboration between the organization and crucial interlocutors in the business environment, such as clients and suppliers (e.g. in the co-design of products).

Aside from this, what is most striking is the inclusive approach of our companies towards their socio-economic stakeholders. Within our sample, there is *a common propensity to bring the experiences, instances, and competences of a series of context stakeholders directly into strategies for creation and construction of products, with decisive effects on the capacity for, and the results of, sustainable innovation.*

In this regard, we may cite the crucial role played, in the innovative activities of Habitech and Innogest, by the development of synergies among resources and among conceptions of local business development deriving, on one hand, from the business world and, on the other, from the institutional sphere. A pattern of *seeking contributions from*, and *constructing service with*, key stakeholders clearly emerges from Vita's conducts. This is manifest first of all in the concrete translation of a desire to give voice to actors of the Italian non-profit sector and to offer a participatory laboratory for stimulating civil society to become a continuing force. But Vita's propensity for openness sustains itself also through its engaging in dialogue with government initiatives (for example, as a contributor to the Italian debate on the status of non-profit organizations) and the for-profit sector (providing customer-tailored consultancy on CSR projects). The case that provides the best example of how stakeholder involvement and partnership building generate resources for production processes and value distribution is Loccioni. Here, the mission to integrate ideas, people, and technology in innovative ways involves extremely distinctive networking approaches, where the technical and professional component intertwines with elements of interpersonal knowledge and trust to create opportunities for the participants (individual growth, the acquisition and use of know-how, and new commercial prospects). This is apparent in a number of its practices, from the 'Nexus' network to the 'Leaf Community' project, but the most striking example—a form of constant reinvention of corporate boundaries—is offered by the practice of supporting the start-up of independent activities of its former employees, and mediating their entry into Loccioni's business circuits, to the benefit of both the firm and the spin-off.

Among the ties created by these companies is to be noted *a pervasive process of interdependence with the territory*. As has been seen (e.g. Becattini et al. 2009; Trigilia and Burroni 2009), being rooted in a territory is a distinctive feature of the Italian entrepreneurial system and of the development processes of the Italian socio-economic texture generally. Therefore, it is no coincidence that the sense of belonging to distinctive local communities and the pressure to participate in their evolution tend to be mirrored by the culture and practices of the organizations we studied and to influence their sustainable business models. This is, of course, highly visible in the action of those organizations that work for the economic and business development of a given area as a part of their mission, such as Habitech and Innogest. Nevertheless, the role of these social ties is clear in the whole sample and across its size and sector categories. For instance, we find it in the numerous and diversified forms of collaboration between MEG and public institutions and universities in Piedmont, as well as Ferrero's continuous investment in the creation of workplace well-being for its employees (mostly residing in the local community). Once again, the experience of Loccioni is particularly instructive. We have already mentioned the influence of local culture, based on small-farmer values of mutual trust and risk sharing, on the business vision of Loccioni's founder and on a set of practices (support for the independent careers of employees, recruiting candidates from local schools, etc.) which generate significant benefits above all for individuals in the local community. Also to be re-emphasized is a creative form of participation in community life, with very immediate consequences, in which the company has been engaged for some years. This is the project tellingly titled 'Land of Values', which entails the regular involvement of hotels and restaurants representing the eno-gastronomic culture of the Marche Region in Loccioni's 'care' activities for customers and visitors. In this case, mutual benefits of the collaboration with actors from the company's surrounding community (improvement of the company's reputation, acquisition of new customers for the partners) are directly created by valorising local culture, professionalism, and assets.

This interdependence with the local dimension also exists, and has significant organizational effects, where the company's trajectory is associated with processes of geographical business expansion. In these situations, in fact, it is possible to find *a direct connection between the local matrix of certain styles of action and the organization's ability to operate effectively in more global arenas*. This is the case not only of Habitech and Innogest, which are progressively extending beyond regional borders services and practices already successfully applied in

Trentino and Piedmont, but also of other organizations such as Loccioni itself. For this company, reference to typical elements of its home territory (including values, relationships, and concepts of quality work) has translated into a competitive resource for its internationalization strategies. To reverse—or, rather, complete—the well-known adage that one has to think globally to know how to act locally, we might say that the sustainable innovation capacity of some of these organizations also works in the opposite direction ('think locally, act globally').

Ultimately, the building of interdependencies with social and economic stakeholders shows a constant push for inclusion, to be mainly understood as incorporation of their voice (needs and interests, competences, and contributions) into the company. But the aspect leading to the depiction of the second dynamic of sustainable innovation as a virtuous balance between 'centrifugal' and 'centripetal' forces is, more generally, the following: in most of the companies analysed, *these mechanisms for attuning to the instances and resources of the context have concurred to strengthen the already existing orientation to sustainability, thereby integrating its self-propulsive elements (rather than diminishing them)* and promoting its evolution in phases of organizational change.

The Third Dynamic: The Emerging Link Between Citizenship Behaviours *in* and *of* the Company

The third dynamic which appears to have facilitated the creation of shared value across all the companies in the sample consists of *a series of mechanisms for the management and involvement of organizational personnel.*

Firstly, the essential role of human resources dynamics in the pursuit of sustainability appears to be evident from the fact that these organizations' members are undoubtedly among the main beneficiaries of their social responsibility practices. We observe this in the presence of a positive organizational climate based on respect for individual needs and autonomy, and on collaborative relational models which have a significant impact on the quality of life in the workplace. The most interesting findings, however, concern key processes in the personnel management cycle. This occurs in the recruitment stage, where—contrary to current rhetoric on the benefits of employment flexibility for both companies and individuals—open-ended contracts, with their connotations of stability and reciprocal commitment over the long term, are usually conceived and used as a fundamental prerequisite for operating consistently with organizational strategy. An eloquent aspect of the integration of human resource management practices into the framework of sustainable innovation strategies is the propensity to use selection criteria which assess the skills and potential of candidates with direct reference to the organizational mission, so that evaluation of a person as a whole, taking account of all of her characteristics (such as values, openness to relationships and change, and previous—not only professional—experience), often prevails over the job description concerning the work to be carried out. This holistic approach to the evaluation of job candidates and newcomers not only characterizes the two organizations oriented to political and social ends (Vita and WIS), as would be expected, but is also evident in other cases, such as MEG and Loccioni. Another domain cultivated by our organizations, and frequently at an advanced stage (especially in the more structured companies), is that of employee development, career, and incentive schemes. A good example, in this regard, is offered by the continuing investment made by companies such as Loccioni and Ferrero; today, the latter occupies a front-line position in Italy with its corporate wellness system which provides a wide range of services for the work–life balance needs of its employees (from on-site corporate childcare to health services).

On the other hand, in the enterprises studied, *human resources are not only among the stakeholders subject to sustainability actions; they also tend to represent an essential means for the attainment and success of those same practices.* In most of the companies, the valorization of personnel appears to be indispensable for the production of context value because everyday organizational reality moves from *impacts on the human resource* (which are themselves crucial for sustainability) to *impacts of the human resource*, by which we mean outcomes generated by the employees' active role in the management of core and support processes (production of goods and services, delivery of services to clients, etc.), in the dynamics of stakeholder engagement, and, even further upstream, in the internal dissemination of a sustainability-driven culture. In this regard, going back to our previous discussion on the development of a distinctive cultural 'character' within these organizations, we would stress that the majority of them are able to generate, sustain, and utilize employees' personal commitment to organizational goals and in particular to the elements most closely tied to sustainable innovation. Considering this from an internal branding perspective (Barrow and Mosley 2005; Rosethorn et al. 2009), we may emphasize the capacity of the companies examined to serve various symbolic needs and aspirations that individuals bring into working life.

In sum, the pervasiveness of sustainability instances in the organizational life of these companies leads us to discern *a direct relationship between 'citizenship of a company'* (another well-known concept for expressing an organization's engagement with context values) *and 'citizenship in a company'*, which refers primarily to the cultural mechanisms of involvement and participation of employees in the priorities pursued by the company (Podsakoff et al. 2000). Our case studies decisively suggest that citizenship 'inside a company'—if oriented to socially significant goals—is essential for generating reliable sustainability conducts towards the company's 'external context'. More generally, by stressing that one of the organizational capacities crucial for the institutionalization of sustainable business strategies lies in the interpersonal communication of cultural priorities, we offer support to the ideas of scholars (such as Cohen 2010) who call for an end to the tendency—in theory and in practice—to treat personnel management and the area of corporate sustainability as parallel rather than strictly interdependent domains.

Concluding Remarks: Business Innovation as the Integration of Social Responsibility and Social Opportunity

The case studies of the research outlined in this chapter appear to indicate the feasibility, all the more in the current scenario, of business models revolving around sustainability-oriented strategies that generate integrated value. These are models in which sustainability, in its various dimensions, and above all in how they intertwine, can drive innovation. In a nutshell, the primary condition that has characterized these companies' development path is *their 'activation' of sustainability principles in terms not only of social responsibility, but also of social opportunity, making these the cornerstones of their market positioning*. As Butera (2009) suggests, this profile should not be understood as an ideal-typical condition, but as something that can be pursued by normal companies built to last.

The experiences that we have investigated confirm, firstly, the role of certain drivers of sustainable innovation which the most recent literature has already begun to examine. At least three of them can be inferred from our discussion: (1) the 'precociousness' and long-term orientation of investments in sustainability strategies; (2) the collective (re)production of relatively shared interpretations, starting from the decision-making levels of the organization, on the significance of sustainability for corporate objectives and operations; and (3) the structural incorporation in organizational core processes of innovative tools and solutions generated by a commitment to sustainable action.

The chapter has especially considered certain dynamics underpinning these organizations' capacity to generate integrated value which satisfies business requirements and the emerging needs of society. We have dwelt in particular on the key role of *three mechanisms or sources of creative tension* (given some apparently paradoxical features), at the basis of the sustainable innovation patterns observed. Such processes are those oriented to reach a delicate balance between, respectively: (a) valorization of organizational tradition and openness to exploration; (b) centripetal and self-propulsive forces (protecting and strengthening distinctive competences developed within the company) and centrifugal forces (leading to the acquisition of competences and stimuli for action from the relationship with external stakeholders); and (c) involvement of organizational members as, on the one hand, 'targets' and, on the other, 'vehicles' of sustainable action.

As a whole, the organizational experiences we have outlined lend support to the idea that in the near future CSR practices can, or must, be linked to business practices as never before. To return to one of our initial points, it is quite certain that the strategic implications of CSR are largely still unexplored (McWilliams et al. 2006; Porter and Kramer 2006) and that new ways must be found to address the challenges raised by the need to regard social issues as constituent elements of corporate activity. In this respect, our concept of integration of different well-being values is consistent with the stress placed by Porter and Kramer (2011) on the possibility/necessity of creating economic value through the creation of social value: an opportunity that they view as capable of transforming capitalism and becoming 'one of the most powerful force driving growth in the global economy' (p. 15).

From this standpoint, these organizations' socially oriented strategies are innovative not only because they seem able to generate an overall value added by the creation, integration, and distribution of multiple values, but also because this condition is the result of two interconnected elements: a recognition of the wider and more dynamic context factors which affect the company's performance and which can determine whether it is successful in the long term; and the organization's capacity to concentrate its efforts on 'divergent' CSR practices (Misani 2010), which represent a true source of competitive advantage due to their marked differentiation and scant imitability in the market (unlike 'convergent' practices, modelled on conventional forms of CSR that are easily accessible to others).

Finally, all this leads inevitably to move the focus of analysis on the actions and the involvement of other actors such as public institutions and civil society groups. Allowing and expecting an active role of these stakeholders in the process of creating integrated business value means calling explicitly into question the company-centric approach of the initial phases of CSR, that is, the assumption that companies' solitary efforts (however enlightened) in addressing social problems can generate lasting positive impacts. After all, such an expanded vision is fully in line with the original systemic connotation of sustainability as an inherently pluralistic challenge which, as noted in the trajectories of our organizations, necessarily extends beyond the sphere of businesses' autonomous initiative.

References

Alvesson M, Sveningsson S (2008) Changing organizational culture: cultural change work in progress. Routledge, Milton Park

Barrow S, Mosley R (2005) The employer brand. Wiley, Chichester

Becattini G, Bellandi M, De Propris L (eds) (2009) A handbook of industrial districts. Elgar, Cheltenham

Butera F (2009) L''impresa integrale': teoria e metodi. L'integrazione fra gestione economica e responsabilità sociale nelle 'imprese costruite per durare': un percorso governabile. Sviluppo & Organizzazione October–November–December, pp 18–38

Christianson C (1997) The innovator's dilemma: when new technologies cause great firms to fail. Harvard Business School Press, New York

Cohen E (2010) CSR for HR: a necessary partnership for advancing responsible business practices. Greenleaf Publishing, Sheffield

Couper B, Crawford J, Young C (2009) Strategic sustainability consulting. In: Galea C (ed) Consulting for business sustainability. Greenleaf Publishing, Sheffield, pp 65–85

Eisenhardt KM (1989) Building theories from case study research. Acad Manag Rev 14:532–550

Epstein MJ (2008) Making sustainability work: best practices in managing and measuring corporate social, environmental, and economic impacts. Greenleaf Publishing, Sheffield

Esty DC, Winston AS (2009) Green to gold: how smart companies use environmental strategy to innovate, create value and build competitive advantage. Wiley, Hoboken

Gagliardi P (1986) The creation and change of organizational cultures: a conceptual framework. Org Stud 7:117–134

Gherardi S (2005) Organizational knowledge: the texture of workplace learning. Blackwell-Wiley, London

Hatch MJ (2004) Dynamics in organizational culture. In: Poole MS, Van de Ven AH (eds) Handbook of organizational change and innovation. Oxford University Press, New York, pp 190–211

Laszlo C (2008) Sustainable value: how the world's leading companies are doing well by doing good. Greenleaf Publishing, Sheffield

Lehni M (2000) Eco-efficiency: creating more value with less impact. World Business Council for Sustainable Development, Geneva

Louche C, Idowu SO, Filho WL (eds) (2010) Innovative CSR: from risk management to value creation. Greenleaf Publishing, Sheffield

Lubin DA, Esty DC (2010) The sustainability imperative. Harv Bus Rev 88:42–50

March J (1991) Exploration and exploitation in organizational learning. Organ Sci 2:71–87

McWilliams A, Siegel DS, Wright PM (2006) Corporate social responsibility: strategic implications. J Manag Stud 43:1–18

Misani N (2010) Convergent and divergent corporate social responsibility. In: Louche C, Idowu SO, Filho WL (eds) Innovative CSR: from risk management to value creation. Greenleaf, Sheffield, pp 62–83

Nidumolu R, Prahalad CK, Rangaswami MR (2009) Why sustainability is now the key driver of innovation. Harv Bus Rev 87:56–64

Pfeffer J (1981) Management as symbolic action: the creation and maintenance of organizational paradigms. In: Staw BM, Cummings LL (eds) Research in organizational behavior. JAI Press, Greenwich, pp 1–52

Podsakoff PM, MacKenzie SB, Paine JB, Bachrach DG (2000) Organizational citizenship behaviors: a critical review of the theoretical and empirical literature and suggestions for future research. J Manag 26:513–563

Porter ME, Kramer MR (2006) Strategy and society: the link between competitive advantage and corporate social responsibility. Harv Bus Rev 84:78–92

Porter ME, Kramer MR (2011) Creating shared value: how to reinvent capitalism and unleash a wave of innovation and growth. Harv Bus Rev 89:2–17

Ricotti P (2010) Sostenibilità e green economy: quarto settore. Angeli, Milan

Rosethorn H, Members of Bernard Hodes Group and Contributors (2009) The employer brand. Gower, Farnham

Schein EH (1983) The role of the founder in creating organizational culture. Org Dyn 12:13–28

Sethi SP (1975) Dimensions of corporate social performance: an analytical framework. Calif Manag Rev 17:58–64

Shaetegger S, Burritt R, Petersen H (2004) An introduction to corporate environmental management: striving for sustainability. Greenleaf, Sheffield

Smith CN, Lenssen G (eds) (2009) Mainstreaming corporate responsibility. Wiley, Chichester

Stark D (2009) The sense of dissonance: accounts of worth in economic life. Princeton University Press, Princeton

Swanson D (1999) Toward an integrative theory of business and society: a research strategy for corporate social performance. Acad Manag Rev 24:506–521

Trigilia C, Burroni L (2009) Italy: rise, decline and restructuring of a regionalized capitalism. Econ Soc 38:630–653

Walker P (2009) Supporting the change agents: keeping ourselves effective on the journey of change. In: Galea C (ed) Consulting for business sustainability. Greenleaf, Sheffield, pp 27–46

Yin RK (2009) Case study research: design and methods, 4th edn. Sage, Thousand Oaks

Chapter 28
Sustainability and the Firm: From the Global to the Corporate Ecological Footprint

Luisa Soares and Cristina Chaves

Abstract Nowadays, sustainability is a topic of fundamental importance, having its roots in a large body of literature. The concerns of overpopulation putting pressure on scarce natural resources are not new, but the search for appropriate indicators to assess the performance of a geographical space or entity is becoming more and more urgent. The ecological footprint is one of the most widely used sustainability indicators on a global scale, extending its influence down to local firm level. In this section, we briefly review its origins, main features, strengths, and weaknesses. Companies are increasingly seeking metrical sustainability measures as a means of differentiating in competing markets. The account of a corporate (ecological) footprint may help the firm to find gaps and opportunities for enhancing its behavior, reducing internal and external costs, while improving its market image. However, results from this tool must be analyzed carefully and may lead to some misunderstandings. We illustrate the vulnerabilities of relying solely on the outcomes from the corporate footprint with the case study of a firm operating in Portugal, using a method which has been recently applied to some firms on the Iberian Peninsula.

Keywords Sustainability · Carrying capacity · Ecological footprint · Corporate footprint's accounting method · Portuguese company

Introduction

The question of overpopulation as previously envisaged by Malthus in his essay on the principle of population emphasizes the problem of overconsumption in a world of limited resources. The barrier of limited natural capital which is not able

---

L. Soares · C. Chaves (✉)
Faculty of Economics, University of Porto, R. Dr. Roberto Frias, 4200-464 Porto, Portugal
e-mail: cristina.chaves@fep.up.pt

L.W. Zacher (ed.), *Technology, Society and Sustainability*,
DOI 10.1007/978-3-319-47164-8_28

to support an exponentially growing population raises a fundamental question about growth and sustainability.

Concerns about how to deal with overconsumption and/or overproduction in a context of increasing depletion of our natural resources have resulted in an extensive literature relating to sustainability indicators. Traditional monetary indicators, such as GDP, proved to be unsuitable for measuring these complex variables. The need to look for new tools and means of calculation that are able to capture the different dimensions of sustainability (social, economic, and environmental) has become urgent. Moreover, sustainability has increasingly become a key issue at every level: the national, regional, urban, individual, and last but not the least, the corporate level.

The *Ecological Footprint* (*EF*), first proposed by Wackernagel and Rees, plays a central role in the quest to find ways of dealing with our current economic behavior patterns. With concrete advantages as well as some recognized negative aspects, the extent of the EF's calculations went beyond its original main scope, analyzing and scoring the situation of some nations. It extended its influence to various applications, namely to the behavior of companies, which constitutes our main focus here.

Being aware of its increasing importance, we want to emphasize that the calculation of the *Corporate Ecological Footprint* (*CEF*) may have a double advantage. It may be beneficial for the company's image, being seen as a sign of goodwill toward the central national and international powers, both faced with commitments regarding the sustainability issue. It is becoming a vital tool for presenting suppliers, investors, customers, and other competition partners with the desired social and environmental performance. For the internal sphere of the firm, it may constitute a basis both for furthering improvements in efficiency and also for allowing for a mitigation of eventual negative effects of the company's operation on social, environmental, and thus economic results, especially in the long run. Technological advances may be a key issue in this regard, helping to solve the problem. After analyzing its CEF in detail, a firm is able to come up with a structured plan of how to reduce its footprint, by investing in greener technologies and even developing new ones.

However, external factors beyond the firm's control may offset the effects of the internal will to adopt accurate measures. Therefore, given both points of view, if the CEF is seen primarily as a work tool, it is worth pointing out that it may also be viewed as a source of uncertainties and misunderstandings, and thus its cautious use and interpretation are recommended.

We begin with a general framework, defining some basic and crucial concepts related to a view of sustainability suitable as background for the ecological footprint approach. The "tricky" notion of carrying capacity will be emphasized as playing an essential role in the definition of the EF. After presenting and discussing the methodology of the EF, pointing out its aims, components, strengths, and weaknesses, we turn to the application of its traditional scope to the particular sphere of the firm. We then focus on the CEF, its definition, and goals. Afterwards, we briefly present a possible methodology for calculating the CEF (the so-called

MC3 method), which we see as the most appropriate, and which is being expressly applied in some organizations in the Iberian Peninsula. Thereafter, some possible general internal measures to reduce the CEF are listed. A brief case study of a Portuguese firm is presented, to illustrate some points of view and to strengthen the standpoint that external factors are also a key to understanding the value obtained via the calculations of the CEF.

A Path to Sustainability: Roots and Main Concepts

During the last two centuries, a large body of economics literature focused mostly on a somewhat "simplified" and "optimistic" view of the future of mankind. "Mainstream economics,"[1] based on the strong assumptions of the "homo economicus," sees the economic system as one in which the increase in the intensity of use of production factors (mainly limited to labor and capital) ensures an increase in production output.

According to this view, the environment may not be an issue. Even if it were, scarcity of resources and pollution are not properly addressed as constraining factors of economic growth. This approach points in a different direction from the classical school of economics. Thus, Malthus' "pessimistic" message confronted the issue of a food supply which was supposed to increase following an arithmetical progression, whereas population growth was expected to increase geometrically (for more details, see Kula 1998).

Some authors in the twentieth century distanced themselves from the mainstream, "revisiting" the Malthusian perspective, but also making some notable individual contributions, which have broadened the discussion about the scarcity of resources in a finite world.

Georgescu-Roegen (1976) also provides a well-known contribution, criticizing the way in which mainstream economics sees the economic process as an isolated, self-sufficient, and ahistorical process. The bioeconomic framework of his work points out that on the basis of the entropic process, there is an escalating extraction of natural resources and production of wastes deriving from human economic activity.

Furthermore, the allegories used by Kenneth Boulding of "cowboy economy" (an open system with no concerns about resource limits), contrasting with a "spaceman economy" (a closed system representing the earth) draw special attention to the importance of constructing a new way of thinking about mankind. As he states: "The closed earth of the future requires economic principles which are somewhat different from those of the open earth of the past" (Boulding 1966, p. 7). In his

[1]We call "mainstream economics" the body of literature which developed from the works of Adam Smith, and which specifically refers to features of the neoclassic school of thought.

words, "in the spaceman economy, the throughput is by no means a desideratum, and is indeed to be regarded as something to be minimized rather than maximized" (Boulding 1966, p. 8). This approach conflicts with the mainstream theoretical constructions, where growth (in production and consumption) is seen as possible, positive, desirable, and "unlimited."

The widely known Report "Limits to Growth" (Meadows et al. 1972), although seen as controversial, reinforces the idea that the future of humanity may be threatened by scarce resources, focusing on the to some extent arguable relationship between growth and sustainability (concerning the topic of the "de-growth" concept see Juknys et al. 2014). However, the decisive turning point in the literature and in terms of "common sense" relating to this issue came with the Report of the United Nations World Commission on the Environment and Development (WCED 1987). A "development that meets the needs of the present without compromising the ability of future generations to meet their own needs" constitutes the main and most popularized definition of sustainable development (see Pezzey 1992, Appendix 1, for an interesting survey of sustainability definitions in the literature). Furthermore, concerns about social equity and the "limitations imposed by the state of technology and social organization on the environment's ability to meet present and future needs" were explicitly stressed by the Commission as guidelines. Some mention was made in the publication regarding the "potential population-supporting capacity of land" or "population carrying capacity of the Earth" (see WCED 1987, Chap. 4, paragraph 47), a vital concept within the meaning of the ecological footprint.

Carrying Capacity: The Takeoff to the Ecological Footprint

Malthus' essay on the limited capacity of land to feed a growing population may be considered as providing a basis for the notion of carrying capacity. Much research was undertaken in several fields concerning this emerging concept. It was first theorized by Verhulst in 1838 and later consolidated by Pearl and Reed in 1920 (Seidl and Tisdell 1999; Manning 2007), whose works gave rise to the so-called logistic growth equation. This rationale was based on the theory that populations begin to grow slowly, and then increase faster until they reach an inflection point associated with achieving the environment limits. Afterward, they grow more slowly, tending to an asymptote, which represents the carrying capacity, an ultimately limiting factor in the environment (Manning 2007), also termed "saturation level" (Monte-Luna et al. 2004).

Despite all efforts to clarify the definition of carrying capacity, defining it as suitable for any level—populations, communities, ecosystems, and the biosphere (Monte-Luna et al. 2004)—the concept remains vague (Catton 1987).

Carrying capacity is generally seen as the maximum number of individuals of a given species that can be supported by a particular level of resources, without impairing the productive capacity of the occupied habitat in the future (Hanley et al. 1999; Rees 2002; Ayllón et al. 2012). In the words of Hardin (1986), the "carrying capacity of a territory is defined as the maximum number of animals that can be supported year after year without damage to the environment". In short, it states how much we can use an environment without spoiling it (Manning 2007).

The application of this concept to human beings has raised concerns about the maximum human population that a given environment can support, taking account of the available finite resources (such as food and water), that is to say, its "logistic growth" (Monte-Luna et al. 2004). In this perspective, we are measuring, so to speak, the number of persons supported per habitat/area.

An alternative to characterizing carrying capacity is to "reverse" that definition, emphasizing our concern relating to the needs of area per person. Bearing in mind the need to avoid progressive damage to the bioproductivity and ecological integrity of relevant ecosystems, the carrying capacity may be defined as the maximum rate of resource harvesting and waste discharge that can be sustained indefinitely, independently of the location of those supporting ecosystems (Rees 1996). This maximum persistently supportable load, including the support of future generations, refers not only to the number of users of that environment but "to the total demands they make upon it" (Catton 1987).

The "human load" imposed on the ecosphere by the people who live there underlies the concept of ecological footprint. It "is the product of population size times average per capita resource consumption and waste production. The notion of 'load' recognizes that human carrying capacity is a function of both population size and material/energy throughput" (Rees 2002). Furthermore, sustainability as the basis of the construction of the EF comprises two vital factors which are seen as constraints: overpopulation and human lifestyle. As Catton states: "The world is being required to accommodate not just more people, but effectively 'larger' people" (Catton 1986 cited in Rees 1996, p. 197).

Likewise, the load imposed by the population on a certain habitat is not static; it changes constantly with the available resources and with the needs of that population (Catton 1987; Ayllón et al. 2012). Furthermore, trade and advances in technology are often seen as a means of counteracting the danger of an excessive human load. However, according to Wackernagel and Rees (1996), this argument is refutable, acknowledging that technological innovation does not increase the carrying capacity, but only the efficiency of resource use. With regard to the possible gains from trade, the authors consider them illusory. Despite the redirection of production/supply between countries, there is no apparent reason for a reduction in consumption/demand for resources, a topic which supports the definition of ecological footprint.

Ecological Footprint: General Framework, Discussion, Strengths, and Weaknesses

In the 1990s, William Rees and Mathias Wackernagel proposed and developed a measure of sustainability based on the concept of the ecological footprint (Rees and Wackernagel 1994; Wackernagel and Rees 1996), which constitutes an unavoidable tool as regards sustainability issues. "The ecological footprint is a measure of the 'load' imposed by a given population on nature. It represents the land area necessary to sustain current levels of resource consumption and waste discharge by that population" (Wackernagel and Rees 1996, p. 5). Likewise, according to Kitzes and Wackernagel (2009, p. 812), the EF "measures the amount of biologically productive land and water area required to support the demands of a population or productive activity."

The method underlying the calculation of the EF was developed as an accounting tool to assess the relationship between nature and humans, given the fact that each person requires an area that provides essential goods and services, including waste assimilation (Nunes et al. 2013).

The measuring unit of the EF in the majority of the ecological footprint accounts is the *global hectare* (*gha*), meaning a standardized average productive hectare, i.e., representing an equal amount of biological productivity (Wackernagel et al. 2005; Moore and Rees 2013). This metric, which represents the global average biological world's productivity, was meant to facilitate the comparison of EFs between countries (Wackernagel et al. 2005; Wiedmann and Lenzen 2007).

Like the EF, the *biocapacity* or *biological capacity* (*BC*) is measured in gha. It may be defined as the regenerative capacity of the existing natural capital (Wackernagel et al. 2005), and therefore as the ability of ecosystems to produce useful biological materials and to absorb waste, keeping up with human demand (Moore and Rees 2013), according to current management schemes and extraction technologies (GFN 2012b). "Natural capital is …the totality of nature - resources, plants, species and ecosystems - that is capable of providing human beings with material and non-material utility" (Neumayer 2003, p. 8).

Biocapacity and ecological footprint also have in common the sum of five types of components (Fig. 28.1): cropland,[2] grazing land,[3] forest,[4] fishing grounds[5], and

[2]Type of land devoted to agriculture activities yielding vegetables, fruits, oil, coffee, cereals for animals, cotton, etc.

[3]Area intended to produce meat, milk, wool, leather, and so on.

[4]Land occupied by forests which produce timber resources used in the production of goods, as well as wood for fuel.

[5]Biologically productive marine surface exploited by humans for fish and other marine food products.

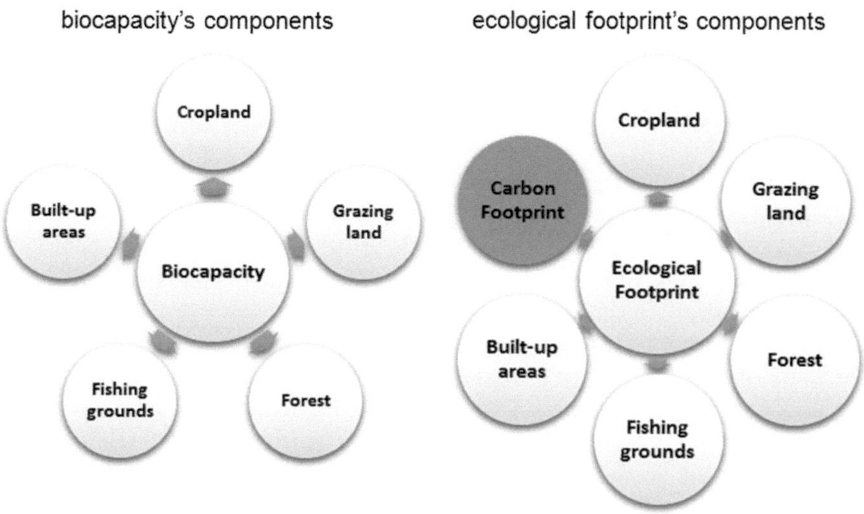

Fig. 28.1 Components of the biocapacity and the ecological footprint

built-up areas.[6] But an extra component in the calculation of the EF, in fact the most important one, is the so-called carbon footprint (CF), also termed "CO_2 area," "CO_2 land," or "energy land," consisting in the demands on the bioproductive area required to sequester the carbon dioxide emissions from fossil fuel combustion (Wackernagel and Rees 1996; Kitzes and Wackernagel 2009; Malghan 2011; GFN 2012a). It is measured in global hectares, as in the case of the other components. This corresponds to the "earlier" concept of "carbon footprint" embedded in the ecological footprint calculations (Cranston and Hammond 2012).

Meanwhile, the expression "carbon footprint" has gained a somewhat different meaning (Muthu et al. 2011; GFN 2012a). Cranston and Hammond (2012) point out this differentiation, because as a component of the EF, a carbon footprint would normally be measured in spatial units, namely global hectares, being actually presented in kilograms of carbon dioxide emitted per person or activity. "Carbon footprints represent the amount of carbon (or carbon dioxide equivalent) emissions associated with a given activity or community, and are closely related to ecological or environmental footprints. But unlike the latter, they are generally presented in terms of units of mass or weight (kilograms per functional unit), rather than in spatial units (such as global hectares). These carbon footprints have become the 'currency' of debate in a climate-constrained world. They are increasingly popular ecological indicators, adopted by individuals, businesses, governments, and the media alike" (Cranston and Hammond 2012, p. 91). To avoid ambiguity, we will adopt the first meaning (measured in gha).

---

[6]areas occupied by all buildings and other infrastructure related to housing, transportation, and industrial production. Part of the area is paved, and other areas remain bioproductive as gardens or parks (Costa 2008; Pereira 2008; GFN 2010)

In short, EF and BC are "two sides of the same coin," with the same metric, and using both the global yield and equivalence factors for their calculation (for a critique of the use of these measures, see Wiedmann and Lenzen 2007). The yield factor compares national average yield per hectare to world average yield in the same land category, i.e., it is the ratio between national and world average yields (GFN 2010). The equivalence factor translates the area of a specific land-use type (e.g., world average cropland or grazing land) into units of world average biologically productive area—global hectares (GFN 2010). At the core of this balance is the key concept of overshoot, a negative outcome in the comparison between the demand and supply of productive ecosystems. When the EF exceeds the BC, the area faces a depletion of productive ecosystems.

On the issue about the role of trade in solving this deficit, it seems obvious that not all countries can have a positive EF and be net importers of biocapacity, but those who are importing transfer their EFs to the exporting countries (Hanley et al. 1999; Wackernagel et al. 2006). In fact, there is a trade-off between importing and exporting regions, with a transfer of pressure on the local resources from the former to the latter. Hence, a change in the global EF does not necessarily ensue, being only dependent on a variation in the consumption/demand for resources. In fact, the importing countries are consuming areas of land/water outside their own ecosystems, but independently of their location they remain the consumer party in this trade; therefore, the potential damage to the ecosystems' regenerative capacity is still being assigned to them. On the other hand, trade is a temporary solution. Thus, if local overshoots are expected to occur, global ones cannot continue indefinitely (Moore and Rees 2013), leaving humanity "in a state of 'overshoot' living, in part, by depleting accumulated stocks of 'natural capital' … and degrading critical ecosystems" (Rees 2010, p. 18). Society is then seen to be heading along an unsustainable path.

In short, a global EF which is higher than the global biocapacity indicates unsustainability. On the other hand, a value of the EF lower than the global biocapacity is a necessary but not sufficient condition to attain sustainability (Moffatt 2000; Wackernagel et al. 2006; Kitzes et al. 2009; Kitzes and Wackernagel 2009).

Despite the apparently intuitive concept and its seemingly "unchallenged" use, the notion of ecological footprint has led to controversial debates and several criticisms in the literature (see for instance van den Bergh and Verbruggen 1999).

In order to provide a brief albeit broad picture of the debate surrounding the use of the EF, some of its most frequently quoted strengths and weaknesses, those considered most relevant for the present study, are summarized in Tables 28.1 and 28.2.

This short summary of advantages and shortcomings presented in Tables 28.1 and 28.2 points out the need to be aware of the weaknesses and misleading aspects of the calculation of the EF as well as the need to refine and improve the concept and its scope, a process which has already begun.

However, the demand for a measure of sustainability with the above-identified features has grown steadily, broadening its scope and persuading the economic agents, especially firms, of its importance.

Table 28.1 Strengths of the ecological footprint

Strengths	Comments	References
Aggregation in a single measure	It aggregates in a single measure the environmental pressures on the ecosystems	Wackernagel et al. (2004), White (2007), Fiala (2008), Bagliania and Martini (2012), Gondran (2012)
Balance between complexity and simplicity	Making abstractions of complex functions of the ecosystems makes it workable	Wackernagel and Rees (1996)
Clear and simple communication	Public opinion may become aware of the relationship between humanity and the ecosphere	Holland (2003), Lenzen et al. (2007), White (2007), Wiedmann and Lenzen (2007), Carballo-Penela et al. (2008), GFN (2010), Bagliania and Martini (2012), Browne et al. (2012), Gondran (2012)
Availability of data	Data to calculate the EF are mostly available in national statistics, international organizations, or firms' accounting records	Browne et al. (2012)
Application to different levels	It may be employed at global, national, local, corporate, and individual levels	White (2007), Mostafa (2010)
Consistency with the laws of thermo dynamics	In the opinion of some authors, the EF is consistent with the thermodynamics laws and is appropriate for measuring strong sustainability, considering the complementarily between natural and manufactured capital	Carballo-Penela et al. (2008)
Ability to measure strong sustainability		Neumayer (2003), Dietz and Neumayer (2007), Carballo-Penela et al. (2008)
Comparison between environmental impacts	It allows the comparison of the environmental impacts at several levels: countries, families, or organizations	Mostafa (2010)
Emphasis on the distance to ecological overshoot	The analysis of an EF time series can reveal whether the population under study is getting closer or further away from the ecological overshoot	Barrett and Scott (2001), Wackernagel et al. (2004), Doménech (2006), Bagliania and Martini (2012), Browne et al. (2012)
"Fairness" of distribution of responsibility concerning climate change	Most of the effort demanded by the Kyoto Protocol falls on the manufacturers. With the EF, everyone (families, countries, all types of organizations) became aware of their share of responsibility and can take action to reduce it	Doménech (2006)

(continued)

Table 28.1 (continued)

Strengths	Comments	References
Identification of critical issues and possible answers	Despite being not normative (it only accounts for the demand of environmental resources in various areas), this measure allows recognition of some of their critical spots and the proposition of better measures	Wackernagel et al. (2004)
Recognition of critical questions concerning the (dis) advantages of trade	The establishment of a series of regional ecological accounts may help countries to estimate their true ecological load and monitorize their ecological trade balances	Wackernagel and Rees (1996), Browne et al. (2012)
Helpfulness in designing policies to reduce the environment's overexploitation and monitor the progress toward sustainability	The EF analysis recommends that the carrying capacity stays below the Earth's biocapacity, which means that it is essential to implement an alternative strategy to "business as usual"	Wackernagel and Rees (1996), Barrett and Scott (2001), GFN (2010), Peters (2010), Browne et al. (2012), Galli et al. (2012)

Table 28.2 Weaknesses of the ecological footprint

Weaknesses	Comments	References
Oversimplistic vision of complex systems	The result of the EF used in an isolated manner can create a too simplistic vision of complex systems	Kitzes and Wackernagel (2009), Browne et al. (2012)
Linear relationship	It assumes a linear relationship between the ecological impact and its effects. But, in reality, ecological systems are complex and nonlinear	Holland (2003)
Difficulty in making comparisons	Data needed for the calculations are very different, and there is an intrinsic uncertainty in the application of the methodologies	Nunes et al. (2013)
Weak capacity for showing the dimensions of sustainability	It is unable to reveal the sociopolitical, economic, and eco-justice dimensions of sustainability, constituting more a measure of unsustainable overshoot than of ecological sustainability. Therefore, it should be complemented with other indicators	Moffatt (2000), Holland (2003), Wackernagel et al. (2004), Nourry (2008), Kitzes and Wackernagel (2009), WWF (2010), Browne et al. (2012)
"False concreteness"	The EF represents hypothetical land area, but there is a serious danger that it will be interpreted as representing realistic land use	van den Bergh and Verbruggen (1999)
Inconsistencies in converting hectares to global hectares	Loss of locally important information about the management of natural resources	Wiedmann and Lenzen (2007)

(continued)

Table 28.2 (continued)

Weaknesses	Comments	References
Focus on stock measurement	It does not take into account the inflows to and outflows from an area	Moffatt (2000)
Static calculations	It only takes a "picture" of the relationship between the economy and the land at a given moment in time	Wackernagel and Rees (1996), Moffatt (2000)
Difficulty in identifying where the environmental degradation takes place	Due to international trade, the EF is spread all over the planet. It is accounted for in the country where the goods are consumed, but the use of resources occurs in the exporting countries	Costa (2008)
Desire for "ecological autarky"	Calculating the ecological deficit through the comparison between the footprint and biocapacity implies the desire for "ecological au-tarky" because the more the country imports, the bigger is its EF, so that each country prefers to use natural resources that are locally available	van den Bergh and Verbruggen (1999), Pearce (2000), Ayers (2000) *in* White (2007)
Disregard for technological changes	The EF ignores the role of technological changes; if they were considered, the EF could decrease	Costa (2008), Fiala (2008)
Provision of misleading signals to policy makers	The issue of bioproductivity and metric used can be elusive, when for instance an increase in biocapacity takes place at the expense of the damage induced in biodiversity and the health of ecosystems	Lenzen et al. (2007)
Lack of political adequacy	On the one hand, it is only used to describe the human demand on natural resources. On the other hand, in the local and regional sphere, there are difficulties in calculation: (a) the smaller the area and population analyzed, the bigger the difficulty in obtaining correct consumption data; (b) the use of different methods and data makes the comparability difficult	Moffatt (2000), Wiedmann et al. (2006), White (2007), Nourry (2008)
Difficulty of comparison with a country's physical area.	The obstacles in comparing the EF with the physical area of a country relate to the fact that the borders are environmentally irrelevant	van den Bergh and Verbruggen (1999), Fiala (2008)
"CO_2 land" in carbon footprint calculations	This is one of the most controversial issues because it usually concerns the biggest portion of the EF. It is doubtful that the amount of land-intensive forestry for sequestering CO_2 is the most appropriate measure for calculating the fossil fuel footprint	van den Bergh and Verbruggen (1999), Pearce (2000), Neumayer (2003), White (2007), Browne et al. (2012)

(continued)

Table 28.2 (continued)

Weaknesses	Comments	References
No distinction between sustainable and unsustainable use of land	A big footprint might be more sustainable than a smaller one, depending on how the land is used. If the population uses the land ineffectively but without destroying it, the system maybe sustainable. On the other hand, the destruction of land and the need to transfer the production to other areas may lead to unsustainability	van den Bergh and Verbruggen (1999), Fiala (2008), GFN (2010), Browne et al. (2012)
Solely valorization of the land with human value	This tool excludes deserts, oceans, and subsoil resources. This has two consequences: the underestimation of Earth's biocapacity potential and the omission of the human impacts caused in these "nonproductive" although crucial areas	Moffatt (2000), Browne et al. (2012), Hopton and White (2012)
Treatment of areas as exclusive	The different areas are treated as exclusive, regardless of the multifunctional possibilities of an area	van den Bergh and Verbruggen (1999), Browne et al. (2012)
Omission of a large spectrum of environmental pressures and impacts	EF takes no notice of a large part of the environment's problems	Hanley et al. (1999), Costa (2008), GFN (2010), Bagliania and Martini (2012), Browne et al. (2012)
Impossibility of considering all aspects of economic activities	This is due to the impossibility to convert most aspects into physical units	Nourry (2008)
False easiness in measuring wastes	The EF considers that it is easy to measure waste flows and convert them into land area, which in reality is quite complex	Pearce (2000), Costa (2008)

From the General Footprint to the Level of Firm: Corporate Footprint as a Trademark of Sustainability

The development of tools for analyzing the environmental performance of organizations is essential in order to evaluate how much of the biosphere is required for a company to maintain its business activity without necessarily impairing the sustainable use of resources and to help contribute toward a global sustainable development (Carballo-Penela et al. 2008; Doménech 2009). Besides, it is unthinkable that governments will achieve sustainable development without the support of organizations and the public in general (Barrett and Scott 2001).

It is a fact that many organizations see sustainability as a means of differentiating, which is vital for raising productivity and competitiveness (Carballo-Penela et al. 2009; Doménech 2009; Lee 2011). The corporate footprint (CEF) allows companies to establish clear and concrete aims relating to environmental sustainability, providing a method for the support of decision making and for monitoring the effectiveness of the implemented policies (Barrett and Scott 2001; Doménech 2006; Gondran 2012; Branco 2012). Accordingly, having a single measure of the firm's ecological impact means owning a tool (Holland 2003; Doménech 2006; Carballo-Penela et al. 2008, 2009; Peters 2010; Lee 2011), which:

- Identifies unsustainable demands of the biosphere, seeking alternative resource use;
- Allows the identification of unnecessary costs and unexploited opportunities at the internal management level;
- Attempts to facilitate external communication and improves image, with possible economic benefits;
- Enables forecasting, by identifying the products which have a bigger impact on the ecological limits;
- Encourages companies to create ecolabels and consumers to choose the ones that are most environmentally friendly.

In consequence of the exposure to carbon markets, the analysis of the CEF may also help the firm to recognize opportunities to reduce carbon emissions, diminishing risks, and costs associated with that exposure. The scrutiny of the components resulting from the calculation may also be paramount for success in identifying areas for a possible reduction in energy consumption costs, or where it is reasonable to do so, in adopting new greener technologies (Lee 2011).

The use of an indicator expressed in hectares of productive surface could be seen as inappropriate for use at the level of the firm, but the conversion of different units (energy consumption, waste generated, etc.) into a common metric will provide relevant information to companies (Carballo-Penela et al. 2008). Moreover, hectares can mostly be converted into CO_2 tones, as in the case studies which are presented in the following sections.

Calculation of the Corporate Footprint Applying One Possible Methodology: Some Case Studies from the Iberian Peninsula

The "Composed Method on Financial Accounts", abbreviated as MC3, also called "method composed of financial statements" (Carballo-Penela and Doménech 2010) is an organization-product-based-life-cycle assessment type of methodology (Cagiao et al. 2011; Carballo-Penela et al. 2012). It was first developed by Doménech (Carballo-Penela et al. 2008; Carballo-Penela and Doménech 2010) based on the Composed Method developed by Wackernagel and Rees (Carballo-Penela et al. 2012). Even if we acknowledge that there are several important methodologies for calculating firms' footprint, we consider that the scope of this section is not to survey them. We therefore decided to focus on the MC3 methodology, not only because its quality is widely acknowledged, but also because we find it appropriate for this case study. The "Método Compuesto de las Cuentas Contables"—MCCC, as it was called by its authors, was first developed by Doménech between 2000 and 2002 (Carballo-Penela et al. 2012) and ever since has been improved to embrace different sectors of activity.

The footprint can be expressed in both land area (global hectares) and CO_2 tones (Doménech 2006; Carballo-Penela and Garcia-Negro 2008; Carballo-Penela et al. 2008; Carballo-Penela and Doménech 2010).

The authors of MC3 created a consumption land-use matrix similar to the one used in the calculation of countries' footprints, which enables the registration of the company's main consumption categories (Carballo-Penela et al. 2009). The matrix lines show each product's category footprint, and the columns include, among other things, the different land types which constitute the footprint (Carballo-Penela et al. 2008, 2009).

Supported by the developments based on the MC3 methodology (the so-called version MC3 V.2.0), a new picture of the calculation matrix was summarized (Table 28.3). Consumption data are mostly obtained from company accounting, which makes it applicable to any organization on any scale. Yet, the difficulties are noteworthy, starting with the fact that several data on the basis of the calculation using the MC3 methodology are difficult to obtain for the populations and periods analyzed, bringing about the need to make several proxies and assumptions, which directly influence the results. Furthermore, we can mention the difficulty of matching the accounting information with the consumption categories. Several problems arise, such as finding information about equivalence factors for the year under study, productivity factors for the specific year and country, and the composition of energy mix, just to name a few.

Some advantages were pointed out by the followers of this approach. It is considered a complete, transparent, flexible, and technically feasible method, which allows researchers to add or update the factors used in the calculations, customizing the tool to the specificities of the company (Carballo-Penela et al. 2008, 2009; Carballo-Penela and Diz 2011). Further, it has a mixed approach: (a) bottom–up for the input products (all the firm's consumptions), and (b) top–down for output products. In other words, the footprint is distributed among the firm's products (goods and services). It allows us to calculate the organizations' and products' footprint(s) simultaneously (Doménech et al. 2010; Carballo-Penela et al. 2012).

The application of MC3 has given rise to several calculations of the CEF for different Iberian Peninsula organizations belonging to various sectors. An illustration of some of the results obtained is shown in Table 28.4. Attempting to understand to what extent a value obtained for the CEF would be "high" or "low" yields unsuccessful results, as we would be "trying" to compare data from different industries and companies, even though the methodology is supposed to be the same or very similar. In fact, even in the same sector, no comparison is possible, because of the large number of distinct specific assumptions which are needed to make the calculations feasible. This constitutes a pitfall in the interpretation of the CEF, the problem we turn to now.

Table 28.3 MC3, V.2.0

Column composition	Consumption categories and subcategories	Annual consumption					Emission factor[a]/conversion factor[b]		CCF by ecosystem type, in tCO2/CEF by ecosystem type, in ha[c]						Total footprint and counter-footprint	
Units	–	In units (L; kWh; m^3; ha)	In euros	In tonnes	Energy intensity (GJ/t)	GJ/year	Natural productivity (t/ha/year)	Energy productivity (GJ/ha/year)	Fossil energy tCO_2; ha	Cropland	Grazing land	Forest	Buildup land	Fishing grounds	tCO_2; ha	tCO_2; ha
1. Direct emissions	1.1. Fuel															
	1.2. Other direct emissions															
2. Indirect emissions	2.1. Electricity															
	2.2. "Other indirect emissions"															
3. Materials	3.1. Flow materials (merchandise)															
	3.2. Non-redeemable materials															
	3.3. Redeemable materials (generic)															
	3.4. Redeemable materials (construction)															
	3.5. Use of public infrastructures															
4. Services and contracts	4.1. Low mobility services															
	4.2. High mobility services															

(continued)

Table 28.3 (continued)

Column composition	Consumption categories and subcategories	Annual consumption					Emission factor[a]/conversion factor[b]		CCF by ecosystem type, in tCO2/CEF by ecosystem type, in ha[c]						Total footprint and counter-footprint	
Units	–	In units (L; kWh; m^3; ha)	In euros	In tonnes	Energy intensity (GJ/t)	GJ/year	Natural productivity (t/ha/year)	Energy productivity (GJ/ha/year)	Fossil energy	Cropland	Grazing land	Forest	Buildup land	Fishing grounds	tCO_2; ha	tCO_2; ha
									tCO_2; ha							
	4.3. Passenger transport services															
	4.4. Merchandise transport services															
	4.5. Use of public infrastructures															
5. Agricultural and fishing resources	5.1. Clothing															
	5.2. Agricultural products															
	5.3. Restaurant services															
6. Forestry resources																
7. Water footprint	7.1. Consumption of drinking water															
	7.2. Consumption of non-drinkable water															
8. Soil use	8.1. On land															
	8.2. On water															

(continued)

Table 28.3 (continued)

Column composition	Consumption categories and subcategories	Annual consumption					Emission factor[a]/conversion factor[b]		CCF by ecosystem type, in tCO2/CEF by ecosystem type, in ha[c]						Total footprint and counter-footprint	
Units	–	In units (L; kWh; m^3; ha)	In euros	In tonnes	Energy intensity (GJ/t)	GJ/year	Natural productivity (t/ha/year)	Energy productivity (GJ/ha/year)	Fossil energy	Cropland	Grazing land	Forest	Buildup land	Fishing grounds	tCO_2; ha	tCO_2; ha
									tCO_2; ha							
9. Waste, discharges and emissions	9.1. Non-hazardous waste															
	9.2. Hazardous waste															
	9.3. Radioactive waste															
	9.4. Discharges in effluents															
	9.5. Emissions															
	9.5.1. Kyoto Protocol GHG															
	9.5.2. Other GHG or precursors															
	9.5.3. Other atmospheric emissions															

[a]It allows to convert data in gigajoule (GJ) into tCO_2/GJ, obtaining the carbon footprint: [tCO_2] = [GJ/year] × emission factor [tCO_2/GJ]

[b]It allows to convert the accounting data in Euros into tonnes (t): [t/year] = [€/year]/conversion factor [€/t]

[c]For each consumption category, the calculation formula slightly varies. For example, to calculate the material footprint, first of all it is necessary to obtain the data from the accounts of the company. By entering this information on the worksheet, the calculations are made automatically, following these steps

1. Conversion of Euros to tonnes of product: [tonnes/year] = [€/year]/conversion factor [€/tonnes]
2. Conversion of tons to gigajoules consumed during the life cycle: [GJ/year] = [tonnes/year] × energy intensity [GJ/tonnes]
3. Carbon footprint ("energy land"): [tCO_2] = [GJ/year] × emission factor [tCO_2/GJ]
4. Ecological footprint ("energy land"): [ha] = carbon footprint [tCO_2]/absorption factor [tCO_2/ha]

Table 28.4 CEF of several firms, calculated using MC3—Iberian Peninsula

Firm	Year	Activity	Country	CEF (gha)	CCF (tCO_2)	References
"*Trans*"	2011	Perishable goods transportation	Portugal	2,136.3	5,915.0	Soares (2013)
"T"	2011	Perishable goods transportation	Portugal	5,736.1	15,860.9	Soares (2013)
"E"	2011	Electric equipments producer	Portugal	32,430.6	39,493.2	Branco (2012)
APG	2004	Port services	Spain	5,297.9	30,426.2	Doménech (2009)
"Gamma"	2006	Wine producer	Spain	–	152.7	Carballo-Penela et al. (2009)
"Alfa"	2007	Mussel producer	Spain	11.0	52.2	Carballo-Penela and Garcia-Negro (2008)
"Anónima"	2006	Car dealership	Spain	4,945.9	22,547.7	Moncho et al. (2008)
"BI"	2006	Fishing	Spain	1,083.5	1,678.2	Carballo-Penela et al. (2008)
"B2"	2006	Fishing	Spain	540.2	1,026.4	Carballo-Penela et al. (2008)

Measures to Reduce the Corporate Footprint: The Pitfall of Interpreting Variations Based Solely on Internal Measures: An Illustration

The difficulties in the interpretation of a specific value of the CEF raise some questions about how important it is to know companies' CEF. An obvious benefit is that firms want to win recognition from their stakeholders, and the CEF is an additional tool which contributes to this aim. In addition, from an internal perspective, it is no less relevant, because it gives an indication of what can be done in order to reduce the footprint, which is normally associated with increasing cost efficiency. Thus, depending on the specific activity of the firm under consideration, there are various measures that can be applied in order to reduce the CEF. These are controlled by the firm and allow it to come closer to being "sustainable," while at the same time becoming "more efficient" in reducing costs.

We can briefly refer to some well-known general internal steps concerning several fields of action. Broadly speaking, it is appropriate to apply the "3 Rs" (reduce, reuse, recycle) strategy, to demand its application by the suppliers and promote it among clients (Doménech 2009; Muthu et al. 2011; Pasqualino et al. 2011). Concrete measures consist of, e.g., (1) fuel use reduction through the utilization of new technologies or more efficient ones, such as hybrid or electric vehicles (light or heavy); (2) application of "green energy" and renewable energy, such as solar thermal installation in the roofs of buildings (Doménech 2009); (3) purchasing policy with preference for certified suppliers and "green" materials (Doménech 2006, 2009).

These measures focus on internal actions of the company, several of them demanding technological advances and new production and consumption paradigms.

However, the firm works in a market and belongs to a chain in society as well as constituting a "part" of the environment, being subject to changes in all these systems. Factors which are *external* to the company can strongly influence the CEF of that organization, sometimes tending in the same direction as internal measures, and thus contributing to greater reduction in the CEF, but they may also militate against this purpose, neutralizing or even counteracting the effects achieved at the level of the firm. These factors may be due to market structure, regulations, and to more "indirect" factors such as climate conditions among others.

By way of illustration, we refer to a case study based on a Portuguese firm here designated as *Trans*, whose activity is transportation of perishable goods. The deliveries are made in lorries, some belonging to the company and some being subcontracted.

The Illustration: Trans' Corporate Footprint in 2006 and 2011

Trans collects goods from different producers/distributors in an area, placing them all in a single vehicle and delivering it to their destination.

The *Trans'* CEF was calculated for two different years, using the MC3 methodology. The results by ecosystem and by consumption category are presented in Table 28.5 and Fig. 28.2. According to these results, from 2006 to 2011 the CEF of *Trans* hardly altered, showing an insignificant increase, apparently resulting from the increase in the main component (services and contracts), due to the firm's requirements for its activity. However, the causes of that behavior of the CEF's

Table 28.5 *Trans'* total CEF and by type of ecosystem

Year	Footprint by type of ecosystem (gha)						Total footprint (gha)		
	Energy land	Cropland	Grazing land	Forest area	Built-up area	Fishing grounds	Total CEF	Counter-footprint[a]	Net CEF
2006	1983.9	2.4	0.1	128.0	1.0	1.7	2117.0	1.4	2115.6
2011	1981.2	0.5	0.1	153.4	1.0	0.4	2136.6	0.3	2136.3

[a]The counter-footprint consists of a commitment assumed by the firm, which involves buying or taking care of forest or garden areas, contributing to reducing part of its footprint. These investments in natural capital can be made primarily by supporting ONG's nature conservation projects (Doménech 2006; Carballo-Penela et al. 2008; Moncho et al. 2008; Doménech 2009; Carballo-Penela et al. 2009)

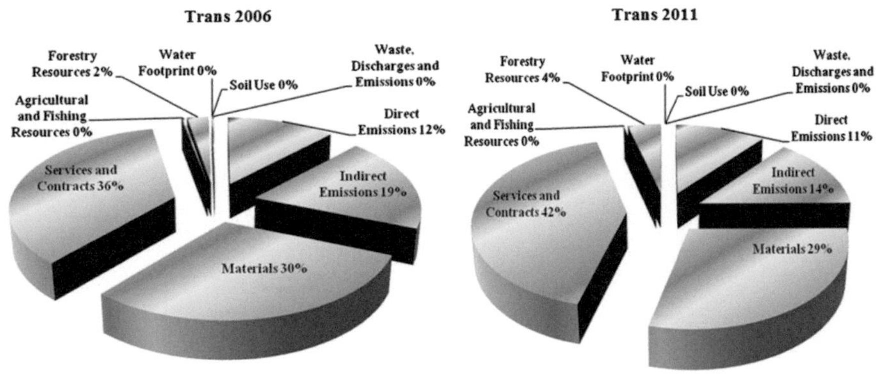

Fig. 28.2 CEF of *Trans*, by category

value can and should also be sought in external factors. To justify that statement, we will explore some simulations relating to the situation of *Trans* in the market and in social interaction.

To achieve the values of CEF, although the MC3 spreadsheet and its features were used and adapted to the case study, several assumptions (which were seen as reasonable) had to be made. The calculations were made for 2011 (based on the most recently available data) and for 2006, a previous selected year, to attempt a temporal comparison of the results (Soares 2013).

Simulation of Alternative Scenarios and Its Impact on the Global EF

Being aware that *Trans* is a firm that delivers food products, we conducted a simulation with two possible situations—Scenarios 1 and 2 (Table 28.6), with two distinct ways of delivering goods to the two major food distribution groups in Portugal—*S* and *J*.

Because *Trans* provides a service which causes a high concentration of transportation and goods, the main question here is whether this service provided by *Trans* (Scenario 1) represents a noticeable contribution to the overall environmental improvement and global sustainability in comparison with an abstract situation in which the firm would not exist, and the deliveries to *S* and *J* would be made by the producers/distributors themselves (Scenario 2). The comparison is made through the simulation of the amount of kilometers and fuel consumption that would result if the deliveries were made individually by each of the *Trans* customers in 2011, using their own fleets.

As a consequence of the services provided by *Trans*, there is a decrease in lorries using the roads and the number of kilometers is cut by half. Consequently, the fuel consumption and CO_2 emissions decrease, which is beneficial for the environment and contributes to reducing the global footprint.

Table 28.6 Calculation results for Scenarios 1 and 2

Scenarios			
1.	Deliveries are made by the transportation services of Trans, collecting the products from its clients and distributing them to S and J. This corresponds to the real market and firm situation in 2011		
2.	Deliveries are made by means of transportation of Trans' clients, using their own fleets, to S and J. This situation corresponds to a simulation of deliveries data in 2011		
Items	Results		
	Scenario 1	Scenario 2	Scenario 1 versus scenario 2 (%)
Total (km)	825,730	1,912,367	−57
Total fuel (L)	325,489	650,205	−50
CO_2 emissions (t)	908.1	1814.1	−50

Impact of Market Concentrations on CEF

As mentioned above, there are some external factors which affect the firms and influence their CEF. Bearing that in mind, we propose to make two simulations, assuming either partial or total concentrations in the destinations of *Trans* deliveries.

In the first simulation, we depart from the real fact of the occurrence of a partial concentration in the food supply market. During 2007, a major change took place in the food distribution market in Portugal, which had a strong impact on *Trans*' business: *S* group acquired *C* (n stores), concentrating *Trans*' deliveries at two *S* central purchasing units. Drawing on that fact, we made a simulation for the year 2006 (scenarios A and B, Table 28.7), assuming that the concentration of *S* and *C* had already occurred, estimating the distance in kilometers covered by the transport of goods. If the market change had occurred prior to 2006, *Trans* would have seen a decrease in the number of kilometers by about 41 %, because of the reduction in the number of delivery destinations. As a consequence, there would have been a decrease in the fuel consumption of the fleet, as well as in the subcontracting cost, and the CEF would have diminished to 2106.8 gha. This favorable variation in the CEF occurs with no proactive intervention from the firm, only stemming from market changes.

In the second simulation, we try to calculate the impact on *Trans* of a market change which would involve the *complete* centralization of the deliveries at the two major groups (*S* and *J*). At present, *S* and *J* centralize most, but not all, of their purchases on central platforms. In 2011, the total *Trans* deliveries at the central purchasing unit were 97 % in the case of *S* and 87 % in the case of *J*, the rest being transported to individual stores. Scenarios I and II (Table 28.7) show two

Table 28.7 Impact of concentrations of the food supply market

Concentration of *S* and *C*: simulation for the year 2006								
Scenarios		Results						
A	The deliveries are made in 2006 by *Trans* at all C stores (real situation)	Delivery location	Total kms	Own fleet		Subcontracted services		Total footprint
				Own fleet kms (17 %)	Consumption (fuel liters)	Subcontracted kms (83 %)	Subcontracting cost	(gha)
		C stores	95,791	16,285	6225	79,507	62,104	2115.6
B	Simulation for *Trans* in 2006, assuming that the deliveries are made at the two purchasing centrals of *S*	Delivery location	Total kms	Own fleet		Subcontracted services		Total footprint
				Own fleet kms (17 %)	Consumption (fuel liters)	Subcontracted km (83 %)	Subcontracting cost	(gha)
		S central purchasing	56,415	9591	4028	46,824	36,575	2106.8

Total centralization of S and J: Simulation for the year 2011								
Scenarios		Results						
I	*Trans* deliveries are made mostly at the central purchasing and a small part at the stores (real situation in 2011)	Delivery location	Total kms	Own fleet		Subcontracted services		Total footprint
				Own fleet kms (17 %)	Consumption (fuel liters)	Subcontracted kms (83 %)	Subcontracting Cost	(gha)
		S and J stores	18,979	1708	595	17,271	15,302	2136.3
II	*Trans* deliveries are entirely made at the *S* and *J* central purchasing platforms	Delivery location	Total kms	Own fleet		Subcontracted services		Total footprint
				Own fleet km (17 %)	Consumption (fuel liters)	Subcontracted km (83 %)	Subcontracting cost	(gha)
		S and *J* central purchasing	9656	869	365	8787	7785	2134.5

In order to make these calculations, the MC3 method was again adopted modifications that influence the *Trans'* CEF: fuel consumption reduction and subcontracting cost reduction, as in the previous situation, which generate a CEF of 2134.5 gha, smaller than the actual value.

Although the CEF reduction is very small (in both scenarios), these examples show the importance of considering the significance of the impact of external factors on the footprint, namely market factors. The impact in these cases was not particularly significant because the volume of goods affected by these changes was also minor. Bigger changes would, predictably, have had more substantial impacts.

As a consequence of several market changes occurring in that period, in this case "external factors" played their role, though not a very significant one. In conclusion, we can state that *Trans'* CEF in the period 2006–2011 underwent an evolution that was not entirely within the company's control and was not confined to its decisions and will alone.

Conclusion

It is becoming increasingly difficult to ignore that the future of mankind is dependent on a biocapacity which is not infinitely expandable. In fact, in recent years it has been increasingly recognized that the world is becoming biologically "overloaded," in contrast to the idea that the earth is providing and will be able to provide everything that humanity demands without "limits".

Many sustainability methodologies are being developed in the literature. We have chosen the ecological footprint, which has achieved outstanding popularity among the scientific community as well as institutions, governments, consumers, and organizations.

Even though it is regarded as a "young" instrument, the EF has experienced a rapid growth in attractiveness, being endorsed as a unique indicator of sustainability with different levels of application, where the corporate field constitutes one of the current main applications.

With reference to the current position of companies in a strong competitive global world, sustainability is becoming an unavoidable issue, being an important factor in helping the firm to be a first mover and to capture a higher share of the market. In line with this reasoning, the corporate ecological footprint is becoming more widely used and demanded by stakeholders and society in general. It is not only a tool for measuring a firm's impact on the environment, along with its share of responsibility for the degradation of common natural resources, but also a benchmark for the firm's image in the competition with its peers and a sign of "good behavior" relating to the accomplishment of national and international norms.

Nevertheless, the CEF's assumptions and calculation methodology are engendering several criticisms. In this section, we have surveyed some of the most pertinent advantages and limitations of this instrument, the outcome of which is that it must be analyzed with some precautions, revealing some pitfalls related to its use.

With the illustration presented, a case study applied to a Portuguese firm, the main aims were first to show the advantage of computing CEF for the firm as well as for social well-being, and second to interpret the results obtained with a view to rethinking its procedures. It should however be noticed that corporate and global footprints may interact.

Technology plays an important role, supporting measures to reduce the footprint and achieve more efficiency, two results which can often go hand in hand.

However, there can be a significant influence of exogenous factors, external to the firm's decisions whether to support or counteract its internal action to reduce its footprint, an issue which demands further research.

References

Ayllón D, Almodóvara A, Nicolab GG, Parra I, Elvira B (2012) Modelling carrying capacity dynamics for the conservation and management of territorial salmonids. Fish Res 134–136:95–103

Bagliania M, Martini F (2012) A joint implementation of ecological footprint methodology and cost accounting techniques for measuring environmental pressures at the company level. Ecol Ind 16:148–156

Barrett J, Scott A (2001) The ecological footprint: a metric for corporate sustainability corporate environmental strategy. Corp Environ Strategy 8(4):316–325

Boulding KE (1966) The economics of the coming spaceship earth. In: Jarrett H (ed) Environmental quality in a growing economy. Resources for the Future/Johns Hopkins University Press, Baltimore, pp 3–14

Branco C (2012) A pegada ecológica das organizações: uma aplicação do método MC3 ao estudo de caso Efacec. Dissertation, Faculty of Economics, University of Porto

Browne D, O'Regan B, Moles R (2012) Comparison of energy flow accounting, energy flow metabolism ratio analysis and ecological footprinting as tools for measuring urban sustainability: a case-study of an Irish city-region. Ecol Econ 83:97–107

Cagiao J, Gómez B, Doménech JL, Mainar SG, Lanza HG (2011) Calculation of the corporate carbon footprint of the cement industry by the application of MC3 methodology. Ecol Ind 11:1526–1540

Carballo-Penela A, Diz JLC (2011) Gestión ambiental de las cadenas logísticas: concepto y caso de estudio. Revista Galega de Economía 20(1):1–29

Carballo-Penela A, Doménech JL (2010) Managing the carbon footprint of products: the contribution of the method composed of financial statements (MC3). Int J Life Cycle Assess 15(9):962–969

Carballo-Penela A, Garcia-Negro M (2008) Hacia el desarrollo sostenible de organizaciones y empresas: la huella ecológica corporativa y su aplicación a un productor de mejillón en Galicia (España). Revista Luna Azul 27:8–26

Carballo-Penela A, Garcia-Negro MC, Doménech JL, Villasante CS, Rodríguez GR, Arenales MG (2008) La huella ecológica corporativa: concepto y aplicación a dos empresas pesqueras de Galicia. Revista Galega de Economía 17(2):1–29

Carballo-Penela A, Garcia-Negro M, Doménech JL (2009) A methodological proposal for corporate carbon footprint and its application to a wine-producing company in Galicia, Spain. Sustainability 1(2):302–318

Carballo-Penela A, Mateo-Mantecón I, Doménech JL, Coto-Millán P (2012) From the motorways of the sea to the green corridors' carbon footprint: the case of a port in Spain. J Environ Plan Manag 55(6):765–782

Catton WR (1987) The world's most polymorphic species: carrying capacity transgressed two ways. Bioscience 37:413–419
Costa A (2008) Desenvolvimento de uma metodologia expedita de cálculo da pegada ecológica de uma cidade – o caso de Lisboa. Dissertation, Technical University of Lisboa
Cranston GR, Hammond GP (2012) Carbon footprints in a bipolar, climate-constrained world. Ecol Ind 16:91–99
Dietz S, Neumayer E (2007) Weak and strong sustainability in the SEEA: concepts and measurement. Ecol Econ 61(4):617–626
Doménech JL (2006) Guía metodológica para el cálculo de la huella ecológica corporativa. Centro Argentino de Estudios Internacionales—Programa Recursos Naturales & Desarrollo. Paper presented at the 3rd symposium on sustainable development and population, Universidad de Málaga
Doménech JL (2009) Huella ecológica y desarrollo sostenible. 3rd edn. Asociación Española de Normalización y Certificación (AENOR), Spain
Doménech JL, Carballo A, Herrero J, Cruz JL (2010) Estándares 2010 de huella de carbono MC3. Paper presented at the Conama10—Congreso Nacional del Medio Ambiente
Fiala N (2008) Measuring sustainability: why the ecological footprint is bad economics and bad environmental science. Ecol Econ 67(4):519–525
Galli A, Wiedmann T, Ercin E, Knoblauch D, Ewing B, Giljum S (2012) Integrating ecological, carbon and water footprint into a "footprint family" of indicators: definition and role in tracking human pressure on the planet. Ecol Ind 16:100–112
Georgescu-Roegen N (1976) The entropy law and the economic process. Harvard University Press, Cambridge
GFN (2010) Ecological footprint atlas 2010. http://www.footprintnetwork.org/images/uploads/Ecolo-gical_Footprint_Atlas_2010.pdf. Cited 5 Oct 2012
GFN (2012a) Carbon footprint. http://www.footprintnetwork.org/en/index.php/GFN/page/carbon_footprint/. Cited 3 Feb 2014
GFN (2012b) Glossary. http://www.footprintnetwork.org/en/index.php/gfn/page/glossary/. Cited 1 Feb 2014
Gondran N (2012) The ecological footprint as a follow-up tool for an administration: application for the Vanoise National Park. Ecol Ind 16:157–166
Hanley N, Moffatt I, Faichney R, Wilson M (1999) Measuring sustainability: a time series of alternative indicators for Scotland. Ecol Econ 28(1):55–73
Hardin G (1986) Cultural carrying capacity: a biological approach to human problems. Bioscience 36(9):599–606
Hardin G (2001) Carrying capacity as an ethical concept. Soc Contract XII(1):48–57
Holland L (2003) Can the principle of the ecological footprint be applied to measure the environmental sustainability of business? Corp Soc Responsib Environ Manag 10(4):224–232
Hopton M, White D (2012) A simplified ecological footprint at a regional scale. J Environ Manag 111:279–286
Juknys R, Liobikienė G, Dagiliūtė R (2014) Sustainability of catch-up growth in the extended European Union. J Clean Prod 63:54–63
Kitzes J, Moran D, Galli A, Wada Y, Wackernagel M (2009) Interpretation and application of the ecological footprint: a reply to Fiala (2008). Ecol Econ 68(4):929–930
Kitzes J, Wackernagel M (2009) Answer to common questions in ecological footprint accounting ecological indicators. Ecol Econ 9(4):812–817
Kula E (1998) History of environmental economic though. Routledge, London
Lee K (2011) Integrating carbon footprint into supply chain management: the case of Hyundai Motor Company (HMC) in the automobile industry. J Clean Prod 19:1216–1223
Lenzen M, Hansson C, Bond S (2007) On the bioproductivity and land-disturbance metrics of the ecological footprint. Ecol Econ 61(1):6–10
Malghan D (2011) A dimensionally consistent aggregation framework for biophysical metrics. Ecol Econ 70(5):900–909

Manning RE (2007) Parks and carrying capacity: commons without tragedy. Island Press, Washington
Meadows DH, Meadows DL, Randers J, Behrens WW (1972) The limits of growth. Universe Books, New York
Moffatt I (2000) Ecological footprints and sustainable development. Ecol Econ 32(3):359–362
Moncho A, Esteve M, Gascón A, Ribera S, Doménech JL, Arenales M (2008) La huella ecológica corporativa de los materiales: aplicación al sector comercial. Observatorio Iberoamericano del Desarrollo Local y la Economía Social 1(4):1–21
Monte-Luna P, Brook BW, Zetina-Rejón MJ, Cruz-Escalona VH (2004) The carrying capacity of ecosystems. Glob Ecol Biogeogr 13:485–495
Moore J, Rees W (2013) Getting to one-planet living. In: Assadourian E, Prugh T et al (eds) State of the World 2013: is sustainability still possible?. The World Watch Institute, Island Press, Washington, pp 39–50
Mostafa M (2010) A Bayesian approach to analyzing the ecological footprint of 140 nations. Ecol Ind 10(4):808–817
Muthu SS, Li Y, Hu JY, Mok PY (2011) Carbon footprint of shopping (grocery) bags in China, Hong Kong and India. Atmos Environ 45:469–475
Neumayer E (2003) Weak versus strong sustainability: exploring the limits of two opposing paradigms. Edward Elgar Publishing, USA
Nourry M (2008) Measuring sustainable development: some empirical evidence for France from eight alternative indicators. Ecol Econ 67(4):441–456
Nunes LM, Catarino A, Teixeira MR, Cuesta EM (2013) Framework for the inter-comparison of ecological footprint of universities. Ecol Ind 32:276–284
Pasqualino J, Meneses M, Castells F (2011) The carbon footprint and energy consumption of beverage packaging selection and disposal. J Food Eng 103:357–365
Pearce D (2000) Public policy and natural resources management—a framework for integrating concepts and methodologies for policy evaluation. Economics, prepared for DGXI, European Commission
Pereira L (2008) Síntese dos métodos de pegada ecológica e análise emergética para diagnóstico da sustentabilidade de países—o Brasil como estudo de caso. Dissertation, University of Campinas
Peters GP (2010) Carbon footprints and embodied carbon at multiple scales. Curr Opin Environ Sustain 2:245–250
Pezzey J (1992) Sustainable development concepts: an economic analysis. World Bank environment paper; no. 2
Rees W (1996) Revisiting carrying capacity: area-based indicators of sustainability. Popul Environ J Interdiscip Stud 17(3):195–2015
Rees W (2002) Carrying capacity and sustainability: waking Malthus' ghost. In: Bell D, Cheung Y (eds) Introduction to sustainable development, Encyclopedia of life support systems. EOLSS Publishers, Oxford, pp 393–422
Rees W (2010) What's blocking sustainability? Human nature, cognition, and denial. Sustain Sci Pract Policy 6(2):13–25
Rees W, Wackernagel M (1994) Ecological footprints and appropriated carrying capacity: measuring the natural capital requirements of human economy. In: Jansson A, Hammer M, Folke C, Costanza R (eds) Investing in natural capital: the ecological economics approach to sustainability. Island Press, Washington
Seidl I, Tisdell C (1999) Carrying capacity reconsidered: from Malthus' population theory to cultural carrying capacity. Ecol Econ 31:395–408
Soares L (2013) Pegada ecológica corporativa: discussão metodológica e aplicação à STEF. Dissertation, Faculty of Economics, University of Porto
van den Bergh JCJM, Verbruggen H (1999) Spatial sustainability, trade and indicators: an evaluation of the 'ecological footprint'. Ecol Econ 29:61–72

Wackernagel M, Monfreda C, Moran D, Wermer P, Goldfinger S, Deumling D, Murray M (2005) National footprint and biocapacity accounts 2005: the underlying calculation method. Global Footprint Network, Oakland

Wackernagel M, Monfreda C, Schulz N, Erb K, Haberl H, Krausmann F (2004) Calculating national and global ecological footprint time series: resolving conceptual challenges. Land Use Policy 21(3):271–278

Wackernagel M, Moran D, White S, Murray M (2006) Ecological footprint accounts for advancing sustainability: measuring human demands on nature. In: Lawn P (ed) Sustainable development indicators in ecological economics. Edward Elgar Publishing, USA, pp 246–267

Wackernagel M, Rees W (1996) Our ecological footprint—reducing human impact on the earth. New Society Publishers, Canada

WCED (UN World Commission on Environment and Development) (1987) Our common future: report of the World Commission on Environment and Development, Switzerland

White T (2007) Sharing resources: the global distribution of the ecological footprint. Ecol Econ 64(4):402–410

Wiedmann T, Lenzen M (2007) On the conversion between local and global hectares in ecological footprint analysis. Ecol Econ 60:673–677

Wiedmann T, Minx J, Barrett J, Wackernagel M (2006) Allocating ecological footprints to final consumption categories with input–output analysis. Ecol Econ 56(1):28–48

WWF (2010) Living planet report 2010. http://wwf.panda.org/about_our_earth/all_publications/living_planet_report/living_planet_report_timeline/2010_lpr2/. Cited 1 Nov 2012

Chapter 29
Sustainability and Public Finances in the Time of Austerity

Noel Gomes

Abstract The scope of this study is to underline, within the current European context of economic and financial crisis, the importance and relevance of the sustainability of public finances, understood not only as an end in itself but also as a means of reaching that end, enabling the further development of society in general. Methodologically, because the subject has taken on a supranational significance, the present work will pay special attention to the EU legal framework in the first stage and then analyse the measures that have been taken in at-risk countries such as Portugal, in order to repair public finances at the local and national level. Concerning these measures, a critical attitude will be adopted, not only to highlight the importance of the increase in financial control and fiscal responsibility, but also to analyse the compatibility of some such measures with the national legal order.

Keywords Sustainability of public finances · Economic and financial crises · Austerity · Financial control and fiscal responsibility

Sustainability of Public Finances: Conceptual Precision, Timeliness and Importance

In recent years much argument has arisen at a global level surrounding the topic of public finances and their sustainability. Such debate has gained particular importance in Europe, largely stemming from a serious crisis (at the economic and financial level and also in the area of public debt), wreaking havoc (and is still doing so) on this continent and, in particular, on European countries with

---

N. Gomes (✉)
Porto School of Economics, University of Porto, R. Dr. Roberto Frias, 4200-464 Porto, Portugal
e-mail: ngomes@fep.up.pt

L.W. Zacher (ed.), *Technology, Society and Sustainability*,
DOI 10.1007/978-3-319-47164-8_29

problems in terms of the consolidation of public accounts, as is the case with the so-called PIGS (Portugal, Italy, Greece and Spain).

Nevertheless, before delving into this theme, it is necessary to define the meaning and reach of public finances. As such, the theme of sustainability of public finances will be considered in two parts: on the one hand, in regard to public finances (as an object), and on the other hand, in regard to sustainability (as an adjective, related to the first definition, of public finances).

Starting with the concept of public finances, notwithstanding its multiple meanings, it can be defined as the "*economic activity of a public body inclined to allocate assets to satisfy the needs with which it has been entrusted*" (Franco 2001, p. 3). Here, then, is a concept which, as a result of its own labelling, is grounded in the public domain, whether from an objective viewpoint on emerging activity for the allocating of goods in order to meet certain needs, or from an organic perspective as the public body responsible for managing available resources to satisfy social needs, keeping in mind the broadest sense of this concept, in order to encompass the financial situation of national bodies as well as all subnational levels of government such as regions, municipalities and even the public business sector.

Abstractly defining what the sustainability of public finances consists of is not such an arduous task. Generically speaking, sustainability translates as the "*ability of a government to assume the fiscal burden of its debts in the future*" (European Commission 2012, p. 1). More concretely, it is possible to say that "*fiscal policy is not sustainable if it implies an excessive accumulation of government debt over time and ever-increasing debt service*". In the same way it can be said that "*sustainability means avoiding an excessive increase in government liabilities—a burden on future generations—while ensuring that the government is able to deliver the necessary public services, including the necessary safety net in times of hardship, and to adjust policy in response to new challenges*". The task of defining the limits of sustainability in a precise and universal manner, that is to say, the line that separates sustainability from unsustainability of public finances, is much more problematic and difficult. Accordingly, from our point of view since a limit is at stake that should not be defined abstractly, or with the least amount of precision, it thus requires being evaluated minutely case by case, since its boundaries are dependent on a wide set of variables, especially in terms of time and space.[1]

[1]According to the European Commission (2012, p. 11), "*limits to sustainability differ across countries and over time. The capacity to run high debts depends inter alia on the degree of development of financial markets, perceived risks, and trust in the capacity of a government to implement structural reforms and consolidate deficits. It also depends on the degree of global risk aversion and the attractiveness of investments alternative to government bonds*". Despite this understanding, we must not fail to point out that austerity policies were largely shaped by two Harvard economists, Reinhart and Rogoff (2010, p. 2), in which they argue for a relationship between public debt and growth, particularly when public debt exceeds 90 % of the GDP. According to these authors, "*whereas the link between growth and debt seems relatively weak at normal debt levels, median growth rates for countries with public debt over 90 % of GDP are roughly one percent lower than otherwise*".

In light of this, we can see that the sustainability of public finances, besides being a current topic of discussion, is also of relevance outside of its own scope, that is to say, beyond public administration (understood *lato* sensu) and its own financial concerns. In reality, the goal of sustainability of public finances is not merely an end in itself, that is, sustainability of public finances for the sake of sustainability, but rather should be seen as a means of reaching a concrete end, which is economic and social development (cf. point 1 of Council Regulation (EC) No. 1466/97, of 7 July 1997). What is meant by this is that there is a link between public finances and development in general, in that healthy and balanced public finances both enable and/or reinforce favourable conditions for growth as well as social and economic development, while at the same time the opposite is true—which, as will be shown later, is recognized by the European Union (EU) and embodied in its own legal orders.

Sustainability of Public Finances and European Union Law

The goal of sustainability of public finances, in line with its importance, naturally constitutes a national plan shared by different States and is incorporated in the legal orders of each one. Nevertheless, the importance of public finance sustainability goes much further than a strictly national scope, constituting a supranational plan. It is exactly this that is observed at the level of the EU, whose law adopted the rule and the objective of sustainability of public finances.[2]

As we know, the EU can be seen as a "*two-speed legal system*": (1) on the one hand we have spaces or areas of centralization and uniformity such as that of monetary policy, with a single legal framework, a single currency (the Euro) and a single authority (the European Central Bank—ECB), and (2) on the other hand, we have spaces of decentralization and mere coordination, as in the case of financial policy. Here, in view of the great difficulty (or even impossibility) of establishing a common legal structure and a supranational financial authority, the maximum point reached is the attempt to approach the different legislations of each member state (MS). So, for now it is impossible to talk about a common public finance policy, a common budget, a common tax system or a common control system and, consequently, each MS approves and executes its own measures concerning public money.

However, not only theory but also reality provides some evidence that the Euro zone cannot survive without a substantial degree of financial alignment, because

---

[2]See article 126.° of the Treaty on the Functioning of the European Union, the Protocol no. 12 on the excessive deficit procedure, the Council Regulation (EC) No. 1466/97 of 7 July 1997 on the strengthening of the surveillance of budgetary positions and the surveillance and coordination of economic policies, and, more recently, the Treaty on Stability, Coordination and Governance in the Economic and Monetary Union and the Council Directive 2011/85/EU of 8 November 2011 on requirements for budgetary frameworks of the member states.

the stability and solidity of the whole depends in large measure on the stability and solidity of the parts, which, when isolated, are not always good performers.

It was in this context that, early on, the EU took measures aimed at self-control and budgetary discipline within member states or, in other words, that each and every MS strives to avoid excessive budgetary deficits. Put simply, we can say that the restrictions put into place—built upon two criteria (budgetary deficit and public debt)—fundamentally break down into the following: (1) budget deficits may not exceed 3 % of the gross domestic product, and (2) public debt may not surpass 60 % of the gross domestic product.[3]

In this respect, it is worth pointing out that the EU did not restrict itself to establishing these limitations, also having foreseen the process of excess deficits and, for this purpose, the sanctioning of measures in case such rules are not respected.

This procedure essentially consists of two phases:

(i) It begins with a supervisory moment by the Commission (the Guardian of the Treaties), which must follow the financial situation of each MS closely in order to identify possible deviations. If this is the case, a report must be made and the MS in question is notified and the Counsel is informed[4];
(ii) If that MS persists in its deviation and does not carry out the necessary measures to correct the situation, the procedure continues with the intervention of the Counsel. Here, a *multistage voting procedure* takes place, beginning with the declaration of the excessive deficit situation, after which the MS is notified to carry out appropriate measures and finally, if the situation still persists some "sanctions" can be applied (for example, publishing additional information before issuing bonds and securities, to make a non-interest-bearing deposit of an appropriate quantity for the Union or to pay fines of an appropriate amount).[5]

Despite the approval of this measure and this sanctioning strategy, the truth is that critical appraisal of such measures is far from positive, namely in the light of:

(i) the bureaucratic nature of the procedure;
(ii) the vagueness and subjectivity of the rules concerning sanctions;
(iii) the lack of enforceability as a consequence of the protection agreements between MS;
(iv) the risk of treatment disparities.

More recently, as a result of recognizing that insufficient implementation had been achieved, further accentuated by the European public debt crisis, a tendency to reinforce the economic and monetary pillars of the EU has since been observed,

[3]The budget deficit and the public debt must be faced *lato* sensu (as prescribed in the European System of accounts 1995—ESA 95), including not only the financial situation of the national bodies (specifically, the State itself and the central administration) but also all subnational levels of Government, as Regions, Municipalities and even the Social Security system.

[4]See, article 126. no. 3 and 5 of the Treaty.

[5]See article 126, ns. 6 to 11 of the Treaty.

through the adoption of a set of rules aimed at, above all, promoting budgetary discipline. Additionally, there is specific emphasis not only on "*the need for governments to keep their public finances healthy and sustainable as well as to avoid excessive deficits*", but also on the importance that this represents for achieving the objectives of the EU. This reinforcement, especially as it pertains to the "*golden rule*" of limiting and restricting public debt, is clearly visible in the recent Treaty on Stability, Coordination and Governance in the Economic and Monetary Union, from which the following essential aspects arise:

(i) Demand that national budgets remain in a balanced or in a positive state[6]; this rule is considered to be adhered to if the annual structural balance of the general government is at its country-specific medium-term objective, as defined in the revised Stability and Growth Pact, with a lower limit of a structural deficit of 0.5 % of the gross domestic product at market prices[7];
(ii) The rule of budget balancing has to be incorporated into national law within one year of the entry into force of the treaty, using provisions that are guaranteed to be adhered to throughout national budgetary processes.[8]
(iii) In the event of significant observed deviations from the medium-term objective or the adjustment path towards it, a correction mechanism shall be triggered automatically. The mechanism shall include the obligation of the Contracting Party concerned to implement measures to correct the deviations over a defined period of time.[9]
(iv) The contracting parties whose currency is the Euro commit themselves to adopting Council decisions within the framework of the excessive deficit procedure unless opposed by a qualified majority (article 7.°).[10]

The Sustainability of Public Finances and Economic and Financial Assistance Programmes

The goal of sustainability of public finances is also deeply ingrained in the economic and financial assistance programmes signed by Greece, Ireland and Portugal.

---

[6]See article 3.°, n.° 1, al. a.

[7]See article 3.°, n.° 1, al. d. Article 3.° also stipulates "*where the ratio of the general government debt to gross domestic product at market prices is significantly below 60 % and where risks in terms of long-term sustainability of public finances are low, the lower limit of the medium-term objective specified under point (b) can reach a structural deficit of at most 1 % of the gross domestic product at market prices*".

[8]See article 3.°, n.° 2. In case of non-compliance of this duty, the EU Court of Justice may apply sanctions on MS, at a level no less than 0.1 % of their GDP (article 8.°, n.° 2).

[9]See article 3.°, n.° 1, al. e.

[10]See article 7.°.

In a period of little more than a year, these three countries (each a MS of the EU), for not entirely parallel reasons [on the causes that form the basis for requests for external financial assistance of each one of these countries, see European Commission (2010, pp. 3–9; 2011a, pp. 6–18; 2011b, pp. 9–15); for further developments about the Portuguese case, see Reis (2013)], all turned to external financial assistance. Each entered into agreement with the so-called *Troika* (made up of representatives of the European Commission, the European Central Bank and the International Monetary Fund), at different moments in time, agreeing to a Memorandum of Understanding on Specific Economic Policy Conditionality.[11]

Under these documents, as a condition of receiving financial assistance, these countries took on a broad and meaningful set of obligations to international creditors, in a wide variety of areas:

(i) public finances and budgetary policies;
(ii) public administration (central, regional and local);
(iii) public health and education systems;
(iv) legal systems;
(v) regulation and supervision of the financial sector;
(vi) public services of transports, telecommunications, energy and national postal service; and the
(vii) job market.

It must be pointed out that the financial assistance that these three States received was made available in instalments, whose disbursements were made under the condition of observing accepted commitments. Thus, those States subjected themselves to a tight set of control schemes and financial reporting that has taken place during regular evaluations of accordance over the course of the programme.

Despite the diverse grounds which form the basis of requests for external financial assistance by each of these countries, a comparative analysis of the three *Memoranda* allows for observing that the approved model of intervention has a common matrix based on adopting austerity policies. [12], [13] In accordance with the

[11]Greece's *Memorandum* dated 3 May 2010, Ireland's 3 September of the same year and, finally, Portugal's was ratified on 17 May 2011.

[12]Austerity is defined by Blyth (2013, p. 2), as "*a form of voluntary deflation in which the economy adjusts through the reduction of wages, prices, and public spending to restore competitiveness, which is (supposedly) best achieved by cutting the state's budget, debts, and deficits.*"

[13]The similarity (materially and formally) between the three aforementioned *Memoranda* is not limited, however, to the adoption of a common matrix, going so far as to incorporating the text used. According to the notice published in the Portuguese newspaper "Público" 8 August 2013, "*the agreements signed by Ireland and Portugal share 75 % of* [the same] *text, those signed by Greece and Ireland coincide on 77 % and those signed by Greece and Portugal 82 %*". Despite the similarities (which in fact exist), the three documents also display some differences between each other, namely in regard to the dimension of the measures aimed at the financial sector, which are more prominent in the *Memoranda* of Ireland and Portugal than in that of Greece.

model set forth in the *Memorandum*, the way of reaching the objectives, especially as far as consolidating public finances is concerned, is through adopting a policy of deflation, based on reducing deficit, public debt and spending, in addition to increasing tax revenue. According to the defenders of the policy of austerity, it was believed that adopting that type of measure (especially in the public sector) would prevent the "*crowding out*" effect and, at the same time, would have the positive effect of creating confidence in the private sector, all of which would together lead to expansion and economic development (Krugman 2013, p. 6; Cavero and Poinasamy 2013, p. 38).

Focusing on the Portuguese case as a reality now under way, it is possible to note that there have been several measures for reducing expenditures and for increasing tax revenue during the programme of financial and economic assistance and in compliance with the commitments made to the *Troika*, bearing in mind the ultimate goal of consolidating public finances and, similarly, the slimming of the State. As it stands, in regards to reducing expenditures, the Portuguese government has achieved a very significant set of cuts, namely in public salary benefits, retirement benefits, social assistance benefits as well as in the sectors of education, culture and sports.

Just as in the case of revenues, a broad set of measures was implemented which including putting State property up for sale and privatizations, an exponential increase in the tax burden, largely through tax rate increases (of personal income tax, VAT and of fees related to delivery of public services such as, for example, co-payments in the national health system) or through the reduction of tax benefits—although later these were somewhat alleviated by a progressive reduction in the tax burden on corporate bodies.

The application of these measures has generated a heated discussion within Portuguese society:

(i) not only regarding the effects they have triggered (which will be noted shortly) and, related to this, the correction of policies that have been adopted, but also on
(ii) their democratic legitimacy, given that they have been imposed by a series of non-elected parties, of doubtful compatibility with two fundamental constitutional principles, those of democracy and sovereignty, and finally,
(iii) from a strictly legal point of view, since measures restricting legal rights, freedoms and guarantees are at risk, in their compatibility owing to constitutional limits in place—to the extent that it becomes questionable whether a process of dismantling the Social State is not under way.

Regardless of the debate that has arisen, nearly three years after the agreement to the Memorandum the outcome of applying deflationary measures—mainly aimed at budgetary consolidation—will undoubtedly be negative. And it will be negative first and foremost due to the social impact that these measures have, especially in increased unemployment, as well as the reduction in disposable income available to people which, as a consequence, brings about a decrease in the standard of living. The same conclusion can be reached in analysing the outcome arising from public finances, which are far from corresponding to those outcomes initially foreseen. In fact, since 2011 to the present, an increase in public debt has been observed and, moreover, an increase in this compared to the GDP, such that these both are much higher than the objectives initially established.

Nonetheless, according to the *Memorandum*, the purpose of budgetary consolidation cannot be reached simply through cutting expenses and increasing revenue. Alongside such measures, mechanisms for implementing and strengthening a system of ongoing control and accountability of decision-makers have been prioritized, spanning across all sectors and administration, contributing to these goals in one way or another.

Internally speaking, in accordance with the *Memorandum*, the implementation and strengthening of financial control has translated into the adoption of various measures, of which the most salient ones are:

(i) Firstly, and as a way of fighting so-called *de-budgetization*, expanding the reach of public administration through introduction of so-called national accountancy rules, including entities that traditionally were excluded from the public administrative sector but whose accounts have relevance for public finance;
(ii) Secondly, dissipation in terms of control functions brought about through the creation of new institutional control structures (such as the Council of Public Finances, the technical surveillance units both for Public and Private Partnership Projects as well as for the Monitoring of the Public Business Sector), which were attributed special functions in "*areas of budgetary risk*";
(iii) Thirdly, a new trend has been detected in regard to the moment of exercising control—concretely, to a trend of exercising this control in advance. In fact, whereas previously a posteriori inspection of offenders[14], was preferred, it is currently possible to say that this practice is being called into question and replaced with and/or complemented by another that, without neglecting posterior inspection and properly holding financial decision-makers responsible, values control during the period prior to taking on and/or incurring expenditures. This new trend is justified for reasons that have to do with questions of efficiency of control itself—and naturally the inefficiency of the previous model—and a certain lack of (self) confidence in public administration;
(iv) Fourthly, with the objective of making financial control more efficient, there has been a marked increase in information disclosure duties placed upon varying sectors and administrations (on the importance of financial information for multiple purposes, including fiscal control, Lundqvist (2013); for more information on the Portuguese local financial system, see Freitas da

[14]A trend which, it should be noted, also was found in other legal orders—in French law, see Bouvier et al. (2008, p. 867), Bouvier (2010, p. 229), Levoyer (2007, p. 109); in Spanish law, see García-Alos (2009, p. 105); for further reading on the general tendency, see Bilbao (2009, pp. 248–250).

Rocha (2011, pp. 455–488)), coupled with an increase in the level of penalties, focussed on the importance and instrumental role that information plays in exercising control.

Nevertheless, as was previously mentioned, it is not only at the level of financial control that this trend towards reinforcement has been noted. At the same time, there has been considerable movement in the direction of greater demands on and accountability for financial decision-makers, translating on the one hand into a significant broadening of duties which various administrations are assigned, as well as into applicable penalties on the other.

It is worth noting that when saying that a tendency towards strengthening of control and financial responsibility has been witnessed, this is not to say that there is truly a new trend evolving, in the sense that prior to the memorandum, financial control and responsibility were undervalued realities or even ignored. Such a viewpoint does not correspond to reality, since, even though on different levels, prior to the memorandum there was already a sense of the weakness of national public finances and, related to this, a sense of the need and importance of strengthening those control mechanisms and of financial responsibility itself. What the *Memorandum* did, was because of necessity and the external link taken on by the Portuguese State and as a condition for ensuring the financing necessary for the national economy to accelerate, intensify and, perhaps, improve this very movement.

From all that has been shown, it is quite apparent that financial accountability and control are current and essential realities for the reinstatement of sustainability of public finances.

Conclusions

The sustainability of public finances is clearly a question of great concern and relevance in the present day. The timeliness of the topic is further accentuated by the European public debt crisis and, in particular, due to the difficulties experienced by several European states in securing financing. Far from being confined to the public sector, the goal of sustainability of public finances should be seen as a way of attaining social and economic development.

The goal of sustainability in public finances, besides being deeply rooted in EU law, is present in the economic and financial assistance programmes entered into by Greece, Ireland and Portugal. Through these programmes as a counterpart to financial assistance, those States agreed under the terms of their creditors to follow a vast set of obligations on economic policy, especially on the matter of budgetary consolidation. In all of these programmes, despite some differences, the model of intervention adopted was based on a common matrix—austerity. In accordance with the defined model, restoring public finances can be carried out through a deflationary policy based on the reduction of the deficit, of public debt, the reduction of spending and the increase of tax revenues.

In the Portuguese case, the balance we have made in terms of public finance and its sustainability reveals negative aspects as well as some positive ones.

The negative aspects are many and significant: bearing in mind the social impact of austerity measures and the suffering of citizens as a consequence of these measures, the fact is that, despite certain measures adopted in the area of revenue (especially in terms of taxes), an increase in public debt has been observed (in addition to an increase in relation to the GDP), meaning that the increase was greater than was foreseen in the programme. This phenomenon is, from our point of view, a direct result of the strict application of a programme centred on the successive implementation of austerity measures (austerity on top of austerity), which by not being accompanied by measures aimed at stimulating growth has lead to a pronounced economic recession. To make our standpoint completely clear, we are not against the adoption of austerity measures. We believe that in the Portuguese case, much could be (and should be) done at the level of rationalizing the State and public spending. What went wrong, from our point of view, was not having tried to strike a balance between austerity and growth.

Nevertheless, the implementation of the economic and financial assistance programme, in the area of budgetary consolidation, did not bring out only negative aspects. Among them, several positive aspects should also be noted. One of these, as we pointed out earlier, concerns the strengthening of control mechanisms and of financial accountability, which, in their essence, seek to introduce greater strictness, discipline and also responsibility in public financial management. However, beyond this aspect which has a normative nature (that is, it results fundamentally from legislative changes), another stands out which is much more appealing and which we can label as arising from a collective civil conscience related to public finances. From this we can say that now more than ever, as a product of the financial and economic crisis which Portugal has been experiencing and the costs which have come with it, people are concerned with public finances and the need for these to remain at a sustainable level, not only for the present generation (a self-centred concern) but also for future generations (an altruistic concern). In fact, when considering public finances people express a range of viewpoints, among which are: (i) firstly, a sense of belonging and of ownership of public finances; (ii) secondly, as a result of raised awareness of collective "ownership" of public finances, a sense of responsibility has emerged, made clear by the fact that citizens have progressively resisted distancing themselves from public financial management but rather perform an active role of continuous surveillance (and being increasingly demanding, even from a qualitative point of view, which goes beyond mere compliance with law, calling for good administration of public money through effectiveness, efficiency and economy in the measures put into place) in the hands of those agents upon whom the management of public monies is bestowed.

References

Bilbao G (2009) El control de las cuentas públicas. Thomson Reuters-Civitas, Navarra
Blyt M (2013) Austerity—the history of a dangerous idea. Oxford University Press, New York
Bouvier M (2010) Les finances locales. LGDJ, Paris
Bouvier M, Esclassan M, Lassale JP (2008) Finances publiques. LGDJ, Paris
Cavero T, Poinasamy K (2013) A cautionary tale. The true cost of austerity and inequality in Europe. Oxfam Brief Pap 174:1–38
European Commission (2010) Economic adjustment programme for Greece. Occas Pap 61:3–9
European Commission (2011a) Economic adjustment programme for Ireland. Occas Pap 76:6–18
European Commission (2011b) Economic Adjustment Programme for Portugal. Occas Pap 79:9–15
European Commission (2012) Fiscal sustainability report 2012. http://ec.europa.eu/economy_finance/publications/european_economy/2012/pdf/ee-2012-8_en.pdf. Accessed 1 Feb 2014
Franco S (2001) Finanças Públicas e Direito Financeiro—vol I. Almedina, Coimbra
Freitas da Rocha (2011) Portugal. In: Biziole G, Sacchetto C (eds) tax aspects of fiscal federalism: a comparative analysis. IBFD, Amsterdam
García-Alos Vacas L (2009) La responsabilidad contable de los gestores públicos y de los perceptores de subvenciones en la jurisprudencia. Claves de la Jurisprudencia, Madrid
Krugman P (2013) How the case for austerity has crumbled. http://www.nybooks.com/articles/archives/2013/jun/06/how-case-austerity-has-crumbled/. Accessed 1 Feb 2014
Levoyer L (2007) Finances locales. Hachette Livre, Paris
Lundqvist K (2013) Financial information of good quality, in time and for multiple uses—the Swedish experience. http://www.bportugal.pt/pt-PT/OBancoeoEurosistema/Eventos/Documents/Paper_Financial_information_of_good_quality20sep2013.pdf. Accessed 1 Feb 2014
Reinhart C, Rogoff K (2010) Growth in a time of debt. http://nber.org/papers/w15639.pdf?new_window=1. Accessed 1 Feb 2014
Reis R (2013) The Portuguese slump and crash and the Euro Crisis. Brookings papers on economic activity. Spring, 143–193

Index

L.W. Zacher (ed.), *Technology, Society and Sustainability*,
DOI 10.1007/978-3-319-47164-8

F

G

H

I

K

L

MIX
Papier aus verantwortungsvollen Quellen
Paper from responsible sources
FSC® C105338

If you have any concerns about our products,
you can contact us on
ProductSafety@springernature.com

In case Publisher is established outside the EU,
the EU authorized representative is:
Springer Nature Customer Service Center GmbH
Europaplatz 3, 69115 Heidelberg, Germany

Printed by Libri Plureos GmbH
in Hamburg, Germany